DOWNSTREAM PROCESSING OF NATURAL PRODUCTS

DOWNSTREAM PROCESSING OF NATURAL PRODUCTS

A PRACTICAL HANDBOOK

Edited by
Michael S. Verrall
SmithKline Beecham Pharmaceuticals, Surrey, UK

JOHN WILEY & SONS
Chichester · New York · Brisbane · Toronto · Singapore

Library of Congress Cataloging-in-Publication Data
Downstream processing of natural products : a practical handbook /
   edited by Michael S. Verrall.
      p.   cm.
   Includes bibliographical references and index.
   ISBN 0-471-96326-7 (alk. paper)
   1. Biomolecules—Separation.   2. Natural products.
3. Biotechnology.   I. Verrall, M. S., 1937–   .
TP248.25.S47D695   1996
660'.6—dc20   95–42364
                                                     CIP

British Library Cataloguing in Publication Data

A catalogue record for this book is available from the British Library

ISBN 0 471 96326 7

Typeset in 10/12pt Times by Dobbie Typesetting Limited, Tavistock, Devon.
Printed and bound in Great Britain by Bookcraft (Bath) Limited.
This book is printed on acid-free paper responsibly manufactured from sustainable forestation, for
which at least two trees are planted for each one used for paper production.

CONTENTS

LIST OF CONTRIBUTORS   xv

PREFACE   xvii

1 GENERAL STRATEGY: THE NATURE OF THE PROBLEM   1
   Michael S. Verrall

1.1   Introduction   1
1.2   Strategy for Isolation   4
       1.2.1   Analytical methods   4
       1.2.2   Stability data   5
       1.2.3   Isolation methods   5
       1.2.4   Scale of operation   6
       1.2.5   Purification procedures   6
1.3   References   9

2 BROTH CONDITIONING AND CLARIFICATION   11
   Duncan Mackay

2.1   Introduction   11
2.2   Broth Clarification Unit Operations   12
       2.2.1   Dead end filtration   12
       2.2.2   Tangential flow filtration   18
       2.2.3   Centrifugal sedimentation   20
2.3   Broth Conditioning   21
       2.3.1   Temperature variation   22
       2.3.2   Filter aid addition   23
       2.3.3   Coagulation and flocculation   23
       2.3.4   Modification of the fermentation   26
2.4   Design Guidelines   27
       2.4.1   Selection of separation technique   28
       2.4.2   Testwork to measure separation performance   28
       2.4.3   Testwork to measure separation performance
               improvement after broth conditioning   36

		2.4.4	Scale-up of clarification process	38
2.5	Summary			40
2.6	References			40

3   CELL DISRUPTION: A PRACTICAL APPROACH   41
Eli Keshavarz-Moore

3.1	Introduction		41
3.2	Choice of Unit Operation		41
	3.2.1	Bead mills	41
	3.2.2	High-pressure cell disruption	42
	3.2.3	New developments	42
3.3	How to Assess Cell Breakage		43
3.4	Parameters Affecting Cell Disruption Kinetics		44
	3.4.1	Pressure and number of passes	44
	3.4.2	Application of high pressures	45
	3.4.3	Location of the product	45
	3.4.4	Process fluid variables	46
	3.4.5	Unicellular organisms	46
	3.4.6	Cell concentration	46
	3.4.7	Effect of temperature	47
	3.4.8	Conditions of cell growth	48
3.5	Interaction Between Cell Disintegration and Other Downstream Operations		48
	3.5.1	Cell debris formation	48
	3.5.2	Viscosity of homogenate	49
	3.5.3	Comparison between different equipment	49
	3.5.4	Storage conditions	49
	3.5.5	Resuspension fluid	50
	3.5.6	Effect of recombinant DNA technology	50
	3.5.7	Pretreatment of cells	50
3.6	Future Methods		50
3.7	Concluding Remarks		51
3.8	References		51

4   TWO-PHASE AQUEOUS SYSTEMS FOR THE RECOVERY OF INTRACELLULAR MICROBIAL PRODUCTS   53
Jonathan G. Huddleston and Andrew Lyddiatt

4.1	Introduction		53
4.2	Properties of PEG–Salt ATPS		55
	4.2.1	The phase components	55
	4.2.2	Choice of phase-forming salts	57
	4.2.3	Choice of PEG molecular weight	57
	4.2.4	Increase of solute exclusion and salting-out	58
	4.2.5	Electrostatic effects	58

4.3	Strategic Selection of Systems for the Recovery of Intracellular Proteins	59
4.4	Recovery of Protein Products from Phase Systems	63
4.5	Recovery of Phase-forming Components	66
4.6	Conclusion	68
4.7	References	68

5 SOLVENT EXTRACTION OF FERMENTATION BROTH — **71**
Laurence R. Weatherley

5.1	Introduction	71
5.2	Drop Behaviour	74
5.3	Mass Transfer	80
5.4	Contacting	83
5.5	Future Developments	89
5.6	References	90

6 CHEMICALLY ASSISTED SOLVENT EXTRACTION — **93**
C. Judson King

6.1	Introduction		93
	6.1.1	Incentives	93
	6.1.2	Limitations	94
	6.1.3	Likely uses of chemically assisted extraction	95
6.2	Types of Chemical Interactions		95
	6.2.1	Acid–base interactions	95
	6.2.2	Ion-pair extraction	97
	6.2.3	Affinity interactions	98
6.3	Regeneration Alternatives		98
6.4	Extraction at $\mathrm{pH} > \mathrm{p}K_a$		99
6.5	Toxicity and Contamination		100
	6.5.1	*In situ* separation	100
	6.5.2	External recycle	100
	6.5.3	Sequential processing	101
6.6	Reactive Solid Sorbents		101
6.7	References		102

7 PRODUCT MODIFICATION TO ASSIST ISOLATION (FACILITATED PROCESSING) — **105**
Alan R. Thomson

7.1	Introduction		105
7.2	Tail Types		108
	7.2.1	Enzymes	108
	7.2.2	Polypeptide-binding proteins	111
	7.2.3	Antigenic epitopes	112

7.2.4   Carbohydrate-binding domains   112
7.2.5   Poly(amino acid) tails   113
7.2.6   Biotin-binding domains   116
7.3   Tail Cleavage   116
7.4   Present Status   116
7.5   Process Design Strategy   118
7.6   Conclusions   119
7.7   References   119

8   PROCESS INTEGRATION IN BIOTECHNOLOGY   123
Juan A. Asenjo and E. W. Leser

8.1   Introduction   123
8.2   Integrating Upstream   126
8.3   Integrating Fermentation   127
8.4   Integrating Downstream   129
8.5   Process Synthesis and Integration   132
8.6   Integration and Environment   135
8.7   References   137

9   MEMBRANE PROCESSING IN BIOTECHNOLOGY: A CASE HISTORY   139
Michael A. Cook

9.1   Introduction   139
9.2   The Nature of the Problem   139
9.3   Initial Experiments   140
9.4   Process Design   142
9.5   Membrane Fouling   145
9.6   Equipment Configuration   146
9.7   Summary   146

10   DISPLACEMENT CHROMATOGRAPHY   147
Denise M. Wallworth

10.1   Introduction   147
10.2   Overload Elution Chromatography   148
10.3   Displacement Chromatography   149
10.4   Equipment Required for Displacement Chromatography   151
10.5   Method Development   151
10.6   Retention and Selectivity Optimization   152
10.7   Adsorption Isotherm Measurement and Choice of Displacer   154
10.8   Self-displacement   155
10.9   Displacement Runs   156
10.10   Conclusions   156
10.11   References   157

11   NON-IONIC ADSORBENTS IN SEPARATION PROCESSES   159
Hiroaki Takayanagi, Junji Fukuda and Eiji Miyata

11.1   Introduction   159
11.2   Basic Properties of Synthetic Adsorbents   161
    11.2.1   General aspects and pore characteristics   161
    11.2.2   Polystyrene resins   161
    11.2.3   Modified polystyrene resins   163
    11.2.4   Polymethacrylate resins   163
    11.2.5   Other properties   166
    11.2.6   Stability of synthetic adsorbents   170
11.3   Practical Operations   170
    11.3.1   Pretreatment   170
    11.3.2   Testing methods   171
    11.3.3   Operating conditions   172
    11.3.4   Cycle tests   173
    11.3.5   Regeneration and rejuvenation   173
    11.3.6   Preservation of adsorbents   174
11.4   Examples of Applications   174
    11.4.1   Separation of cephalosporin derivatives   174
    11.4.2   Rejuvenation of a synthetic adsorbent (HP20)   174
    11.4.3   Other applications in the literature   175
    11.4.4   Application of smaller particle adsorbents   177
11.5   References   177

**12   ION EXCHANGE CHROMATOGRAPHY AND
SECONDARY ADSORPTION EFFECTS   179**
Peter R. Levison

12.1   Introduction   179
12.2   Feedstock Preparation   179
12.3   Media Preparation   181
12.4   Column Packing   181
12.5   Media Fouling   183
    12.5.1 Media discolouration   184
    12.5.2 Capacity reduction   184
    12.5.3 Chromatographic performance tests   185
12.6   Column Regeneration   190
12.7   Summary   190
12.8   References   190

13   AFFINITY CHROMATOGRAPHY   193
Steven J. Burton

13.1   Introduction   193
13.2   Preparation of Affinity Adsorbents   195

13.2.1   Affinity ligands   195
13.2.2   Support matrices   196
13.2.3   Coupling chemistries   197
13.3   Use of Affinity Chromatography in Downstream Processing   199
13.3.1   Adsorbent selection   200
13.3.2   Binding and elution   201
13.3.3   Examples   202
13.4   Future Prospects   205
13.5   References   205

14   THE USE OF RADIAL FLOW CHROMATOGRAPHY
     IN THE PURIFICATION OF BIOMOLECULES   209
     Denise M. Wallworth

14.1   Introduction   209
14.2   Column Requirements   209
14.2.1   Sample introduction   211
14.3   Radial Flow Column Design   211
14.3.1   Flowrates   212
14.3.2   Pressure drop–flowrate relationship in radial flow columns   215
14.4   Packing Radial Flow Columns   216
14.5   Applications for Radial Flow Columns   218
14.5.1   Desalting on a radial flow column   220
14.6   Conclusions   221
14.7   References   221

15   HIGH-PERFORMANCE LIQUID CHROMATOGRAPHY   223
     Peter R. Shelley

15.1   Introduction   223
15.2   Sample Treatment   224
15.3   Analytical Packings and Columns   225
15.3.1   Types of stationary phase   225
15.3.2   Stability of HPLC column packings   228
15.3.3   Choice of analytical column dimensions   228
15.3.4   Addition types of stationary phase particle   229
15.4   Detectors   231
15.5   Use of Databases   236
15.6   Preparative HPLC   236
15.6.1   Introduction   236
15.6.2   Types of stationary phase   236
15.6.3   Choice of particle size   236
15.6.4   Choice of column types   237
15.6.5   Application of preparative HPLC to the
         purification of milbemycins   237
15.7   References   240

16   SUPERCRITICAL FLUID EXTRACTION AND
     CHROMATOGRAPHY   241
     Anthony A. Clifford

16.1   Introduction   241
16.2   Supercritical Fluid   242
       16.2.1   The nature of a supercritical fluid   242
       16.2.2   Use of entrainers   244
       16.2.3   Advantages and uses of supercritical fluids   244
16.3   Supercritical Fluid Extraction   245
       16.3.1   SFE on a small scale   245
       16.3.2   SFE on a pilot plant scale   247
       16.3.3   Principles of SFE   250
16.4   Supercritical Fluid Chromatography   252
       16.4.1   A preparative SFC system   254
       16.4.2   Optimizing preparative SFC   254
16.5   Microparticle Formation   255
16.6   Bibliography   257

17   LIQUID MEMBRANES IN DOWNSTREAM PROCESSING   259
     Julian B. Chaudhuri and Paul J. Pickering

17.1   Introduction   259
17.2   Liquid Membrane Separations   259
       17.2.1   Separation principles   259
       17.2.2   Selection of carrier species for liquid membrane
                separations   261
17.3   Liquid Membrane Configurations   265
       17.3.1   Emulsion liquid membranes   265
       17.3.2   Supported liquid membranes   269
       17.3.3   Comparison of configurations   270
17.4   References   272

18   LYOPHILIZATION   275
     John W. Snowman

18.1   Introduction   275
18.2   Description of the Freeze Drying Process   276
       18.2.1   Freezing   278
       18.2.2   Sublimation (primary drying)   280
       18.2.3   Desorption (secondary drying)   283
18.3   Equipment   283
       18.3.1   General description   283
       18.3.2   Formulation and cycle optimization   286
18.4   Formulation   288
       18.4.1   Dosage   288

18.4.2 Fill depth   289
18.4.3 Excipients   289
18.4.4 Non-aqueous solvents   290
18.5 Cycle Development   291
18.5.1 Freezing   291
18.5.2 Primary drying   292
18.5.3 Primary to secondary drying conversion   295
18.5.4 Secondary drying   295
18.6 Good Manufacturing Practice   295
18.7 Leakage   296
18.8 The Cost of Lyophilization   298
18.9 References   299

19 INSTRUMENTATION AND PROCESS CONTROL   301
John Noble and Keith G. Robins

19.1 Introduction   301
19.2 Design of Instruments   301
19.2.1 Design of instrumentation for natural product recovery   302
19.2.2 Ease of cleaning   302
19.2.3 Sterilizability and sanitization   303
19.2.4 Maintenance of asepsis   304
19.2.5 Materials of construction   304
19.3 Selection of Instrumentation for Natural Product Recovery   305
19.3.1 Flow measurement   305
19.3.2 Pressure measurement   306
19.3.3 Temperature measurement   306
19.3.4 pH measurement   306
19.3.5 Conductivity measurement   307
19.3.6 Level, volume and mass measurement   308
19.3.7 Analysis   309
19.4 Process Control in Product Development and Pilot Plant
      Processes   309
19.4.1 Control module concept   309
19.4.2 Flameproof and intrinsically safe instruments   312
19.5 Future Trends   314
19.6 References   315

20 SCALE-UP CONSIDERATIONS   317
John Noble and Robert Davies

20.1 Introduction   317
20.2 Considerations for Scale-up   319
20.2.1 Product evaluation   319
20.2.2 Process selection   319

20.2.3 Equipment selection   321
20.2.4 Building requirements   323
20.2.5 Regulatory requirements   323
20.2.6 Economic factors   323
20.2.7 Final process   324
20.3 Practical Aspects of Scale-up   324
20.3.1 Equipment selection   324
20.3.2 Data transfer   326
20.4 Conclusions   326
20.5 References   327

21 GMP AND QUALITY CONTROL   329
David Sherwood

21.1 Introduction   329
21.2 The Production Facility   330
21.3 Equipment and Services   331
21.3.1 Equipment   331
21.3.2 Services   331
21.4 Personnel   332
21.4.1 Training   332
21.4.2 Gowning procedures   333
21.5 Raw Materials   333
21.6 Validation   335
21.6.1 Design qualification (DQ)   335
21.6.2 Installation qualification (IQ)   335
21.6.3 Operational qualification (OQ)   336
21.6.4 Performance qualification (PQ)   337
21.6.5 PQ activities specific to chromatography-based processes   339
21.6.6 Validation of aseptic operations   340
21.6.7 Validation of computer systems   340
21.6.8 Ongoing validation   340
21.7 Documentation Systems   341
21.8 Quality Control   342
21.8.1 In-process testing of intermediate products   342
21.8.2 Bulk product testing   343
21.8.3 Method validation   344
21.8.4 Stability studies   344
21.8.5 Environmental monitoring programme   345
21.9 References   346

Index   347

LIST OF CONTRIBUTORS

Juan A. Asenjo	*Biochemical Engineering Laboratory, Department of Food Science and Technology, University of Reading, Whiteknights, Reading, Berkshire, UK*
Steven J. Burton	*Affinity Chromatography Ltd, 307 Huntingdon Road, Girton, Cambridge, UK*
Julian B. Chaudhuri	*School of Chemical Engineering, University of Bath, Claverton Down, Bath, Avon, UK*
Anthony A. Clifford	*School of Chemistry, University of Leeds, Leeds, West Yorkshire, UK*
Michael A. Cook	*SmithKline Beecham Pharmaceuticals, Clarendon Road, Worthing, West Sussex, UK*
Robert Davies	*AMEC Design and Management, Stratford-upon-Avon, Warwickshire, UK*
Junji Fukuda	*Speciality Chemicals Laboratory, Mitsubishi Kasei Corporation, 1000 Kamoshida-cho, Midori-ku, Yokohama 227, Japan*
Jonathan G. Huddleston	*Biochemical Recovery Group, BBSRC Centre for Biochemical Engineering, School of Chemical Engineering, University of Birmingham, Edgbaston, Birmingham, UK*
Eli Keshavarz-Moore	*Advanced Centre for Biochemical Engineering, University College London, Torrington Place, London WC1, UK*
C. Judson King	*Department of Chemical Engineering, University of California, Berkeley, California, USA*
E. W. Leser,	*Fundação Oswaldo Cruz, FIOCRUZ/MS, Av. Brasil 4365, 21045-900 Rio de Janeiro, Brazil*
Peter R. Levison	*Whatman International Ltd, Springfield Mill, Maidstone, Kent, UK*
Andrew Lyddiatt	*Biochemical Recovery Group, BBSRC Centre for Biochemical Engineering, School of Chemical Engineering, University of Birmingham, Edgbaston, Birmingham, UK*

Duncan Mackay   *AEA Technology, Harwell, Oxfordshire, UK*
Eiji Miyata   *Speciality Chemicals Laboratory, Mitsubishi Kasei Corporation, 1000 Kamoshida-cho, Midori-ku, Yokohama 227, Japan*
John Noble   *AMEC Design and Management, Stratford-upon-Avon, Warwickshire, UK (now Foster Wheeler Energy, Station Road, Reading, Berkshire, UK)*
Paul J. Pickering   *University of Auckland, New Zealand*
Keith G. Robins   *SmithKline Beecham Pharmaceuticals, Brockham Park, Betchworth, Surrey, UK*
Peter R. Shelley   *SmithKline Beecham Pharmaceuticals, Brockham Park, Betchworth, Surrey, UK*
David Sherwood   *Celltech Biologics, 216 Bath Road, Slough, Berkshire, UK*
John W. Snowman   *Edwards Freeze Drying, Manor Royal, Crawley, West Sussex, UK*
Hiroaki Takayanagi   *Speciality Chemicals Laboratory, Mitsubishi Kasei Corporation, 1000 Kamoshida-cho, Midori-ku, Yokohama 227, Japan*
Alan R. Thomson   *Alan Thomson and Associates, 135 Amersham Road, Beaconsfield, Buckinghamshire, UK*
Michael S. Verrall   *SmithKline Beecham Pharmaceuticals, Brockham Park, Betchworth, Surrey, UK*
Denise M. Wallworth   *BAS Technicol Ltd, Adcroft Street, Stockport, Cheshire, UK*
Laurence R. Weatherley   *Department of Chemical Engineering, The Queen's University of Belfast, Stranmillis Road, Belfast, UK*

PREFACE

Natural products have been known and exploited by mankind for many millennia. They were formerly used with minimal purification but with the requirement for higher potency and predictability, selective concentration or isolation methods have been undertaken. Following the development of microbiology as a science and the subsequent antibiotic revolution there has been intense interest in the diverse natural products produced by micro-organisms. The majority of the research performed has had a pharmaceutical bias and the contributions in this book are similarly orientated. Many *ad hoc* methods for producing and studying microbial products have been used. With the passage of time procedures have been rationalized in terms of scientific principles, but theory has generally followed behind general practice. Many techniques and observations which are truly 'state of the art' may not appear in conventional scientific literature until a certain amount of supporting theory is available. Thus at any one time there exists an accumulation of wisdom among experienced practitioners which is difficult to access from the usual academic sources. Contributors to this volume have been encouraged to share their practical experience with the reader in order to further the cause of natural product isolation. The isolation of proteins and secondary microbial metabolites are often treated as separate subjects, as are laboratory-scale or process-scale operations. This book aims to unify natural product isolation by finding common themes which run through the various operations from crude feedstock to bulk purified product. Although much of the text describes the processing of microbial products, similar principles can be applied to plant products. The new biotechnology has made available in bulk many previously unobtainable or very scarce proteins and represents a major advance in biochemistry. Rapid advances are also being made in understanding the biosynthetic pathways of secondary metabolites and again genetic engineering can play a part in increasing the activity or specificity of a key biosynthetic enzyme. The science of downstream processing consists of a wide variety of techniques. It is now the task of the downstream processing scientist to make the correct choice and efficiently apply the current state of the art in each technique to match the success of the biotechnologist and fermentation scientist.

The array of potential separation and purification techniques for natural products and their implementation may seem rather bewildering for a newcomer. Even more experienced workers often prefer to use a limited range of familiar procedures. Certain chapters in this book discuss the potential of less common methods such as the use of supercritical fluids and liquid membranes. Other methods such as aqueous two-phase partition are not as widely used as they could be because there are few practical guidelines available.

The feedstocks from which natural products are prepared are complex, variable and poorly defined. The contributors to this volume are well aware of these problems and offer pragmatic methods for handling such streams. Despite the nature of the starting material it is necessary to eventually obtain a pure, defined substance of guaranteed purity and potency. Thus it is essential to have high-resolution purification procedures which are adequately documented. Guidelines are therefore presented for scaling-up, controlling, documenting and generally conforming to the principles of Good Manufacturing Practice (GMP).

Thus it is anticipated that the reader will use this book as a supplement to the academic texts and will find practical guidance and inspiration for all stages of natural product isolation, whether the operation be small or large scale, research or production oriented, or for a protein or secondary metabolite.

M.S.V.

1 GENERAL STRATEGY: THE NATURE OF THE PROBLEM

Michael S. Verrall

1.1 Introduction

What is a natural product? Since the earliest prehistoric era of man's existence he has used naturally occurring materials to advantage. The first conscious selection of a natural product was probably for use as a tool; later various plant products would have been selected on the basis of particular effects rather than purely for survival. In these cases, although selection has taken place, the whole product was used. We can imagine that the first deliberate fractionation process was the isolation of readily obtained metals such as copper and gold or possibly the recovery of salt from seawater. Thus 'natural products' exist as mixtures containing a small portion of valuable substance, the remainder being waste material.

Plants possessing medicinal properties have been known since the dawn of history and it is difficult to ascertain when it was first appreciated that their effects were due to small amounts of active substances. There are references to aqueous extracts, oils and distillates in the earliest written records from the Near East. A long period of trial and error has given mankind a large number of plant-derived pharmaceutical products. It has been known for centuries that pure inorganic substances could be isolated from appropriate minerals but it required the establishment of organic chemistry in the nineteenth century before defined chemical substances were isolated from plants, followed by fungi and bacteria.

Investigations then followed two lines: firstly the fractionation of extracts of plants known to have medicinal properties and secondly the *ad hoc* extraction of lower life forms such as fungi, leading to the isolation of some of their readily crystallizable metabolites such as mycophenolic acid from *Penicillium*

Downstream Processing of Natural Products. Edited by Michael S. Verrall
©1996 John Wiley & Sons Ltd

brevi-compactum [1]. It was only after the introduction and spectacular success of penicillin in the 1940s [2] that systematic large-scale screening of micro-organisms commenced. Since antibacterial activity is easy to detect in unpurified culture extracts screening work was initially focused on this type of agent. Attempts to detect other types of biological activity were generally frustrated by the crude (usually *in vivo*) test methods available. Present-day biochemical test methods have largely redressed the problem and there is now renewed interest in screening both plants and micro-organisms.

It is generally the case that medicinal products, whether produced by plants or micro-organisms, are present in very low concentrations and are species specific with respect to production. The compounds often have no obvious function in the organism's primary metabolism and have been called secondary metabolites. In the case of plants the compound may be found in only one tissue and with micro-organisms production is associated with particular growth conditions. Among the large number of secondary metabolites which have been isolated from plants or microbes the majority have been found to have some kind of biological activity such as antimicrobial, enzyme inhibition or specific binding to receptors of pharmacological significance. Although this is largely due to isolations being performed with the guidance of biological assays, there are many instances of chemically guided fractionation resulting in biologically active products [3,4].

In the early days of natural product research, work was limited by the prevailing technology and directed mainly towards isolating the readily obtainable compounds, followed by structure elucidation using classical degradative methods and a confirmatory formal synthesis. Evaluation of biological activity may have followed. Today the approach is to submit source materials to sophisticated, highly selective and sensitive screening procedures which can detect the presence of very low concentrations (nanomoles/litre) of compounds giving a specific response. It is then the task of the natural product chemist to isolate the compound and elucidate the structure. The latter operation is often possible on a few milligrams of compound in comparison with the multigram quantities required by the early workers. Although more data may now be obtained on small quantities of product, the concentration of product in the feedstock may now be very low as screening methods have become more sensitive. Thus it is often still necessary to process large volumes of starting material such as fermentation broth to obtain the required amounts of product. Nevertheless, the combination of sensitive screens and modern structure elucidation methods is resulting in the discovery of many new natural products each year.

Following the identification of a novel compound, larger quantities may be required for further evaluation. For plant products there is a greater emphasis on total synthesis of the compund. The collection or cultivation of large amounts of plant material is usually too time consuming for either an academic or industrial research project. Certain products have been prepared by plant cell culture but the techniques are difficult and not universal. In contrast, most micro-organisms are easy to grow in submerged culture, have a short doubling time and the process is readily scaled-up. Many commercial fermenters for secondary metabolites have working volumes in excess of $100\,m^3$ and $3\,m^3$ is quite a normal pilot plant scale. Thus the total amount produced may be readily increased, albeit as a dilute solution. Some increases in fermentation titre may be obtained by optimizing the fermentation conditions [5]. The complexity of the biochemical pathways involved in secondary metabolite production often means that expression of the compound cannot be readily increased by a single genetic manipulation. This is a notable difference between secondary metabolites and protein products. Thus the most rapid way to increase output of the former products for research purposes is to scale up the fermentation, giving the isolation chemist the task of recovering possibly only a gram of product from thousands of litres of fermentation broth. However, with the increasing knowledge of the gene clusters involved in secondary metabolite biosynthesis this may become less of a problem in the future. At present initial isolation stages will involve process-scale operations and involve chemical engineers while the final stages will be carried out at bench scale by laboratory chemists. In contrast, developments in microbial genetics have made the isolation and manipulation of protein coding genes a relatively simple matter and the productivity (titre) of a protein producing fermentation can be more readily increased by these means. Consequently, large-volume protein producing fermentations are less common.

Many natural product isolations are developed because of the potential interest of the compound to the pharmaceutical industry. Progression of a new drug substance to the market involves a large number of stages and the compilation of an extensive dossier on all aspects of the compound, particularly data relating to toxicological studies and clinical trials. Product prepared for clinical trials must be of the same guaranteed quality as that for final marketing and involves operation under current Good Manufacturing Practice (GMP) using validated equipment and processes as discussed in Chapter 21.

As mentioned above, improvements in molecular biology have identified many of the macromolecular agents (usually proteins) responsible for a number of biological responses and it is now possible in principle to produce these products by inserting the appropriate genes coding for their production into bacterial DNA. This has given rise to a new branch of biotechnology. In fact, the term 'biotechnology' is often used exclusively to describe the application of recombinant DNA technology whereas it should be used in the broader context of the harnessing of microbes for any useful purpose and this latter definition is used throughout this book.

Compared with microbial secondary metabolites, protein agents have a number of distinct differences which have given rise to parallel developments in isolation techniques. A few years ago some aspects of protein purification lagged behind those of small molecules, e.g. the application of fast high-resolution chromatography. However, recent developments (e.g. perfusion chromatography) have largely redressed the balance in this respect. On the other hand, protein chemists have had for many years the benefit of highly selective ('affinity') adsorbents. More importantly, it is now a simple matter, using genetic engineering, to attach 'tails' with specific properties to proteins to assist isolation (see Chapter 7). Comparable methods are currently difficult to apply to secondary metabolites, but progress in molecular recognition interactions may have an impact in the future. Despite the obvious differences between secondary metabolites and protein products this book aims to identify common features relating to their isolation from source material.

1.2 Strategy for Isolation

The strategy for isolation outlined assumes that no prior information is available on the compound. Where certain properties are known or where an analogue of a well-defined compound is to be isolated the investigations can be started at the appropriate stage. Guidance is provided to enable the researcher to increase the scale of each stage to the level necessary for his or her purpose.

1.2.1 ANALYTICAL METHODS

If the product to be isolated is a defined chemical entity, a reference compound could be available. However, there are many occasions when a novel product is isolated for investigative purposes including structure elucidation. In the latter case the isolation is usually guided by a suitable assay for biological activity. It is important to use a biochemical assay which is reliably quantitative and not unduly influenced by the presence of various solvents, buffer salts, etc., which will be present in samples from successive stages of isolation. At the earliest opportunity this assay should be used to establish correlations with physicochemical properties of the substance under investigation and thus provide a number of alternative detection methods. It is important to remember that the biological assay will detect only total activity; it may be due to a mixture of active components and the response measured may be the result of synergistic or antagonistic effects. The eventual analytical method chosen will depend on the chemistry of the compound under consideration. A direct spectrophotometric assay could be used for compounds with a strong and distinctive chromophore such as the macrocyclic polyene antibiotics (e.g. amphotericin B) and colorimetric methods are also useful, but in both cases there will be no discrimination from closely related compounds. Unless the target compound is known to be the only one of its class in a sample a chromatographic procedure will normally be necessary; automated HPLC (high-performance liquid chromatography) systems can reliably handle large numbers of samples.

In a fermentation broth the desired product may be an intracellular or extracellular product (or both). Some sample preparation may be required to ensure that the total amount of compound is being determined. A secondary metabolite fermentation could be treated with a water-miscible solvent such as propanone to leach putative products from the microbial cells. Methanol is sometimes used for this purpose but it is less suitable as it is more likely to react with certain compounds containing readily esterified or methanolysed groups. A protein producing fermentation will normally require a cell disruption stage (e.g. sonication, homogenization, lytic treatment) to release intracellular product.

Scale-up of isolation also involves optimization with respect to increasing the efficiency of each stage giving rise to increasing demands on the accuracy of the assay system.

1.2.2 STABILITY DATA

When isolating novel compounds it is worth while spending a little time acquiring stability data before embarking on a series of experiments to assess potential isolation methods. Results obtained at later stages are then less likely to be misinterpreted; in particular the parameters of temperature and pH should be investigated. Initial results will be obtained on a crude extract or fermentation broth and it is important to appreciate that in this environment the compound may be more or less stable than in the pure state. Enzymes present in crude extracts or broth may transform the product; the rate will almost certainly be pH and temperature dependent. Some compounds which are unstable when in the pure state may be stabilized in crude extracts by conjugation with other components. Mixtures of active products may give rise to synergistic or antagonistic effects. Thus although these results usually provide a useful guide for subsequent work, the researcher must always be prepared to re-evaluate stability data on purer preparations.

1.2.3 ISOLATION METHODS

With temperature and pH stability data available it is then possible to determine other properties relevant to the initial isolation stages. With respect to applicability of isolation methods secondary metabolites can be broadly classified as either neutral or ionizable compounds. Protein products can be broadly classified in terms of hydrophobicity or the presence of metal ions. The use of non-ionic (hydrophobic) and ion exchange adsorbents can provide useful information on the nature of the active compound. For secondary metabolites this can be supplemented by partition studies using a variety of water-immiscible organic solvents. Conventional ion exchange resins can be used for small molecules. For macromolecules, ion exchangers based on wide pore polymers in a hydrophilic matrix are available. In many cases the required studies may conveniently be performed using prepacked 'solid phase extraction' cartridges.

1.2.4 SCALE OF OPERATION

Following the above, there will be an indication of the type of compound that is to be extracted. The purpose of the isolation will affect the scale of operations that are required. The unit operations described in this book are applicable to process-scale use, but many will also be found useful on the laboratory scale. A greater variety of techniques is available for laboratory-scale operations where cost or recyclability of reagents is less relevant and techniques such as prolonged batch centrifugation are possible. When developing an isolation for process-scale operation the options can be more limited. However, on scale-up certain other techniques become available and may be more convenient. These include continuous flow liquid–liquid or liquid–liquid–solid centrifugal separation, multistage countercurrent extraction [6] and tangential flow membrane separations. As with the scale-up of fermentation processes, it is easier to control operating parameters on the process scale. Although instrumentation and control of downstream processes is less well developed than fermentation, some guidelines are presented in Chapter 19. Unit operations should generally be designed to be robust, i.e. not greatly influenced by small changes in feed solution or operating parameters.

1.2.5 PURIFICATION PROCEDURES

Downstream processing can conveniently be divided into three phases: firstly product release (if required), secondly product capture or concentration and finally product purification. In many cases where small process-scale operations are required for the first two stages, the final stage may conveniently be performed using laboratory facilities.

1.2.5.1 Product Release

Natural products are normally biosynthesized within living or viable cells but many are excreted into the surrounding medium. The majority of highly polar and/or water-soluble secondary metabolites and many enzymes may be isolated from culture filtrate and it is possible to pass directly to the product capture stage. For most small-scale work or in cases where a limited number of batches is to be processed it is preferable to remove cells of the producing organism before product capture. More detailed process studies will often indicate that it may be more economical to capture product from whole broth using fluidized bed adsorption or whole broth solvent extraction (or possibly aqueous two-phase partition in the case of proteins).

Lipophilic secondary metabolites and some proteins are found as intracellular entities and require release before purification. Most secondary metabolites can be leached from the cells using water-miscible organic solvents, but it may be preferable to use whole broth extraction with a water-immiscible solvent as it has a number of significant advantages. Firstly, it is easier to separate and recycle a water-immiscible solvent compared to a miscible solvent. Secondly, product in culture filtrate and cells can be captured in the one operation and finally the stage of broth clarification is eliminated. Macromolecular products are less readily leached from cells and cell disruption or at least cell permeabilization is usually required. Solubilization of inclusion bodies and subsequent protein refolding may also be required. Practical methods for dealing with the resultant crude mixture are discussed in Chapter 4.

1.2.5.2 Product Capture

This stage is designed to provide an extract containing the desired product and closely related compounds but free of the majority of contaminants in the source material. Methods should be designed to be as selective as possible consistent with the economics of operation. The most selective methods are currently available for proteins where relatively inexpensive adsorbents based on specific chemical interactions are available, e.g. transition metal ion containing materials to chelate with histidine residues in proteins or dye-ligand materials designed for generic π-bond interaction (see Chapter 13). Affinity adsorbents based on ligands of biological origin are normally reserved for use on partially purified streams. Less selective but nevertheless effective adsorbents include ion exchangers and hydrophobic interaction materials and are applicable for compounds of any molecular size. Higher efficiency adsorption of small molecules is obtained by the use of more highly crosslinked resins with greater surface area (see Chapter 11). For optimal performance it may be necessary to use a highly clarified feed stream such as an ultrafilter permeate (see Chapter 9). Many extracellular secondary metabolites may be conveniently isolated from fermentation broth by solvent extraction (liquid–liquid partition) using a water-immiscible organic solvent, and the process is in commercial operation for products such as penicillin. Although well established, the process is not fully understood. Fundamental studies of the partitioning process have been undertaken over a number of years and recently greater emphasis has been placed on whole broth extraction systems. The complex interactions are discussed in Chapter 5. Conventional solvent extraction incorporates subsequent precipitation of product from the organic phase or extraction back into a second aqueous strip phase. It is convenient for large-scale operation because it is readily operated on a continuous basis and it also effects a rapid transfer of product from an aqueous environment which may contain degradative enzymes. Proteins are insoluble in, and are often denatured by, organic solvents, but two-phase aqueous partition can be an appropriate partitioning technique.

Liquid–liquid partition, both with organic solvents or by means of aqueous two-phase systems, can be enhanced by the incorporation of a suitable complexing agent in the system to form an ion pair, chelate or other complex with the product, as described in Chapter 6. Separation processes in general, but more particularly those for proteins, may also be enhanced by modifying the product to interact with a particular reagent, as described in Chapter 7. The technique of displacement chromatography can give surprisingly high purification factors (see Chapter 10).

1.2.5.3 Purification

Having prepared a crude extract high-resolution purification will be required as secondary metabolites in particular are normally produced as mixtures of closely related compounds. Established fermentation processes for commercial products will have been developed to minimize coproduction of undesired metabolites and contain a higher proportion of product to impurity, whereas new products require careful purification. The requirements for purity of pharmaceutical compounds is becoming increasingly stringent; it is essential to ascertain that the product is a single defined substance. There may also be an interest in isolating minor components for the establishment of structure–biological activity relationships.

High-resolution purification is performed almost exclusively using chromatographic methods. High-resolution HPLC has been a feature of analytical laboratories for the past 20 years or so but it has taken longer to become accepted as a preparative technique. However, commercial equipment and suitable column packings are now available and some practical aspects are discussed in Chapter 15. There is also some interest in using supercritical fluid chromatography on a preparative scale (see Chapter 16).

Many compounds can be readily crystallized from an appropriate solution. Where this is not possible (usually with highly hydrophilic compounds), a solid product may be obtained by lyophilization (freeze drying). The variables involved in this process are often poorly understood. Chapter 18 unveils much of the mystique involved in this unit operation.

It is not sufficient to prepare a compound that only satisfies the criteria set by the manufacturer; it will also be subject to scrutiny by the regulatory authorities. These bodies have the view that final product analysis alone is insufficient to guarantee quality and consistency, particularly with some macromolecular products where the product purity is partially defined by the production process itself. For this reason the concept of Good Manufacturing Practice (GMP) has evolved. Its present application to biological products and general considerations relating to scale-up are discussed in Chapters 20 and 21. Consistent results for unit operations is aided by the application of process control, the fundamental requirements for which are outlined in Chapter 19.

From the viewpoint of GMP, process control philosophies should be as simple as possible to avoid extensive work in validating their operation.

1.3 References

1. C. L. Alsberg and O. F. Black (1913) *USDA Bur. Plant. Ind. Bull.*, **270**.
2. G. L. Hobby (1985) *Penicillin, Meeting the Challenge*, Yale University Press, New Haven and London.
3. H. P. Fiedler (1993) *Natural Product Letters*, **2**, 119–28.
4. A. Nakagawa (1992) in S. Omura (ed.) *The Search for Bioactive Compounds from Micro-organisms*, Springer-Verlag, Berlin, pp. 263–80.
5. A. T. Bull, T. A. Huck and M. E. Bushell (1990) in R. K. Poole, M. J. Bazin and C. W. Keevil (eds.), *Microbial Growth Dynamics*, IRL Press, Oxford, pp. 145–68.
6. M. S. Verrall (1992) in J. Thornton (ed.), *Science and Practice of Liquid–Liquid Extraction*, Vol. 2, Clarendon Press, Oxford, pp. 194–308.

2 BROTH CONDITIONING AND CLARIFICATION

Duncan Mackay

2.1 Introduction

The recovery of products from fermentation broths is often the major cost in a biotechnological manufacturing process. The first step is broth clarification, i.e. the separation of the biomass from the fermentation liquor. Since the overall yield of product is a function of the operating performance of the various unit operations it is important that the clarification step is performed efficiently. Characteristics of the biomass are given in Table 2.1.

Table 2.1 Characteristics of biomass

Biomass	Shape	Size (μm)	Density (kg/m^3)	Shear resistance
Bacteria	Rods	0.5–3	1050–1080	Good
	Cocci	5–10	1050–1090	Good
Yeast	Spherical	5–10	1050–1090	Good
Fungi	Filamentous	20–>100	1050–1090	Fair
Plant	Irregular	20–100	1050–1090	Poor
Mammalian	Irregular	20–50		Poor
Debris	Varied	<1	1010–1200	Poor

There are various solid–liquid separation techniques that can be used to clarify fermentation broth. Bacterial cells and cell debris are the most difficult biomass to recover because of the small size, low density and compressibility. Mammalian cells, although large, are very shear sensitive and so require careful handling during recovery.

This chapter will describe the separation techniques appropriate to broth clarification, methods for conditioning broth to improve the clarification and design procedures for optimizing the separation process.

Downstream Processing of Natural Products. Edited by Michael S. Verrall
©1996 John Wiley & Sons Ltd

2.2 Broth Clarification Unit Operations

The separation of biomass from the fermentation liquor can be achieved by:

(a) dead end filtration,
(b) tangential flow filtration,
(c) centrifugal sedimentation.

There are other solid–liquid separation techniques (e.g. froth flotation, gravity setting, straining) but these are either inefficient or untested for biomass recovery.

The correct choice of separation technique will depend on factors such as the process limitations, requirements of the separation and the broth properties. This selection process is discussed in Section 2.4.

2.2.1 DEAD END FILTRATION

The fermentation broth is passed through a filter medium which prevents the passage of solids but permits liquid and solubles to pass. The driving force can be vacuum, gas pressure or pump pressure.

The specific types of filter equipment include:

(a) rotary drum vacuum filter,
(b) horizontal pressure leaf,
(c) vertical pressure leaf,
(d) filter press — plate and frame, recessed chamber,
(e) variable-volume filter press.

2.2.1.1 Rotary Drum Vacuum Filter

General features are:

(a) It can accept a wide range of feed biomass concentrations (maximum throughput being achieved at high values).
(b) Performance can be adversely affected by operation at high altitudes or with volatile liquids.
(c) Use of a precoat to improve clarification is simple.
(d) Filter cake washing is easy.
(e) Containment is relatively easy.

Rotary drum filters pick up biomass by partial immersion in a trough as the drum rotates. There are two main types of drum — single compartment and multicompartment.

Single-compartment drums are simple in construction, suitable for fast filtering broths and can be run at high speeds to form thin cakes. Simple washing can be accomplished and, with special types, the washings and mother liquor can be separated.

Multicompartment drums are more complicated mechanically, with valves and many internal pipes which can cause loss of driving force at large flowrates. Simple washing can be accomplished. Mother and wash liquors can be separated. These drums are well suited to the use of precoat.

Drum filters may be totally enclosed to conserve heat to contain volatile vapours or to ensure containment.

The method for discharging the biomass filter cake is an important feature. The main methods are:

(a) Knife discharge — most commonly used. When used with assisted discharge by air blow the liquor can re-enter the cake.
(b) Precision knife discharge — when precoat is used.
(c) Belt discharge — assists sticky cake discharge and where blinding of media tends to occur the belt can be washed when it is off the filter.
(d) Roller discharge — for sticky cakes. The cake must not crumble, but must stick to the roller.
(e) String discharge — for fibrous cakes. String must not cut through the cake; the cake must not crumble and should be fairly thick.
(f) Coil discharge — filter medium is coils contained in grooves on the main drum. Coils are removed as in a belt discharge filter and open up, thereby discharging cake. Useful for slimy, difficult cakes.

The various discharge methods are illustred in Figure 2.1.

2.2.1.2 Horizontal and Vertical Leaf Pressure Filters

General features are:

(a) Categorized on the geometry of the filtering surfaces (Figures 2.2 to 2.4).
(b) Horizontal elements are leaves or discs.
(c) Vertical elements are leaves or candles.
(d) Containing vessels are usually cylindrical with either horizontal or vertical axes.
(e) Vessels can be jacketed to allow control of temperature.
(f) Precoat/body feed can be used.
(g) Cake washing is possible.
(h) Vertical vessels have low floor space requirements but need adequate height.
(i) Horizontal filter elements are best for washing applications.
(j) Vertical filter elements provide greater filtering areas, particularly when mounted in a horizontal vessel.
(k) Biomass cake may discharge prematurely from vertical elements.

The biomass filter cake is discharged from the filter vessel by either

(a) opening the vessel and extracting the filter elements — unsuitable if containment is important and is rather labour intensive — or

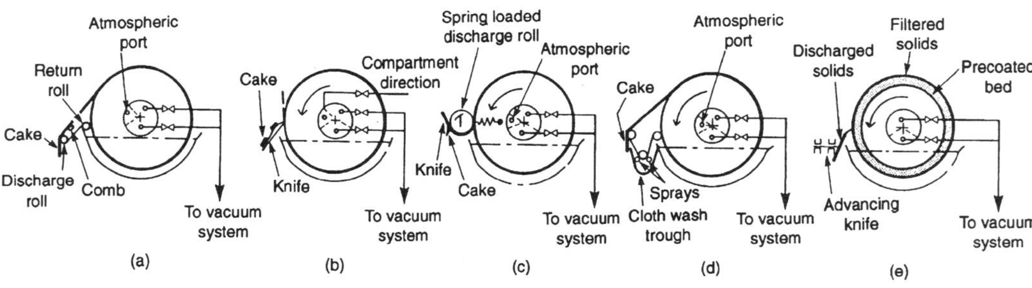

Figure 2.1 Diagrammatic arrangements of cake discharge systems for rotary drum vacuum filters. (a) String discharge; (b) knife discharge with blow-back; (c) roll discharge; (d) belt discharge and belt wash; (e) precision knife discharge for precoat

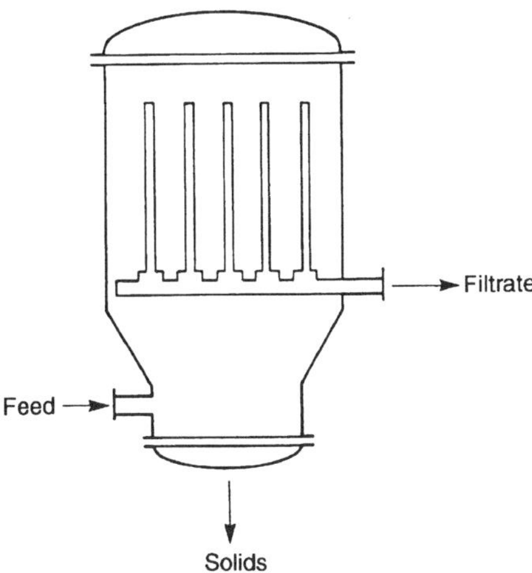

Figure 2.2 Vertical leaf pressure filter

(b) discharging the elements within the vessel and removing the biomass as a cake or slurry. Removal as a cake can be accomplished using vibration (vertical filter elements), centrifugal force (horizontal filter elements) or compressed air (vertical elements). Wet discharge can be carried out by spray or reverse flow of liquid.

2.2.1.3 Filter Presses

The fermentation broth is pumped into a filter press and the biomass filter cake is formed in a series of narrow chambers held closed by mechanical or hydraulic pressure. Cake discharge is achieved by opening the chambers and allowing the solids to fall out.

There are two main types of filter press:

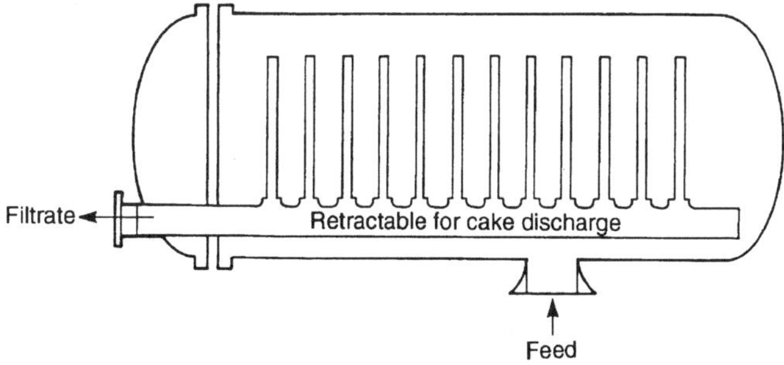

Figure 2.3 Vertical leaf/horizontal shell pressure filter

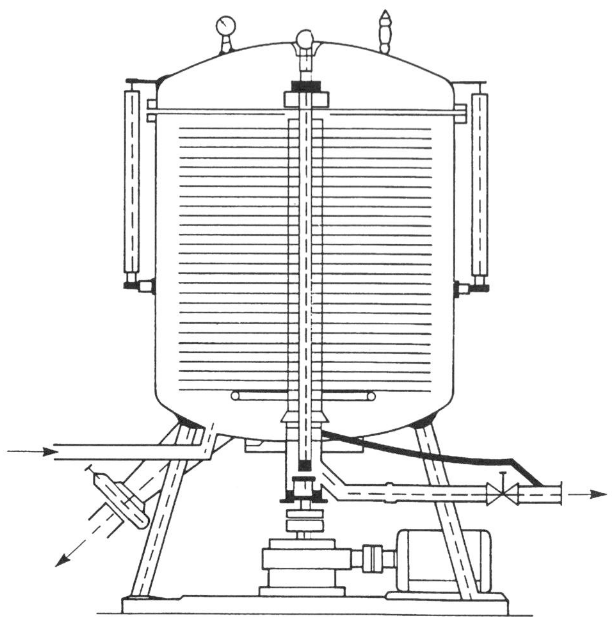

Figure 2.4 Centrifugal discharge horizontal plate pressure filter

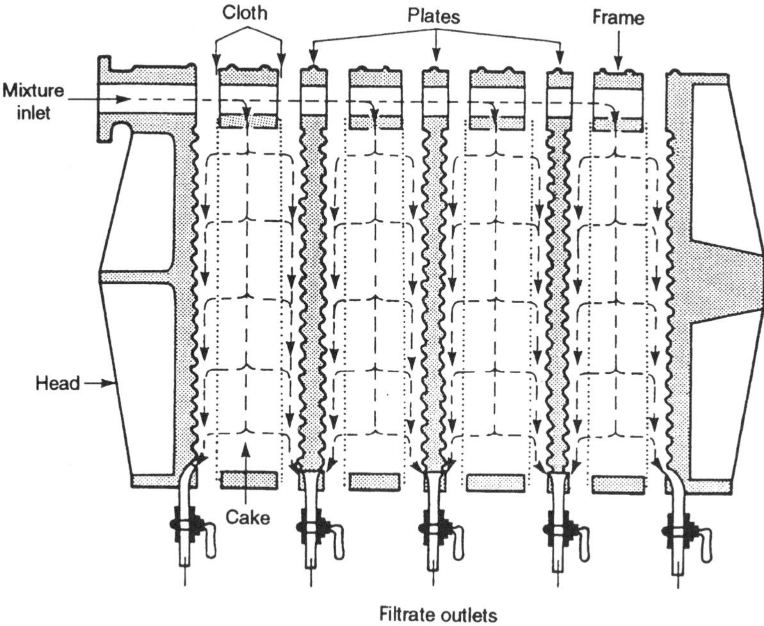

Figure 2.5 Plate and frame press (plain plates)

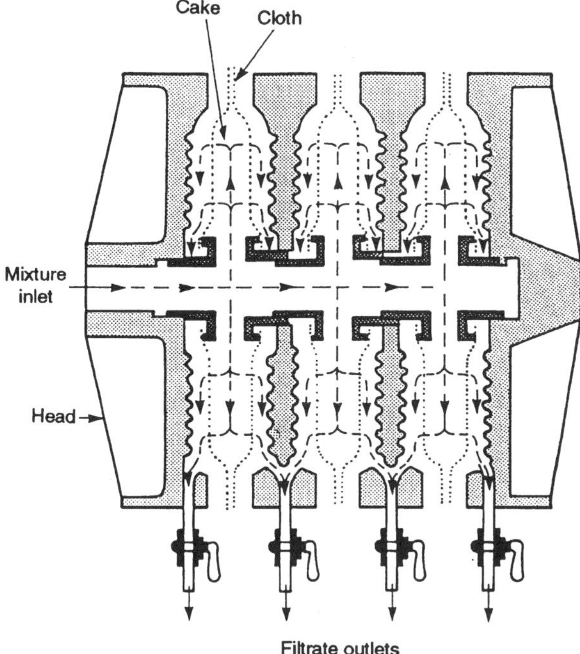

Figure 2.6   Recessed plate press

(a) Plate and frame. The filter medium is placed on the plates and the frames act as the chambers in which the cake builds up (see Figure 2.5).
(b) Recessed plate. The recess in the plate acts as the chamber (see Figure 2.6).

Filter presses are batch operated and require to be fed by pumps capable of maintaining the filtration pressure/flowrate.

Table 2.2   Relative pros and cons of filter press types

Plate and frame	Recessed plate
Pros	
1. Maximum filter medium life and low replacement costs	1. Fewer parts
2. Paper media can be used	2. Smaller number of joints to seal
3. Uniform filter cake thickness	3. Larger feed inlet minimizes blocking
Cons	
1. Capital cost higher	1. Filter medium more strained, leading to shorter life
2. More prone to leakage	2. Uneven filter cake formation
3. Feed inlet can block	3. Limited cake thickness

The advantages and disadvantages of both types of filter press are given in Table 2.2.

The fermentation broth can be fed, and the clarified broth discharged, through several alternative arrangements. The plates and frames can be manufactured with corner ports. Points to consider in the selection of inlet and outlet ports include:

(a) The bottom feed inlet encourages even cake formation.
(b) The bottom filtrate outlet permits good drainage and maximizes the recovery of liquid.
(c) An efficient biomass cake washing can be achieved through appropriate blanking off of certain ports.

2.2.1.4 Variable-volume Filter Press

The biomass filter cake is formed by pumping pressure and is then compressed by a diaphragm to expel more liquid. The compressibility of biomass is an aid to effective deliquoring.

There are three main types of variable-volume filters:

(a) cylindrical mandrel,
(b) rectangular filter plates,
(c) circular filter plates.

Compression pressures up to 16 bar are achievable.

Some filters can be operated in a semi-continuous manner by using an endless length of filter medium which is progressed after each filtration cycle.

Variable-volume filter presses operate on a relatively fast cycle time, thereby giving a high output per unit filtration area. For bacterial broths filtration rates will fall rapidly and so these filters could be more appropriate than vacuum filters, pressure leaf filters or filter presses.

2.2.2 TANGENTIAL FLOW FILTRATION

Fermentation broth can be pumped across the surface of a filter medium during filtration, thereby minimizing the buildup of a biomass filter cake and so maintain a reasonable clarified broth throughput.

Tangential flow (or crossflow) filtration equipment consists of two main parts: the membrane and the housing. The membrane is the filter medium which will generally have a microporous structure with a choice of pore sizes in the range $0.2–5\,\mu$m. The membrane material is either polymeric or inorganic. The most common polymers are cellulose nitrate/acetate and polysulfone. Inorganic membranes tend to be ceramic based, e.g. alumina with a zirconia coating.

The membrane housings are modular in design and are available in four configurations:

(a) tubular,
(b) flat sheet,
(c) hollow fibre,
(d) spiral wound.

Characteristics of these module configurations are given in Table 2.3. All can be used for fermentation broth clarification, although the susceptibility to blockage and the low working pressure limitation of the hollow fibre and spiral wound modules can limit their use.

The separation performance of tangential flow filtration is a function of:

(a) the membrane — pore size, fouling properties,
(b) the fermentation broth — rheology, solute composition,
(c) the hydrodynamics — tangential velocity, laminar/turbulent flow,
(d) the pressure drop.

Table 2.3 Characteristics of crossflow module configurations

Tubular	1. Large channels — less prone to blockage
	2. Crossflow velocity 2–6 m/s
	3. Reynolds number $> 10^4$
	4. Easy to clean
	5. Membrane replacement relatively simple
	6. Low surface area to volume ratio — high space requirement
	7. High liquid hold-up
	8. High energy consumption
Hollow fibre	1. Narrow channels — blockage a problem
	2. Crossflow velocity 0.5–2.5 m/s
	3. Reynolds number 500–3000
	4. Can be back-flushed
	5. High wall shear rates 4000–14 000 s^{-1}
	6. High surface area to volume ratio
	7. Low liquid hold-up
	8. Low energy consumption
	9. Maximum operating pressure limited (approximately 2.4 bar)
	10. Membrane replacement cost high
Sprial wound	1. Narrow channels — blockage a problem
	2. Low capital cost
	3. Surface area to volume ratio high
	4. Reasonably economic
	5. Design susceptible to collapse
	6. Operates best with clean feed, i.e. little or no suspended matter
Flat sheet	1. Crossflow velocity ~ 2 m/s
	2. Membrane replacement relatively simple
	3. Energy consumption lies roughly between spiral wound and tubular designs
	4. Visual observation of permeate from each membrane pair is possible

ience of the closely related technique of tangential flow
s presented in Chapter 9.

RIFUGAL SEDIMENTATION

Enhanced gravity settling is an effective method for broth clarification. The recovery of biomass in a sedimenting centrifuge depends on:

(a) the biomass settling rate,
(b) the biomass residence time within the centrifugal field,
(c) the biomass settling distance.

The settling rate can be increased by increasing the biomass particle size, increasing the density difference between the biomass and fermentation liquor or by decreasing the viscosity of the liquor. The residence time can be increased by reducing the flowrate through the centrifuge. This improves the biomass recovery efficiency. The settling distance can be reduced, thereby enhancing biomass recovery, e.g. in disc stack centrifuges.

The types of sedimenting centrifuge appropriate to fermentation broth clarification are:

(a) disc stack,
(b) scroll decanter.

2.2.3.1 Disc Stack

Disc stack centrifuges are the most commonly used broth clarification equipment. The design of these centrifuges incorporates numerous discs (typically between 30 and 200) which divide the sedimenting bowl into separate settling zones. The different varieties of disc stack centrifuge can be classified according to the method of sedimented solids discharge:

(a) solid bowl,
(b) opening bowl,
(c) nozzle,
(d) solids ejecting (nozzle valve),
(e) solids ejecting (slot).

Table 2.4 gives some characteristics of these machines.
General features are:

(a) Angle of discs 35–50°.
(b) Settling distance 0.5–2 mm.
(c) Solids settle on the lower surfaces of the discs and migrate towards the bowl periphery.

Table 2.4 Characteristics of disc stack centrifuges

Type	Discharge of biomass	Centrifugal force (G)	Feed biomass content (% v/v)	Discharged biomass
1. Solid bowl	Batch	4000–10 000	< 1	Cake
2. Opening bowl	Batch	5000–8000	1–10	Paste
3. Nozzle	Continuous	4000–9000	1–30	Slurry
4. Nozzle valve	Batch	15 000	< 10	Paste
5. Slot	Batch	4000–15 000	0.2–20	Paste

(d) Solid bowl — limited to broths with low biomass content. A removable solids collection basket can be used to reduce the time required for discharge.

(e) Nozzle valve. The discharge valves can be opened for brief periods, less than 0.5 s, to give control of the biomass moisture content.

2.2.3.2 Scroll Decanter

Scroll decanter centrifuges consist of a solid bowl, tapered at one end, and a close-fitting helical screw which rotates at a slight differential speed to the bowl. The screw is thereby able to continuously discharge sedimented biomass from the narrow end of the bowl.

General features are:

(a) The settling distance is varied by altering the height of the pool of clarified liquid.

(b) G-force of 4000–10 000.

(c) Biomass content of feed of 5–80% v/v.

(d) Flow properties of sedimented biomass can cause high torques to develop on the screw.

(e) Clarity of the fermentation broth is not usually as good as with disc stack machines.

This type of machine can be adapted for solvent extraction of whole broth (see Chapter 5).

2.3 Broth Conditioning

The clarification of a fermentation broth can be optimized by:

(a) the correct choice of solid–liquid separation equipment,

(b) adjusting the operating regime, e.g. the driving force (vacuum, pressure, G-force), separation cycle time,

(c) modifying the properties of the broth prior to clarification.

The correct choice of equipment and adjustment of the operating conditions are discussed in Section 2.4.

Conditioning or pretreatment of the broth can be carried out to increase the filtration or sedimentation rates. Appropriate conditioning can improve the biomass recovery efficiency and the deliquoring of the thickened biomass. Such improvements would largely be due to changes in the biomass particle size, the fermentation liquor viscosity and the interactions between biomass particles.

The filtration of fermentation broth is improved if:

(a) The biomass filter cake permeability is as high as possible.
(b) The deliquoring rate of the biomass cake is high, i.e. low liquor viscosity, high permeability.

The centrifugal sedimentation of broth is improved if:

(a) The biomass settling rate is increased.
(b) The broth liquor viscosity is reduced.

Broth conditioning can be achieved in several ways:

(a) temperature variation,
(b) filter aid addition,
(c) coagulation/flocculation,
(d) modification of the fermentation conditions.

Table 2.5 gives some examples.

2.3.1 TEMPERATURE VARIATION

Generally an increase in temperature will lower the viscosity of a liquid and will raise the collision frequency and collision energy of suspended particles. These collisions between particles can lead to aggregation, but the effect of

Table 2.5 Examples of broth conditioning

Broth	Product	Conditioning	Recovery technique
1. *Penicillium chrysogenum*	Penicillin G	Flocculation	Rotary drum
2. *Bacillus subtilis*	Protease	Flocculation	Filter press
3. *Streptomyces griseus*	Mycelial protein	Heating	
4. *Cephalosporium acremonium*	Cephalosporin	Flocculation	Rotary drum
5. *Streptomyces* spp.	Glucose isomerase	Flocculation	Rotary drum
6. *Streptomyces*	Vitamin B12	Flocculation	Rotary drum
7. *Methylophilus methylotrophus*	Single-cell protein	Coagulation	Centrifuge

temperature on the interparticle interaction energy is often unknown and so an increase in coagulation/flocculation may not occur.

A reduction in the viscosity and an increase in the effective biomass particle size are not necessary consequences of heating a fermentation broth.

Thermal instability of solutes may lead to precipitate formation which can either improve or worsen clarification efficiency.

2.3.2 FILTER AID ADDITION

Filtration performance can be improved through the addition of a filter aid to the fermentation broth and/or by precoating the filter medium with a filter aid. A filter aid is a particulate material which forms high-porosity, low-compressibility filter cakes. The two main types of filter aid are:

(a) diatomite — the skeletal remains of aquatic plants,
(b) perlite — volcanic rock processed to give an expanded structure.

Both types are available in a range of particle sizes (typically 3–25 μm) and have a particle density of 2.3 g/cm^3. The permeability of a filter aid cake is in the range of 10^{-13}–10^{-11} m^2. The use of a larger sized filter aid will give greater flowrates but poorer clarity.

As a body feed the filter aid is mixed with the fermentation broth prior to filtration at a concentration of typically 1–5% w/w. It is recommended that the weight of filter aid added be at least equal to the weight of biomass present. However, more filter aid would be necessary if the biomass particle size is significantly lower than that of the aid.

Filter aids, used as a body feed and/or precoat on the filter, offer the advantage of increasing filtration rates and filtrate clarity. However, disadvantages include the possible loss of broth solutes by adsorption to the filter aid. In general, a filter aid should be considered for use in the filtration of whole cells for the recovery of extracellular products or in the filtration of cellular debris. Filter aids are of limited use in improving centrifugal sedimentation performance. Body feed may enhance the discharge properties of the sediment, but their abrasive properties are undesirable.

2.3.3 COAGULATION AND FLOCCULATION

The effective settling size or filtering size of the biomass can be increased by encouraging the biomass to aggregate. The methods of achieving this size increase are:

(a) pH adjustment,
(b) electrolyte addition,
(c) polymer addition.

Aggregation as the result of pH adjustment or electrolyte addition is coagulation. Aggregation as the result of polymer addition is flocculation.

2.3.3.1 pH-Induced Coagulation

The surface chemistry of cells or cell debris is such that the surface charge is a function of the hydrogen ion concentration. A change in pH will therefore alter the electrostatic energy between interacting cells. A reduction in broth pH will reduce the negative charge — a characteristic of the surfaces under normal fermentation conditions — and hence the strength of repulsion between approaching biomass. There is likely to be a pH below which coagulation of the biomass takes place. Optimum coagulation will occur within a relatively narrow pH range of the point of zero surface charge. The latter can be estimated by measuring the zeta potential as a function of pH using microelectrophoresis. The majority of biomass will possess a point of zero charge within the pH range 4.5–6.5.

Because of the buffering properties of many fermentation broths the quantity of acid required to reach the pH range within which coagulation is possible could be significant.

2.3.3.2 Electrolyte-induced Coagulation

Fermentation broths in which the biomass is prevented from aggregating by electrostatic repulsion can be coagulated through the addition of electrolyte.

An increase in the ionic strength reduces the thickness of the electrical double layer and hence the strength of repulsion. The higher the valency of the ion of opposite charge sign to the biomass (usually the cation) the greater the effect. Also the larger the charge per unit ion volume the greater the reduction in the repulsive force.

Typical electrolytes include the nitrates and chlorides of sodium, calcium, magnesium, iron(II), iron(III) and aluminium. However, the use of chlorides is generally undesirable in stainless steel plant.

2.3.3.3 Polymer-induced Flocculation

Polymers can adsorb to the surface of biomass to create the conditions for flocculation. Polymeric flocculants are either natural or synthetic in origin.

Natural flocculants include:

(a) starches,
(b) galactomannans, e.g. guar gum,
(c) celluloses, e.g. carboxymethylcellulose,
(d) alginates,
(e) chitins, e.g. chitosan.

Such flocculants tend to have the following disadvantages:

(a) high doses are often required,
(b) variable quality,
(c) unstable solutions, e.g. hydrolytic reactions,

(d) bacterial degradation (but this can be advantageous for the treatment of
effluent from the process).

Synthetic flocculants are largely based on acrylamide and acrylate.
Classification of the various types is made on the basis of the charge—
nonionic, anionic or cationic. Nonionics consist chiefly of polyols, polyethers
and polyamides. Anionics derive their charge from the dissociation of acidic
groups, e.g. carboxyl, sulfonate or phosphonate. Cationics tend to be co-
polymers of acrylamide and a positively charged monomer based on amino and
imino groups.

The important properties of a polymeric flocculant are:

(a) the molecular weight (typically in the range 50 000 to several million),
(b) the charge density,
(c) the structure and configuration.

A suitable flocculant can usually be selected from the wide range of
commercially available polymers.

Polymeric flocculants can be supplied in a variety of physical forms:

(a) granules,
(b) beads,
(c) spray dried,
(d) concentrated aqueous liquid,
(e) emulsion (water in oil and oil in water).

The choice of the physical form can affect the ease and efficiency of uniform
dispersion within the fermentation broth. The liquid forms are easier to
disperse but are generally only available for the lower molecular weight
polymers. The solid forms need to be dissolved prior to incorporation in the
broth. Dissolution may be very slow.

For practical purposes polymeric flocculants act by altering the electrostatic
repulsive interaction between, or by bridging between, biomass particles. The
lower molecular weight polyelectrolytes (say less than 500 000) tend to adsorb
and reduce the electrostatic repulsion between particles. Higher molecular
weight polymers, ionic and nonionic, tend to adsorb, leaving loops and tails
extending into the solution which can adsorb on the surface of another particle.

Flocculation performance is a function of many factors, including:

(a) Flocculant molecular weight. The higher the molecular weight the better
the bridging.
(b) Flocculant charge density. The higher the charge density the more
extended the molecular conformation. Bridging and/or charge neutraliza-
tion may be improved.
(c) Flocculant dose. Overdosing can lead to stabilization of the biomass
particles.

(d) Biomass concentration. The higher the particle concentration the higher
    the collision frequency and the higher the rate of flocculation, other things
    being the same.
(e) Hydrodynamics. The presence of shear forces (e.g. through mixing) can
    improve the efficiency of biomass flocculation, but there will be an optimum.
(f) Broth liquor. The pH and ionic strength affect the conformation of the
    flocculant molecules and hence the flocculation process. In practice the
    adverse physiological effects on biomass as the result of pH extremes limits
    the useful range of pH adjustment. The relative changes in the biomass
    surface charge, flocculant charge density and soluble polyelectrolyte charge
    will influence the extent of flocculant adsorption and the resulting
    flocculation of the biomass.

2.3.3.4 Summary

Broth clarification methods can be made more efficient by coagulating or
flocculating the biomass. The increase in the filterability or settling rate
improves the separation performance.

Simple pH adjustment or salt addition and/or polymer addition can be used
to improve the biomass recovery rate.

2.3.4 MODIFICATION OF THE FERMENTATION

The properties of the biomass and broth liquor have a strong influence on the
separation characteristics, in particular:

(a) biomass particle size,
(b) biomass particle density,
(c) biomass surface properties (charge density, hydrophobicity, roughness),
(d) broth liquor viscosity.

The fermentation process exerts an important control on these properties. The
following variables can then be considered to influence the ultimate separation
performance:

(a) Fermenter operating conditions

- Batch, continuous or fed batch
- Mixing/agitation regime
- Mass transfer rate
- Use of antifoam agents
- Biomass growth phase when harvested

(b) Broth liquor characteristics

- Chemical composition
- Viscosity

2.3.4.1 Fermenter Operating Conditions

The fed batch mode of operation can be used instead of the batch mode to reduce the broth viscosity. Continuous fermentation minimizes variability in the properties of the biomass and broth liquor.

The shear to which the biomass and polymeric solutes are subjected can affect the surface properties of the former and the extent of degradation of the latter.

The rate of mass transfer of oxygen during aerobic fermentations often exerts the greatest influence on the growth of biomass. Consequently the very properties that affect the clarification process can be altered by changes in the oxygen mass transfer rate.

Antifoam agents can cause difficulties in the clarification process by virtue of their surface activity.

2.3.4.2 Broth Liquor

The physical and chemical characteristics of the broth liquor affect the properties of the biomass and the liquor viscosity. The nutrient composition and the pH are the main characteristics. These factors influence the growth of the biomass (and hence size and density) and the surface properties. The effect of the growth conditions on biomass particle size has been reviewed by Ratwatte and Wase [1].

The presence of divalent cations, particularly calcium and magnesium, can increase the chance of biomass aggregation. Other solutes can prevent aggregation, e.g. potassium, sugars.

Autoflocculation of the biomass can occur as a response to substrate limitation or selection of a flocculating strain of micro-organism.

2.4 Design Guidelines

In order to achieve good performance in broth clarification it is useful to follow a design procedure. Broadly speaking this should include:

(a) selection of the separation method,
(b) testwork to measure separation performance and to assess the need for broth conditioning,
(c) testwork to measure separation improvement as the result of broth conditioning,
(d) scale-up of the clarification process.

This design process is an important part of setting up a broth clarification step. A brief outline can only be given here.

2.4.1 SELECTION OF SEPARATION TECHNIQUE

Various factors need to be considered:

(a) Process limitations

- Integration with other unit operations, e.g. batch or continuous operation
- Containment requirements
- Scale of operation, e.g. small volumes would not justify using a centrifuge because of the significant hold-up volume and significant throughput of the smallest continuous flow machines available

(b) Fermentation broth properties

- Viscosity
- Biomass concentration
- Biomass filterability
- Biomass sedimentation rate
- Biomass dewatering rate
- Degree of variability in the properties

(c) Requirements of the separation

- Maximum permitted level of biomass particles in the clarified broth
- Range of acceptable moisture content of biomass sediment or cake
- Efficiency of washing stage, if needed
- Required equipment reliability
- Flexibility to vary the operating conditions in order to accommodate changes in the fermentation broth properties

(d) Economics

- Capital cost of equipment
- Running costs (labour, utilities, replacement parts)
- Ancillary equipment costs

Proper consideration of these factors will eliminate various separation techniques/equipment to leave a shortlist. Some laboratory testwork will be necessary during this selection process, e.g. filterability and sedimentation rate, and for the assessment of separation performance.

2.4.2 TESTWORK TO MEASURE SEPARATION PERFORMANCE

Laboratory-scale experiments should be carried out to measure the separation performance. These tests should determine both equilibrium characteristics (e.g. moisture content, sediment volume) and the rate at which these characteristics are reached. Average values from several fermentations should be obtained.

2.4.2.1 Filtration Performance

The filterability of the fermentation broth is determined and the need for conditioning assessed.

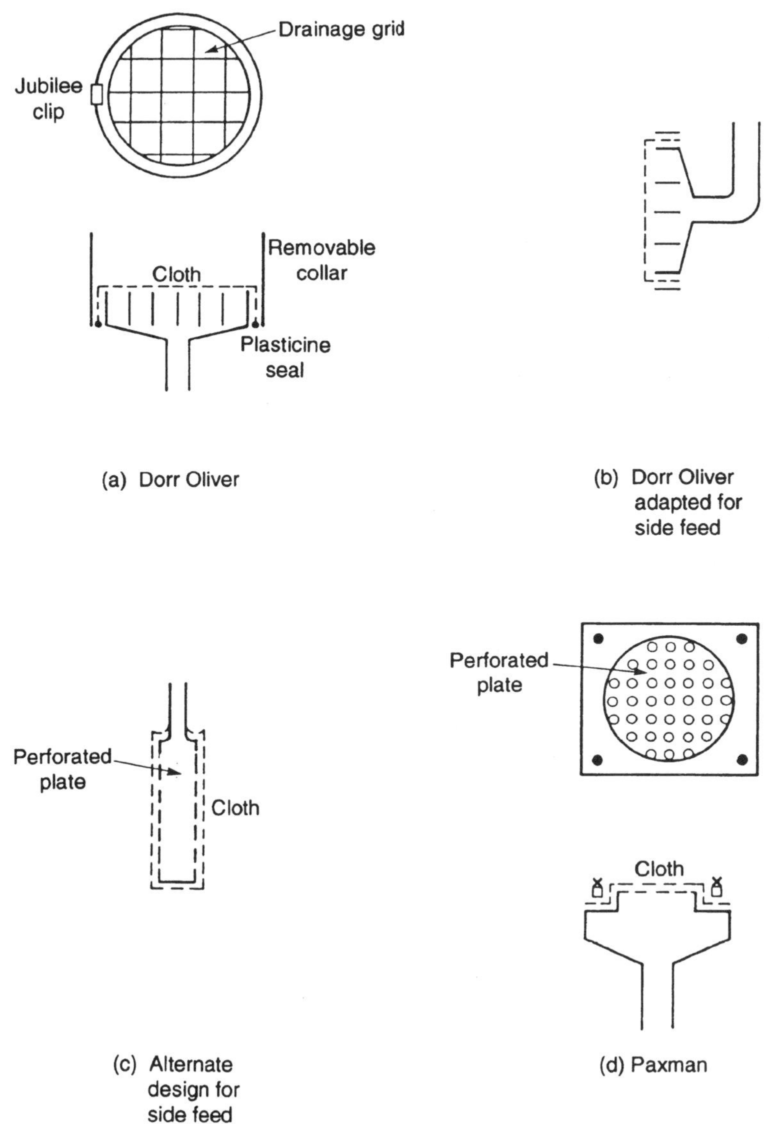

Figure 2.7 Designs of vacuum leaf

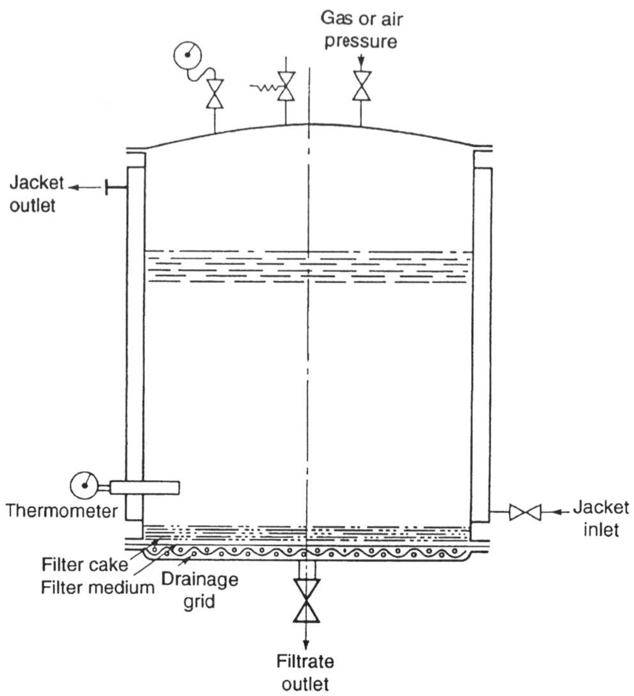

Figure 2.8 Pressure pot with horizontal surface

Apparatus Dead end filtration tests can be performed with a range of small-scale filters:

(a) buchner funnel,
(b) vacuum leaf — simulates rotary drum filters,
(c) pressure pot — simulates pressure leaf filters,
(d) compression cell — simulates variable volume filter presses,
(e) chamber cell — simulates filter presses.

Although a buchner funnel is a simple device it is not reliable for filterability measurement because of the significant resistance to flow created by the drainage surface. Vacuum leafs (Figure 2.7) have an area of typically $0.01\,\text{m}^2$. Pressure pots can be used to simulate horizontal and vertical leaf filters (Figures 2.8 and 2.9). The diameter should be greater than the cake thickness. Small, $10\,\text{cm} \times 10\,\text{cm}$, chamber cells can be used to simulate filter presses with spacer frames altering the cake thickness as required. A hydraulic piston press, with a diameter of approximately $5\,\text{cm}$, is an effective laboratory equivalent of a variable-volume filter press.

Crossflow filtration tests can be performed with:

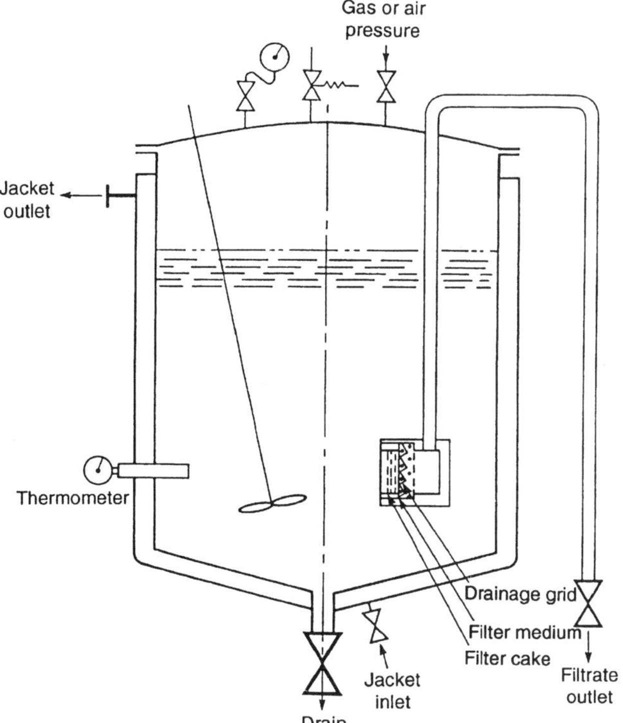

Figure 2.9 Pressure pot with flat vertical surface

(a) stirred cells,
(b) crossflow cells,
(c) small-scale versions of the full-scale crossflow filters.

Test procedure It is important to obtain a representative sample of the fermentation broth and to ensure that sufficient agitation is maintained during the test. For all filter types the broth is forced through the filter medium (cloth, paper, mesh, membrane, sinter) by gas pressure or pump pressure.

(a) *Vacuum leaf.* For horizontal orientation a collar should be fitted and the broth fed on to the filter. For vertical orientation a collar should not be fitted and the leaf should be mounted or hand-held immersed in the broth.
(b) *Pressure leaf.* Broth should be added to the pressure pot and the gas pressure applied as soon as possible.
(c) *Chamber cell.* Broth should be pumped into the filter using the same pump characteristics as would be the case for the full scale.
(d) *Compression cell.* As with the pressure leaf there should be minimal delay between adding the broth and starting to apply the compression pressure.

Table 2.6 Main independent variables in filtration performance testwork

Filter type	ΔP	Q	T	t_F	V_F	v_X	Constant	
							Pressure	Rate
1. Vacuum leaf	✓	—	✓	✓	✓	—	✓	—
2. Pressure leaf	✓	—	✓	✓	✓	—	✓	—
3. Chamber cell	✓	✓	✓	✓	✓	—	✓	✓
4. Compression cell	✓	✓	✓	✓	—	—	✓	✓
5. Stirred cell	✓	—	✓	—	✓	—	✓	—
6. Crossflow cell	✓	—	✓	—	✓	✓	✓	—
7. Crossflow filter	✓	—	✓	—	✓	✓	✓	—

ΔP = applied pressure; Q = filtrate flowrate; t_F = cake formation time; V_F = volume of feed filtered; v_X = crossflow velocity; T = temperature.

(e) *Stirred cell*. It would be preferable to use a cell able to hold the test volume of broth. The need for a separate sample reservoir merely increases problems with biomass settling.

(f) *Crossflow cell*. The feed/retentate flowrate should be set to give a membrane surface crossflow velocity within the range characteristic of commercial crossflow filters.

(g) *Crossflow filter*. Selection of the membrane module configuration is an important step and should be carried out in such a way that the final choice is based on feedback from the laboratory tests.

Test variables   The main variables that can be readily adjusted in fermentation broth filtration performance tests are shown in Table 2.6.

Test data to record   Test data that needs to be recorded for each filtration test is listed below:

(a) applied pressure, as a function of time if constant rate filtration is used,
(b) temperature of fermentation broth,
(c) weight or volume of broth filtered,
(d) weight or volume of filtrate collected as a function of time,
(e) weight of wet filter cake,
(f) weight of dried filter cake, adjusted for weight of solubles if necessary,
(g) filter cake thickness,
(h) retentate flowrate, for crossflow filter and crossflow cell testwork,
(i) cake formation time.

Data interpretation   Analysis of the filtration data provides a measure of the filterability of the fermentation broth [2]. A specific cake resistance, r (units of m/kg), can be determined which is a characteristic of the filtered biomass at a given applied pressure.

For constant pressure filtration A plot of time/filtrate volume against filtrate volume (t/V versus V) should be linear with

$$\text{Slope} = \frac{r\eta C}{2A^2 \Delta P} \tag{1}$$

where η = filtrate viscosity (Pa s)
  C = mass of dry cake per unit volume of filtrate (kg/m^3)
  A = filter medium area (m^2)
  ΔP = applied pressure (Pa)

(If the cake dewaters after it has formed, the volume of filtrate used to calculate C should be corrected.)

The cake resistance, r, is calculated and used to estimate filtration processing requirements, e.g. (assuming filter medium resistance is negligible)

$$\text{Filtration area} = \left(\frac{r\eta C}{2\Delta P} \frac{V^2}{t} \right)^{\frac{1}{2}} \tag{2}$$

(filter area required to separate a given volume of broth in a specified time, t)

$$\text{Filtration time} = \frac{r\eta C}{2\Delta P} \frac{V^2}{A^2} \tag{3}$$

(filtration time required to separate a given volume of broth using a specified filter area, A).

The data plot may exhibit:

(a) Scattered data points
(b) A significant intercept value
(c) Non-linearity due to:
  • Liquid hold-up
  • Incorrect starting time
  • Variable cake resistance
  • Biomass sedimentation
  • Pressure variation

Significant random scatter of the test data reflects a poor experimental technique.

A significant intercept value is indicative of a high filter medium resistance, filter medium blinding or hydraulic restrictions in the drainage from the filter. An alternative filter medium should be considered and/or the drainage improved.

Non-linearity of the data plot can be handled as described below:

(a) Hold-up of filtrate in the test filter. The T/V against V plot exhibits curvature away from the x axis in the initial stages of the filtration. This is dealt with in one of several ways:

- By elimination of hold-up from the filter.
- Determining the actual hold-up volume and adding this value to the filtrate volumes before calculating and plotting t/V against V.
- Start timing when the filtrate is first collected and not when the pressure is applied.
- Using the slope of the plot near to the cake formation time where the data should be linear.
- Using the differential form of the filtration equation and plotting $\Delta t/\Delta V$ against $V + \Delta V/2$. The slope is twice that given by equation (1) and so the denominator does not include the factor 2. Any doubts over the data are best handled by replotting in this way.

(b) Incorrect starting time. The problem of filtrate hold-up arises because $t > 0$ when $V = 0$. Similarly there is a problem if $V > 0$ when $t = 0$. Such a situation should be handled in similar ways.

(c) Variable cake resistance. A variation in the biomass cake resistance during the filtration, as indicated by curvature of the data points, could be due to:

- Sedimentation effects
- Migration of fine particles through the cake
- Cake compressibility

(d) Sedimentation. Changes in the particle size distribution reaching the surface of the forming filter cake case variations in the instantaneous cake resistance — the larger the particles the lower the resistance.

(e) Pressure variations. Curvature of the t/V against V plot will occur if the applied pressure is not held constant during the test. An improved experimental technique would be necessary in such a case.

For constant rate filtration   A plot of ΔP against t should be linear with

$$\text{Slope} = \frac{Q^2 r \eta C}{A^2} \tag{4}$$

where Q is the filtrate volumetric flowrate (m^3/s). Difficulties with the plot can be dealt with as described for constant pressure filtration.

Assessment of need for broth conditioning   The filtration performance tests should provide sufficient information on the filterability of the fermentation broth. The need to improve the filtration rate by broth conditioning can be assessed on the following criteria:

If $r > 10^{13}$ m/kg (dead end filtration)
or   Permeate flux < 50 L/m^{2}h (crossflow filtration)

then conditioning of the broth should be considered.

2.4.2.2 Centrifugal Sedimentation Performance

The sedimentation rate of the biomass can be measured under gravity or enhanced gravity conditions. Because of the small size, low density and liquor viscosity the former is not practical. However, broths that settle under gravity to at least 70% of the total available distance in about 30 min and form a distinct settling interface will be easy to separate in a centrifuge.

Test apparatus A laboratory swing-out rotor centrifuge is the preferred choice at test equipment. An angle rotor would be less satisfactory because the settled biomass would form a sloped sediment.

If a small-scale disc stack or scroll decanter centrifuge is available and there is sufficient fermentation broth (i.e. at least 100 litres) then useful performance data appropriate to a particular centrifuge type can be obtained.

Test procedure The purpose of the tests is to measure the clarity of the supernatant, the sediment volume and the nature of the sediment (moisture content and flow properties) as a function of centrifugation time and G-force. Round bottomed 50 ml centrifuge tubes should be used with the swing-out rotor. A known weight of fermentation broth is added to each tube, the tubes spun at a specified G-force and a tube removed at different times for analysis.

Test variables The main parameters that can be adjusted for the centrifugation tests are:

(a) spinning time,
(b) G-force,
(c) broth flowrate (continuous centrifuges),
(d) settling distance — disc spacing in disc stack centrifuges or pond depth in scroll decanter centrifuges,
(e) residence time — spinning time for centrifuge tubes, broth flowrate for continuous centrifuges,
(f) sediment discharge rate (continuous centrifuges).

Test data to record The minimum data that should be recorded for each centrifugal sedimentation test are listed below:

(a) weight of volume of broth used (tube tests),
(b) broth volumetric flowrate (continuous centrifuge tests),
(c) biomass concentration of broth,
(d) centrifuge bowl/rotor rotational speed,
(e) supernatant or centrate clarity,
(f) sediment volume (tube tests),
(g) sediment moisture content,
(h) qualitative assessment of the flow behaviour of the sediment.

Test data interpretation   Analysis of the test data provides a measure of the sedimentation properties of the biomass. A quantitative description of the solids recovery performance can be made by plotting percentage recovery against *G*-force (for tube tests) or Q/Σ (for continuous centrifuge tests):

$$\text{Solids recovery} = 100\left(1 - \frac{C_s}{C_F}\right) \tag{5}$$

where C_S = solids content of the supernatant
C_F = solids content of the feed, i.e. broth sample
Q = feed volumetric flowrate
Σ = sigma factor (units of m^2) which is a characteristic of a centrifuge

The factor Q/Σ is the settling velocity of the smallest particle that is totally recovered.

These relationships between biomass recovery and the operating conditions allow the separation efficiency to be assessed. By plotting sediment moisture content and/or sediment volume (for tube tests) against *G*-force (for tube tests) or Q/Σ (for continuous centrifuge tests) the extent of biomass thickening can be assessed.

Assessment of need for broth conditioning   The need to improve the settling properties of the biomass by broth conditioning should be assessed in terms of:

(a) the recovery efficiency and
(b) the time required to achieve that efficiency.

Roughly speaking, if the product of *G*-force and spinning time (for tube tests) is greater than 2×10^7 s broth conditioning should be considered. Similarly, if the Q/Σ value is greater than about 10^{-8} m/s for 80% solids recovery (for continuous centrifuges) broth conditioning should be considered.

2.4.3 TESTWORK TO MEASURE SEPARATION PERFORMANCE IMPROVEMENT AFTER BROTH CONDITIONING

Once the decision has been taken to condition the fermentation broth the separation performance testwork described in Section 2.4.2 needs to be repeated. An additional test variable is the extent of conditioning.

2.4.3.1 Temperature Variation

The broth can be heated or cooled immediately prior to filtration or centrifugation, or the broth can be held at a given temperature for a defined time before separation. Several of the test filters can incorporate a heating/cooling jacket whilst the residence time in a continuous centrifuge is sufficiently short so as to minimize temperature variation during separation (however, disc stack and decanter centrifuges can be fitted with jackets). Vacuum filter leafs and swing-out centrifuge rotors cannot be temperature controlled and so care must be taken to monitor the broth temperature throughout the test.

The change in the broth liquor viscosity with temperature must be taken into account when calculating the filtration and sedimentation characteristics.

2.4.3.2 Filter Aid Addition

The conditioning of the fermentation broth using a filter aid introduces additional test variables:

(a) the filter aid properties — type, particle size,
(b) the concentration of filter aid (body feed),
(c) the thickness of precoat.

When used as a body feed with upward facing filter surfaces the filter aid will settle faster than the rate of filtration and so the filter cake will be significantly more permeable in the earlier stages. This effect can be minimized by further additions of filter aid during filtration. For downward facing (e.g. rotary drum filter) and vertical filter surfaces the settling of the filter aid may limit the effectiveness of the latter in improving the filtration rate.

Analysis of the data will be as described in Section 2.4.2.1.

When the filter aid is used as a filter medium precoat, and none is added to the fermentation broth as a body feed, the mechanism of broth clarification could be cake filtration or depth filtration depending on the biomass concentration. If it is the former mechanism, then the previously described data analysis applies, but if the mechanism is depth filtration the data should be plotted as t/V against t.

2.4.3.3 Coagulation/Flocculation

Broth conditioning by biomass aggregation requires a means of dosing coagulant or flocculant into the broth, an effective mixing stage and a conditioning stage for growth/maturity of the coagula or flocs.

Dosing procedure Acid or alkali addition for pH-induced coagulation and salt addition for electrolyte-induced coagulation are best carried out at the laboratory scale using concentrated solutions. Polymeric flocculants are available in various physical forms (Section 2.3.3.3) and should be made up in concentrated stock solutions prior to dosing. The ionic strength of the make-up water will affect the conformation of a polyelectrolyte and hence the flocculation performance. The coagulant or flocculant is best added gradually whilst the broth sample is stirred vigorously.

Mixing and coagula or floc formation procedure Although the uniform dispersion of coagulant or flocculant within the broth sample cannot be separated from the coagulation or flocculation process, in practice the mixing and coagula or floc formation processes can be carried out in defined stages. Depending on the volume of sample required for the separation tests, suitably sized glass beakers with overhead stirrers are the preferred choice. Rapid stirring during coagulant or flocculant addition is followed by slow stirring during coagulum or floc growth.

Additional test variables to consider include:

(a) pH value of broth,
(b) electrolyte concentration of broth,
(c) type and valency of added cation,
(d) molecular weight of flocculant,
(e) charge density of flocculant,
(f) flocculant concentration in broth.

The effectiveness of the coagulation or flocculation is best assessed from the changes to the filterability or sedimentation of the fermentation broth. However, an initial screening of the most appropriate coagulating or flocculating conditions is often required and carrying out separation tests for this purpose would be time consuming. It is therefore recommended that biomass particle size and/or supernatant turbidity be measured and used to rapidly select appropriate conditioning parameters.

For fermentation broths high charge density cationic flocculants are often the best performers. With viscous broths low molecular weight flocculants should be used in order to minimize poor flocculant dispersion.

The filtration and centrifugation test data should be analysed as described earlier.

2.4.3.4 Modification of the Fermentation

The filtration and centrifugation testwork should be performed as indicated in Sections 2.4.2.1 and 2.4.2.2. Each variation in the fermentation parameters can be assessed in this manner.

2.4.4 SCALE-UP OF CLARIFICATION PROCESS

Once a separation technique has been selected and initial laboratory testwork has shown the performance levels achievable it is then necessary to consider how to scale up to production size.

2.4.4.1 Scale-up of Filtration

Dead end cake filtration laboratory testwork can be readily used to estimate filter size and cycle times at the large scale.

The specific cake resistance should be used to extrapolate filtration performance to give filtration areas and times (equations 2 and 3). Care should be exercised since biomass filter cakes are likely to be compressible.

Ideally pilot scale testwork should be carried out. As a rule laboratory filtration rates should be multiplied by a factor of 0.8 for the purposes of sizing.

Crossflow filtration laboratory testwork can be scaled up if crossflow filter modules are used. Allowance should be made for greater pressure losses as the size of the module is increased. Laboratory testwork based on stirred cell experiments should not be relied on for scale-up purposes.

2.4.4.2 Scale-up of Centrifugation

Swing-out centrifuge tube tests can be used to select a continuous centrifuge possessing a similar Gt factor. Small-scale continous centrifuge tests can be used to size a production centrifuge based on the sigma concept. The solids recovery against Q/Σ plot (Section 2.4.2.2) provides an estimate of the sigma parameter for a required recovery level at a specified broth flowrate. The sigma values, Σ, reflect the characteristics of the centrifuge.

Disc stack centrifuge

$$\Sigma = \frac{2\pi n \dot\omega}{3g} (R_2^3 - R_1^3)\cot\theta \tag{6}$$

where $n =$ number of discs
$\theta =$ angle the discs are tilted from vertical
$\dot\omega =$ centrifuge rotational velocity (rad/s)
$R_2 =$ distance from the rotational axis to the outer edge of the discs
$R_1 =$ distance from the rotational axis to the inner edge of the discs

Scroll decanter centrifuge

$$\Sigma = \frac{\dot\omega^2\pi}{g}\left[L_1(\tfrac{3}{2}R_3^2 + \tfrac{1}{2}R_1^2) + L_2\left(\frac{R_3^2 + 3R_3R_1 + 4R_1^2}{4}\right)\right] \tag{7}$$

where $L_1 =$ length of the cylindrical bowl section
$L_2 =$ length of the conical bowl section
$R_1 =$ distance from the rotational axis to the surface of the liquid
$R_3 =$ distance from the rotational axis to the bowl wall

Tubular centrifuge

$$\Sigma = \frac{\dot\omega^2\pi L}{g}(\tfrac{3}{2}R_3^2 + \tfrac{1}{2}R_1^2) \tag{8}$$

where $L =$ length of the bowl
R_3, R_1 have the meanings above

Scale-up between centrifuges of the same type is reasonably reliable using the simple relationship

$$\frac{Q_1}{\Sigma_1} = \frac{Q_2}{\Sigma_2} \tag{9}$$

Scale-up between different types of centrifuge should not be carried out using the sigma approach. Performance and sizing should be performed with small-scale versions of the full-scale centrifuge.

2.4.4.3 Scale-up of Conditioning

The use of filter aid as a precoat and/or body feed necessitates the installation of a filter aid filtration system comprising a precoat tank, body feed tank and pump. The sizing of the system should be based on the laboratory filtration tests.

The scale-up of coagulation/flocculation is difficult because the conditioning is both a mixing and a particle interaction process. However, the coagulation/flocculation can be carried out either as a two-stage or one-stage process:

(a) coagulant/flocculant mixing,
(b) coagula/floc growth.

Stirred tanks and in-line mixers are suitable for either or both stages.

2.5 Summary

The primary step in product recovery is often the separation of the biomass from the fermentation liquor. The main methods to achieve this separation are filtration and centrifugal sedimentation. The performance of these methods can be improved by conditioning the broth, e.g. by heating/cooling, by inducing coagulation or flocculation, by using filters aids [3].

2.6 References

1. A. M. Ratwatte and D. A. J. Wase (1988) *Adv. Biotechnol. Proc.*, **7**, 20–50.
2. L. Svarovsky (1982) *Solid–Liquid Separation*, 2nd edn, Butterworths, Oxford, pp. 242–64.
3. B. Atkinson and F. Mavituna (1982) *Biochemical Engineering and Biotechnology Handbook*, 2nd edn, Stockton Press.

3 CELL DISRUPTION: A PRACTICAL APPROACH

Eli Keshavarz-Moore

3.1 Introduction

In the recovery of intracellular proteins and enzymes, cell disruption is the most significant step. This is because the efficiency at this step of the operation directly affects the subsequent downstream operations and losses occurring at this initial stage cannot be regained. Many products are now the result of recombinant DNA technology. Whilst some are secreted from the cells, others remain intracellular with, in some cases, the added challenge of forming insoluble aggregates which need to be released from the cells and separated from other cell components.

In this chapter, the aim is to bring into focus some of the more practical issues arising from process-scale cell disruption. Only when necessary, reference is made to the theoretical aspects of cell breakage.

3.2 Choice of Unit Operation

Many reviews have addressed different techniques of cell disruption [1–3]. On the whole, mechanical disruption is favoured at the process scale [4]. Bead mill and high-pressure homogenization are the main contenders, although reports indicate that the latter technique is the most widely used [5]. Here, these techniques and their uses are described; however, other potentially useful techniques and possible future developments are also reported.

3.2.1 BEAD MILLS

The design of bead mills depends on the size of the unit and the manufacturer [4,6]. They consist of either a horizontal grinding chamber containing impellers or rotating discs mounted, concentrically or off-centred, on a motor-driven shaft. The grinding action is due to glass or plastic beads typically occupying

Downstream Processing of Natural Products. Edited by Michael S. Verrall
©1996 John Wiley & Sons Ltd

80–85% of the working volume of the chamber. Disruption can be batchwise or continuous.

The beads are retained in the grinding chamber either by a sieve plate or a similar device to retain the beads. The units require high-capacity cooling systems. Horizontal units are preferred for cell disruption to reduce the fluidizing effect in the vertical units.

The parameters involved in the process are numerous [6]. They include bead size and type, loading and density, configuration of disc/impeller and its speed, cell concentration and flowrate of the process feed and residence time. Other factors such as the properties of the cell are also relevant.

The bead mill has been used in the disruption of yeast and both Gram-positive and Gram-negative bacteria. Reports of filamentous organisms and algae being disrupted also exist in the literature [7,8]. The mechanism of disruption is based on shear, although particle impaction is also important.

3.2.1.1 Advantages and Disadvantages of Bead Mills

Given the large number of variables that need to be optimized for disruption of cells in the bead mill, it is not surprising that this technique lags behind high-pressure homogenization. Although the equipment can be run in batch mode, continuous operation is more practical with large fluid volumes. If sampling is required, this should be done with care to avoid back-mixing. Relatively difficult cleaning and initial preparation prior to operation can be drawbacks.

Another factor to consider is the dependency of disruption on the size of the organism. Smaller beads are more effective with bacteria than with yeast cells. However, bead fluidization and fouling of the separator may result. Attrition of beads can also result in fine particle contaminant which may be difficult to remove.

Nevertheless, this method provides a suitable alternative when breakage by other mechanical means are unsuccessful. Also, release of granular material dissociated from peptidoglycan layers and other cell debris may be easily achieved.

3.2.2 HIGH-PRESSURE CELL DISRUPTION

High-pressure homogenizers lend themselves to a relatively user friendly operation. Disruption is essentially achieved by passing the cells at high pressures through a small valve or orifice [9]. Several items of equipment using high pressure as a means of cell breakage are currently on the market. A selection of these is summarized in Table 3.1.

3.2.3 NEW DEVELOPMENTS

In recent years the main emphasis of manufacturers has been on three issues: achievement of higher pressures, better design of the main cell breakage

Table 3.1   Some high-pressure disruption devices for process operations

Equipment	Manufacturer
APV high-pressure homogenizer (e.g. 30 CD)(maximum pressure *ca* 1500 bar)	APV Gaulin (Germany) (part of APV Baker)
Rannie Hyper homogenizers (maximum pressure *ca* 1500 bar)	APV Rannie AS (Denmark) (part of APV Baker)
Ultra high pressure cell disrupter (maximum pressure *ca* 2700 bar)	Constant Systems Ltd (UK)
Microfluidizer (maximum pressure *ca* 1000 bar)	Microfluidics (USA)

component and, in some cases, containment and sterilizability. The first two aspects are aimed at more effective breakage. Where containment has been catered for, it has been aimed at satisfying regulations dealing with organisms falling into stricter containment categories.

Some of the APV Baker homogenizers (encompassing Rannie as well) can achieve pressures of up to 1500 bar and more. Relatively new valve design and material (a composite ceramic) have improved the cell breakage efficiency and reduced wear and tear [9].

Microfluidics design which is based on two impinging process fluid jets can operate at high pressures even though requiring a supply of pressurized air to activate the equipment. It has found a following in the United States although its application in larger than pilot processes is not documented. One of its main alternative applications (with modification) is in emulsion production.

Constant systems are relatively new to the field. The equipment has been tested for containment [10] and can reach very high pressures of up to 2700 bar depending on the model.

3.3 How to Assess Cell Breakage

There are many ways of assessing cell breakage, namely microscopic inspection and counting of cells, measurement of soluble protein and enzyme or particle measurement. With microscopy, it is not always easy to differentiate between whole and broken cells even if staining techniques are employed (as a small fissure at times is enough to release the intracellular material and as staining is not accurate). Nevertheless, it is advisable to carry out visual inspection of samples to better understand the process. Soluble protein measurements are easy and results are reliable, especially if backed by data obtained from a cytoplasmic enzyme which could be used as a marker [11]. Nucleic acids are not commonly used outside small laboratory-scale work as they are easily affected by disruption conditions, as will be seen in a later section. Use of particle measurement techniques [12] require specialist equipment and expertise.

Table 3.2 Comparative breakage rates for different equipment using *Saccharomyces cerevisiae* (90% breakage)

Equipment	Number of passes	Pressure (bar)	Fraction soluble protein release
Microfluidizer	5	600	0.38 [17]
APV homogenizer	5	600	1.0 [16]
Constant systems	1	>1700	1.0 [18]

Maximum releasable protein is the maximum amount of soluble protein that is expected to be released from the cells. As it is empirically derived, problems usually arise from micronization [13], where a given protein assay gives a positive response to non-protein material. Whilst a range is reported in the literature [14,15], an average figure of 30% g/g cell dry weight may be used for yeast. This is in agreement with a theoretical level based on expected protein contents of the cells. Use of a marker enzyme alleviates such problems. Examples of some protein release data are given in Table 3.2.

3.4 Parameters Affecting Cell Disruption Kinetics

Cell disruption in general is governed by, firstly, the mechanism of breakage, i.e. how the actual breakage is effected, and, secondly, by kinetic factors, i.e. how well the disruption has taken place. In the case of high-pressure cell disrupters, although the mechanism of disruption is complex, the operational variables are few and simple to control. Pressure, number of passages through the valve assembly or similar disruption orifices are the main parameters, which are easily set. Machine-related parameters such as the valve/orifice design and flowrate are already defined by the manufacturers.

3.4.1 PRESSURE AND NUMBER OF PASSES

The effect of pressure and number of passes is well documented in the literature on cell disruption [1–4]. Cell breakage follows first-order kinetics [16] with respect to the number of passes, which means that, at a given pressure, breakage is gradual and is dependent on the number of passages through the equipment until all cells are broken. As the pressure increases the amount of cell breakage increases accordingly. It is possible to define a rate constant for cell disruption which gives an indication of the ease of disruption. The pressure dependency can also be quantified.

Data such as those given in Table 3.2 should be interpreted with caution. As will be seen later, the condition of the process fluid is important and relevant to the breakage rate.

Kinetics of cell disruption were first put forward by Hetherington *et al.* [16] for bakers' yeast:

$$\text{Ln } \{R_\text{m}\}/\{R_m - R\} = K N$$

where

R_m = maximum releasable soluble protein
R = amount of soluble protein released at a given pass
K = dimensionless rate constant and represents the rate of cell breakage
N = number of passes through the homogenizer

The exponent of N is found to be unity here and therefore the breakage kinetics are known to be first order. K can be rewritten as $K = kP^a$, i.e. it depends on pressure, where the exponent a indicates the degree of dependency on pressure; the dependence on cell characteristics and process temperature is inherent in k.

Some examples of rate constants are given in Table 3.3.

3.4.2 APPLICATION OF HIGH PRESSURES

With high pressures, the general intention is to reduce the number of passes through the equipment to a minimum, preferably keeping it to a single passage. This will save considerable time and possibly power and may reduce enzyme inactivation resulting from multiple input through the equipment. However, another consequence of homogenization needs evaluation and this is cell debris size formation and whether very high pressures are beneficial or not to the whole process. This is discussed in the section dedicated to particle size formation.

It is possible to achieve at least 95% disruption with bakers' yeast in one pass through a Micron Lab 40 homogenizer at 1200 bar [19].

3.4.3 LOCATION OF THE PRODUCT

One factor in deciding the pressure required for breakage is the location of the product in the cell. A cytoplasmic soluble product is easier to release than a membrane bound enzyme [20] or one located in the vacuole [11]. The nature of

Table 3.3 Protein release from micro-organisms in a high-pressure homogenizer [4]

Micro-organism	Rate constant K^a
Pseudomonas putida	0.41
Escherichia coli	0.39
Bacillus brevis	0.28
Saccharomyces cerevisiae	0.23
Nocardia rhodochrous	0.0085

[a]The rate constant is a measure of cell breakage. Higher values mean easier disruption. In all cases the micro-organisms were harvested near or at the end of growth. Homogenization was at 500 bar, 5 °C in a model 15M APV-Gaulin homogenizer.

the product is also relevant, i.e. whether it is an easily degradable enzyme or a denatured solid particle such as an inclusion body or even a naturally produced particle, as is the case with poly(hydroxybutyrate) (PHB), a biodegradable thermoplastic.

3.4.4 PROCESS FLUID VARIABLES

Process fluid variables pertain to the nature of the micro-organism used and are of crucial importance in disruption. The first major distinction is between unicellular and filamentous organisms. Fundamental differences have been observed between the kinetics of disruption of filamentous and unicellular micro-organisms [13]. Whilst the rate of disruption of cells such as yeast is known to be strongly dependent on pressure, the rate of release of soluble protein from such filamentous organisms as *Rhizopus nigricans* was a weak function of pressure, the breakage being complete after one pass at low pressures (about 100 bar). This, however, is not the only aspect of interest. Filamentous organisms may be septate (e.g. *Aspergillus*) or non-septate (e.g. *Rhizopus*). In the former case there may be a need to disrupt the cells at several points in order to release the contents of the cells, whilst in the latter one fission is adequate. A major obstacle in processing of filamentous organisms is equipment failure due to blockage. This was observed for both *Aspergillus* [21] and *Rhizopus* [13]. The mechanism may be attributed to cell morphology and structural characteristics of the suspension. At high concentrations (e.g. above 22 g/L dry cell weight for *Rhizopus*), cells form closely intertwined structures which easily settled into cellular mats blocking the homogenizer. For pelleted cells (1–2 mm diameter), homogeneous suspensions similar to those of unicellular micro-organisms are formed. Here little interaction exists between the cells; blockage does not occur until relatively high concentrations (30 g/L cell dry weight) [13].

3.4.5 UNICELLULAR ORGANISMS

Wall structures of micro-organisms differ in their composition depending on genetic and environmental factors. However, similarities in overall structural aspects are observed within groups of organisms which may explain some of the trends observed in cell disruption. Where there is no cell wall (e.g. animal cells), cells are highly susceptible to disruption. Gram-negative bacteria are easier to break than Gram-positive bacteria, which in turn are easier to break than yeasts [1,22]. Fungal walls are classified as more resistant to disruption but their rupture depends greatly on the mechanism employed, as seen in the previous section.

3.4.6 CELL CONCENTRATION

High cell concentrations may be processed if unicellular organisms are used. With bakers' yeast no reduction in the cell disruption rate was observed up to

Table 3.4 Cell characteristics and conditions of growth versus pressure exponent (indicator of pressure dependency)

Micro-organism	Source/conditions of growth	Exponent
Bakers' yeast		
Hetherington *et al.* [16]	Commercial yeast. Probably batch or fed batch. Complex medium	2.9
Doulah *et al.* [27]	As above	1.72–1.79
Dunnill and Lilly [28]	As above	2.9
Spent brewers' yeast		
Whitworth [29]	From brewery. Complex medium	1.2–1.3
Engler [26]	From brewery. Complex medium	1.87
Saccharomyces cerevisiae		
Engler [26]	Aerobic, continuous culture specific growth rate $0.1\,h^{-1}$. Synthetic medium	0.86
Candida utilis		
Engler and Robinson [30]	Cyclic batch maximum growth rate $0.5\,h^{-1}$. Synthetic medium	1.17
	continuous culture specific growth rate $0.1\,h^{-1}$. Synthetic medium	1.77
Baldwin and Robinson [31]	A range of specific	1.73–2.98
Candida lipolytica		
Whitworth [32]	Batch. Harvest after $55\,h$. Complex medium	NA
Bacillus brevis		
Augenstein *et al.* [33]	Batch. Late lag phase then frozen at $-20\,°C$. Complex medium	1.8
Bacillus subtilis		
Engler and Robinson [34]	Continuous culture specific growth rate $0.2\,h^{-1}$. Complex medium	1.07

224 g/L cell dry weight [16]. With *Escherichia coli*, it has been reported that the cell fraction disrupted increased with increasing cell concentration when disruption took place in a microfluidizer [23]. This was at variance with results obtained by other investigators using the same equipment [24] and also in earlier work with a high-pressure homogenizer, where no dependence of disruption kinetics on cell concentration was observed [25]. Cell concentrations up to little over 100 g/L of cell dry weight have been disrupted with no reported problem [24].

3.4.7 EFFECT OF TEMPERATURE

An increase in the fluid process temperature results in a higher disruption rate (e.g. 1.5 times higher for a 25 °C increase) [16]. However, this is not desirable due to inactivation/degradation of the product if an enzyme or protein is produced. In general, a rise in temperature of 1 °C per 100 bar is to be expected. This means that cooling of the process fluid, especially in high pressure or multiple pass disruption operations, is essential.

3.4.8 CONDITIONS OF CELL GROWTH

Conditions of growth have a major effect on disruption kinetics. Engler [26] attempted to derive a general kinetics expression for all micro-organisms but concluded that such generalization, even at the level of micro-organisms of the same group, may be misleading. Table 3.4 illustrates the point by showing the variation in the calculated pressure dependency for different yeasts and bacteria under different conditions of growth.

Results obtained for the disruption of *E. coli* cells indicated that batch cultures grown on a synthetic medium were easier to disrupt than those grown on complex medium [25]. In another study [30] it was shown that cells grown at a higher specific growth rate were easier to disrupt than cells grown at a lower specific growth rate. Similarly, cells harvested during the log phase of growth were more susceptible to homogenization than those from the stationary phase. It is well known that cells from the stationary phase commonly have a stronger wall structure than those in the log phase. Furthermore, cells grown at a higher specific growth rate probably direct the available energy towards reproduction rather than synthesis or strengthening of wall structure. Complex media may provide additional nutrients for such synthesis, which are not available in simple synthetic media.

3.5 Interaction between Cell Disintegration and Other Downstream Operations

After cell disintegration, the first major step is to separate the cell debris from the homogenate. Typically this involves centrifugation or filtration. In both cases, the performance of the processes is dependent on the characteristics of the homogenate. These are cell debris size and density, and viscosity of the homogenate. Small particles and high fluid viscosities are detrimental to the separation operation. In the case of centrifugation, the settling velocity is reduced. In filtration, clogging and reduction in flowrate will result. Prior to clarification it may be beneficial to use some of the conditioning treatments mentioned in Chapter 2. Alternatively, it may be possible to separate product from cell debris by aqueous two-phase partition methods (Chapter 4).

3.5.1 CELL DEBRIS FORMATION

For yeast cells, it is shown that smaller cell debris size is associated with higher protein release with each pass through the high-pressure homogenizer [35]. Similar results are obtained for *E. coli* [24]. Use of very high pressures with a single pass may be considered a reasonable alternative as cells are not micronized as a result of multiple passes. However, recent work indicates that similar size distributions may be expected from single-pass homogenization at 1000 bar and with that at around 500 bar and 5 passes [35].

Reduction in cell size may be beneficial in some cases. For instance, if inclusion bodies are to be separated from cell debris it is desirable to reduce the size of the cell debris whilst retaining that of inclusion bodies. This may be used as a tool to help separation in a centrifuge where the operation may be optimized [36].

3.5.2 VISCOSITY OF HOMOGENATE

The reduction in homogenate viscosity is desirable to improve solid liquid separation. The viscosity of the fluid is the result of release of RNA and DNA contents from the cells. Higgins *et al.* [37] reduced the RNA of disrupted *E. coli* suspension by 80% by heat treatment for 20 minutes at 50 °C. Clearly this approach is only applicable when the product is not susceptible to heat denaturation. Another study shows that it is mainly the DNA polymer that contributes to viscosity [24]. DNA polymer is shear sensitive and easily degrades into fragments after several passes through a high-pressure homogenizer, thus resulting in reduced homogenate viscosity.

3.5.3 COMPARISON BETWEEN DIFFERENT EQUIPMENT

Comparison of characteristics of a disrupted *E. coli* cell suspension obtained from a bead mill, a high-pressure homogenizer and a microfluidizer indicates that, for similar levels of protein and enzyme release, the bead mill produces the largest debris size and the high-pressure homogenizer the smallest. However, the viscosity remains highest as well. Overall, higher disruption does not necessarily lead to higher overall product recovery [24].

3.5.4 STORAGE CONDITIONS

It is often desirable to store cells after harvesting, prior to separation and purification. Freezing the cells is a common practice for long storage intervals. It is normally believed that freeze thawing weakens the cells and can even lead to disruption. In a recent study it was shown that whilst this procedure results in some release, freeze thawed yeast cells disrupt less readily than fresh cells when subjected to mechanical disintegration [38]. The reduction in the rate of breakage is dependent on the number of freeze–thaw cycles and the storage temperature (storage at −70 °C is recommended compared to storage at −20 °C). This effect is less pronounced in a bead mill than in a high-pressure homogenizer. This is because damage to the cell wall results in cells becoming 'flaccid'. The impingement mechanism in a high-pressure homogenizer is therefore less effective in breaking the cells than shear and bead collision in a bead mill. Although this phenomenon was not observed for some other micro-organisms [24,33], it is worth noting in order to prevent misinterpretation of results.

3.5.5 RESUSPENSION FLUID

Cells may be processed directly from the fermenter. This may not be desirable if it leads to added steps in removing other broth materials, especially when a complex medium is used or if volume reduction is required. Resuspension in buffer may then be necessary. The buffer should normally have a pH close to the physiological pH of the cells. Addition of proteases such as serine proteases to reduce proteolytic activity is common. It is worth noting that addition of any chemical to the suspension may require removal at a later stage and should be carried out with caution.

3.5.6 EFFECT OF RECOMBINANT DNA TECHNOLOGY

Reports on the effect of gene manipulation on cell breakage differ. Whilst some studies indicate that recombinant strains are easier to disrupt [23,36], others do not [38]. These differences may be partly attributed to the nature of the product (inclusion body versus soluble protein) or resultant changes in morphology of cells.

3.5.7 PRETREATMENT OF CELLS

When cells are particularly resilient to cell breakage, it may be necessary to pretreat the cells either chemically or enzymatically. This is assuming that these additions will not degrade or denature the product. If cells are washed, it should be noted that weak pretreated cells may break in the centrifuge prior to resuspension. In cases where particulate matter is to be extracted from the cells (e.g. PHB from *Alcaligenes eutrophus*), it has been reported that pretreatment allows equivalent or better disruption to be achieved at lower pressures and fewer passes through the homogenizer [39]. Baldwin and Robinson [40] report that if surviving cells of *Candida utilis* that have been enzymatically treated are cultured, these will show increased resistance to disruption in a microfluidizer.

3.6 Future Methods

Whilst the use of lytic enzymes may see a revival in their use in the differential release of cell components [14], other exciting approaches are also being sought. One is the development of strains of yeast by mutation and gene manipulation for the controlled release of intracellular products. Here, ideally, it would be possible to control cell wall integrity and trigger the release of material by a simple mechanism. Some success has been reported in the literature, although their application at large scale requires further investigation [41,42]. Similar works with *E. coli* using different lytic systems are still in their infancy, as in some systems spontaneous lysis only occurs in the mid-exponential phase [43]. Further development at large scale is necessary. A successful mechanism will present an ideal system for downstream processing, as not only no disruption is required but also the resulting cell ghosts are thought to be easy to separate.

3.7 Concluding Remarks

In deciding on technique and conditions for cell disruption, it is essential to remember that this is not a process in isolation but conditions upstream have a significant effect on its success. Equally, the efficiency of disruption should be estimated in relation to other operations further downstream, such as centrifugation, filtration or aqueous two-phase partition (see Chapter 4). Small-scale cell disruption units such as Micron lab 40 (APV Baker) have been shown to give similar results to large-scale homogenizers and can therefore be used effectively to define preliminary process conditions [19].

3.8 References

1. C. R. Engler (1985) in M. Moo-Young (Ed.), *Comprehensive Biotechnology*, Vol. 2, Pergamon Press, Oxford and New York, pp. 305–24.
2. J. A. Chisti and M. Moo-Young (1986) *Enzyme and Microbial Technol.*, **8**, 194–204.
3. H. Schutte and M.-R. Kula (1990) *Biotechnol. Appl. Biochem.*, **12**, 599–620.
4. E. Keshavarz, M. Hoare and P. Dunnill (1987) in M. S. Verrall and M. J. Hudson (Eds.), *Separations for Biotechnology*, published for Soc. Chem. Ind. by Ellis Horwood, Chichester, pp. 62–79.
5. M. D. Scawen and P. M. Hammond (1986) in J. D. Stowell, P. J. Bailey and D. J. Winstanley (Eds.), *Bioactive Microbial Products*, Vol. 3, Academic Press, London, pp. 77–101.
6. J. Limon-Lason, M. Hoare, C. B. Osborn, D. J. Doyle and P. Dunnill (1979) *Biotechnol. Bioengng.*, **21**, 745–74.
7. L. Edebo (1969) *Ferment. Adv.*, 249–71.
8. K. Zetalaki (1969) *Process Biochem.*, **4**, 19–22 and 27.
9. E. Keshavarz-Moore, M. Hoare and P. Dunnill (1990) *Enzyme and Microbiol. Technol.*, **12**, 764–70.
10. D. Foster (1992) *Bio/Technology*, **10**, 1539–41.
11. M. Follows, P. J. Hetherington, P. Dunnill and M. D. Lilly (1971) *Biotechnol. Bioengng.*, **13**, 549–60.
12. A. P. J. Middleberg, I. D. L. Bogle and M. Snoswell (1990) *Biotechnol. Prog.*, **6**(4), 255–61.
13. E. Keshavarz, M. Hoare and P. Dunnill (1990) *Enzyme and Microbiol. Technol.*, **12**, 494–8.
14. R.-B. Huang, B. A. Andrews and J. A. Asenjo (1991) *Biotechnol. Bioengng.*, **38**, 977–85.
15. D. van Gaver and A. Huyghebaert (1990) *Enzyme and Microbiol. Technol.*, **13**, 665–71.
16. P. J. Hetherington, M. Follows, P. Dunnill and M. D. Lilly (1971) *Trans. Instn. Chem. Engrs.*, **49**, 142–8.
17. C. V. Baldwin and C. W. Robinson (1990) *Biotechnol. Techniques*, **4**(5), 329–34.

18. Manufacturer's communication.
19. E. Keshavarz-Moore, unpublished data.
20. B. L. Talboys, and P. Dunnill (1985) *Biotechnol. Bioengng.*, **27**, 1726–9.
21. L. Edebo (1983) in R. M. Lafferty (Ed.), *Enzyme Technology*, Springer-Verlag, Heidelberg, pp. 93–114.
22. R. F. Ramaley (1979) in D. Perlam (Ed.), *Advances in Applied Microbiology*, Vol. 25, Academic Press, New York, pp. 37–55.
23. T. Sauer, C. W. Robinson and B. R. Glick (1991) *Biotechnol. Bioengng.*, **33**, 1330–42.
24. I. Agerkvist and S.-O. Enfors (1990) *Biotechnol. Bioengng.*, **36**, 1083–9.
25. P. P. Gray, P. Dunnill and M. D. Lilly (1972) *Proc. IV IFS Ferment. Technol. Today*, **1972**, 347–51.
26. C. R. Engler (1979) PhD Thesis, University of Waterloo, Ontario, Canada.
27. M. S. Doulah, T. H. Hammond and J. S. G. Brookman (1975) *Biotechnol. Bioengng.*, **17**, 845–58.
28. P. Dunnill and M. D. Lilly (1975) in S. R. Tannebaum and D. I. C. Wang (Eds.), *Single Cell Protein*, Vol. II, MIT Press, Cambridge, Massachusetts, pp. 179–207.
29. D. A. Whitworth (1974) *Compt. Rend. Trav. Lab. Carlsbeg*, **40**, 19–32.
30. C. R. Engler and C. W. Robinson (1981) *Biotechnol. Bioengng.*, **23**, 765–80.
31. C. V. Baldwin and C. W. Robinson (1994) *Biotechnol. Bioengng.*, **43**, 46–56.
32. D. A. Whitworth (1974) *Biotechnol. Bioengng.* **16**, 1399–406.
33. D. C. Augenstein, K. Thrasher, A. J. Sinskey and D. I. C. Wang (1974) *Biotechnol. Bioengng.*, **16**, 1433–47.
34. C. R. Engler and C. W. Robinson (1981) *Biotechnol. Lett.*, **3**, 83–8.
35. E. Keshavarz-Moore (1993) *Sixth European Congress on Biotechnology*, Florence, Vol. 3, WE 023.
36. E. Keshavarz-Moore, R. Olbrich, M. Hoare and P. Dunnill (1991) *Ann. New York Academy of Sciences, Recombinant DNA Technology I*, **646**, 307–14.
37. J. J. Higgins, D. J. Lewis, W. H. Daley, F. G. Mosqueira, P. Dunnill and M. D. Lilly (1978) *Biotechnol Bioengng.*, **20**, 159–82.
38. P. T. Milburn and P. Dunnill (1994) *Biotechnol. Bioengng.*, **44**, 736–44.
39. S. T. L. Harrison, J. S. Dennis and H. A. Chase (1991) *Bioseparation*, **2**, 95–105.
40. C. Baldwin and C. W. Robinson (1992) *Lett. Appl. Microbiol.*, **15**, 59–62.
41. M. Broker (1994) *BioTechniques*, **16**(4), 604–10.
42. P. Alvarez, M. Sampedro, M. Molina and C. Nombela (1994) *J. Biotechnol.*, **38**, 81–8.
43. R. L. Dabora and C. L. Cooney (1990) in A. Fiechter (Ed.), *Advances in Biochemical Engineering/Biotechnology*, Vol. 43, Springer-Verlag, Berlin–Heidelberg, pp. 12–30.

4 TWO-PHASE AQUEOUS SYSTEMS FOR THE RECOVERY OF INTRACELLULAR MICROBIAL PRODUCTS

Jonathan G. Huddleston and Andrew Lyddiatt

4.1 Introduction

The mixing of a solution of a polymer (e.g. poly(ethylene glycol), PEG) with a second solution of a polymer (e.g. dextran) or salt (potassium phosphate) at concentrations above respective critical values yields an aqueous two-phase system (ATPS). PEG-rich top phase and dextran (or phosphate)-rich bottom phase can be dispersed one in the other, and separated in the manner of oil–water mixtures. However, the key characteristic of the high water content of both phases (typically greater than 70%) ensures a biocompatibilty with biological components such as cells, membrane preparations, particulate debris and macromolecules not seen with conventional oil–water solvent systems. Differential partition of mixtures of such components (as in whole fermentation broths, cell disruptates or unclarified biological extracts) has been harnessed to recover target products in simple procedures applicable to both laboratory and production plant (reviewed in Refs. [1] and [2]). In particular, the technique has been widely reported (and is most suited) when applied to the recovery of macromolecular protein products rather than low molecular weight secondary metabolites whose lack of a sensitive tertiary and/ or quaternary structure invite extraction procedures based upon conventional solvent–water systems (see Chapter 5). The report of high recoveries for certain protein products in multistage processes, evidently advantaged by high space–time yield and ease of scale-up (reviewed in Ref. [3]), has encouraged a recognition of the process as a potential substitute for chromatography.

Downstream Processing of Natural Products. Edited by Michael S. Verrall
©1996 John Wiley & Sons Ltd

However, achievable degrees of purification have been less often discussed in tandem with the high recoveries, and the overall selectivity of such systems remains open to question — particularly in the absence of a clear mechanistic understanding of the molecular basis for differential partition.

It is the view of the present authors [4,5] that, with a number of exceptions, selectivity in an ATPS is fundamentally limited for the vast majority of protein products, but can be maximized by rigorous system optimization. Experience indicates that exceptional qualities (extreme isoelectric point, surface hydrophobicity, molecular mass, etc.) will be reflected in unique partition behaviour in *any* two-phase separation (e.g. solvent, chromatography or precipitation based). This may account for both some of the startling reports of individual performances reviewed in the literature [1–5] and also the disappointment of would-be imitators hamstrung by less distinguished target products. However, limitations of product selectivity do not exclude ATPSs from assuming a useful role in the downstream processing of natural products. A common feature of reports of applications of an ATPS (see refs. [1] to [5]) is the transformation of a particulate suspension feedstock (whole broth, cell disruptate, etc.) into a clarified, product-rich feedstream suitable for further purification in conventional operations of precipitation, chromatography or molecular sieving. Thus, solid–liquid separation (conventionally requiring refined centrifugation or microfiltration) and varying degrees of macromolecular fractionation have been achieved in liquid-phase separations enhanced at the most by moderate gravitational fields (500–5000 g). It is therefore as a primary processing tool that ATPS shows immediate promise, and this reflects the tenor of the present chapter. However, improved identification of quantifiable molecular parameters which account for partitioning behaviour [4–6] will advance the selectivity of such operations for all potential products.

Work in our laboratory on the recovery and partial fractionation of intracellular proteins from waste brewers' yeast and commercial bakers' yeast has indicated a practical role for ATPSs assembled from PEG and potassium phosphate [7–9]. Such PEG–salt systems have been identified as the most economic strategy having generic application to problems of protein purification, and form the focus of the present chapter. Additional work [10,11] with variously sourced refractile bodies, protein aggregates and membrane proteins indicates that such a positive indication is neither product or species specific and thus will have generic application to many problems of downstream processing. As a consequence, we outline below proven ATPS strategies in PEG–salt systems for the recovery of products from yeasts in the belief that this will most usefully benefit practitioners interested in applying the method to other, specific separation problems.

However, at the outset, it is worth reviewing the nature of the problem which confronts all practitioners seeking target products in the complex mixtures which generally comprise biological extracts. Those products targeted for purification are commonly individual members of an extensive set of closely similar molecules. Bulk properties (particularly in macromolecules) differ by virtue of only minor changes to the spatial extent and composition of a mosaic of fractional properties (forces) at the phase boundary of their molecular surfaces. Conventional purification technologies attempt to exploit limited groups of such properties (e.g. charge, hydrophobicity, etc.) in a controlled manner, but absolute and specific molecular resolutions are commonly difficult to achieve in a single step. Thus downstream operations require the sequential exploitation of one aggregate force after another in multiple steps targeted towards solids clarification, feedstock dewatering, followed by product fractionation, polishing and formulation.

In contrast, partitioning behaviour in ATPSs exploits the entire molecular surface of phase components in a simultaneous definition of the physical and chemical interaction with, and the distribution of, the target molecule within the two-phase system. Alteration of the solution properties of the phase system influences those molecular forces directing partition, but it is difficult to mechanistically enhance the influence of one force over another. For example, alteration in salt type or concentration will simultaneously effect changes to both electrostatic and dispersion forces.

Although many of those forces that influence partition at the molecular level have been identified, their variously aggregated behaviour is less well understood and permits only a limited degree of predictability in the design and implementation of operational solutions to separations problems. Consequently, the approaches detailed here must necessarily concern themselves with the art of 'fishing' rather than the pursuit of the 'trout' in particular.

4.2 Properties of PEG–Salt ATPS

4.2.1 THE PHASE COMPONENTS

PEG–salt ATPSs form on the admixture of polymers and salts above defined concentrations to generate a PEG-rich top phase and a salt-rich bottom phase. Here, phase separation is related both to salting-out phenomena associated with solutions of strong electrolytes and to the cloud point behaviour of polymers in solution. In this respect, it is phenomenologically different from the phase separation of polymer–polymer solutions (e.g. PEG–dextran [12]). PEG is available commercially, classified in a range of estimated average molecular weights from less than 700 daltons to very much greater than 20 000 daltons. However, the weights quoted almost certainly delimit the useful range applicable in salt-based systems. The higher the PEG molecular weight employed, the lower the concentrations of polymer and salt required for phase formation and subsequently present in the resultant discrete phases so formed. In low molecular weight PEG systems (1450 daltons), the concentration of salt

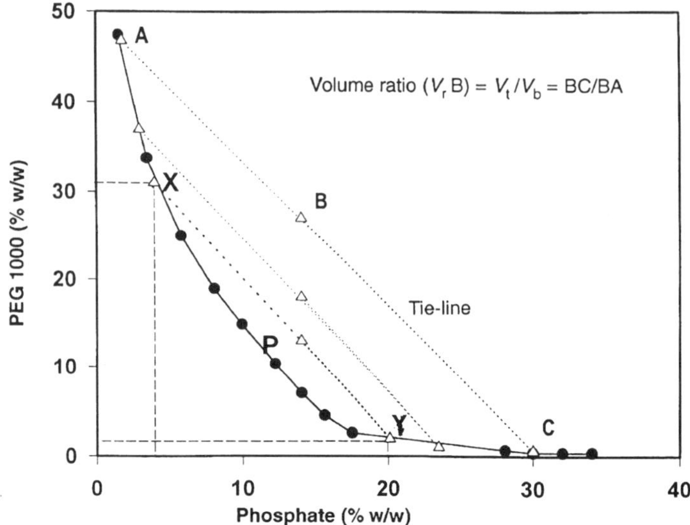

Figure 4.1 Phase diagram for a PEG–phosphate aqueous two-phase system. The curve (A–C) represents the binodal, to the left of which systems are homogeneous solutions and to the right systems form two-phase systems. Phase systems form at lower solute concentrations with increased molecular weight of PEG (binodal moves towards bottom left of phase diagram). Straight lines A–C and X–Y are tie-lines which individually link systems of constant phase composition but varying volume ratio. For example, the PEG and phosphate composition of the top and bottom phases on tie-line X–Y are defined by points X and Y on the binodal, whilst tie-line lengths (TLL) are determined geometrically and are commonly expressed in terms of % w/w PEG. The volume ratio of system B (Vr_B) on the tie-line A–C is given by the length ratio BC/BA. The critical point P is defined by those concentrations of PEG and phosphate that just yield a two-phase system with a volume ratio of 1 (minimal TLL). Partition coefficients (K) of individual solutes in such systems are defined by the ratio of solute concentrations in the top (PEG-rich) and bottom (phosphate-rich) phases

(even in systems close to the critical point) may approach values sufficient to precipitate many globular proteins.

The critical point lies on the binodal curve (which describes the conditions of phase formation; see Figure 4.1) and defines the minimum concentrations of PEG and salt required to initiate a biphasic system with a volume ratio equal to unity [1]. Higher molecular weights of PEG initiate phase formation at lower concentrations of salt, such that the salting-out tendency of the salt-rich lower phase is reduced close to the critical point.

Poly(ethylene glycols), by virtue of their strong association with water molecules, will also promote macromolecular precipitation — the so-called exclusion effect. In contrast to the salting-out effects of agents such as phosphate, the exclusion effect is most pronounced with high molecular weight PEG. As a result, certain proteins are very soluble (and are demonstrably stabilized) in very high concentrations of low molecular weight PEG ($>30\%$ w/w).

In the light of these two counterposed tendencies to precipitate or exclude solutes in high salt concentrations or high PEG concentrations respectively, solute partition may be usefully (if approximately) viewed as the balance in relative solubilities between the salt in the bottom phase and the polymer in the top phase. Useful comparative data and approximations of phase behaviour may be obtained from chromatographic analysis of solutes and mixtures on hydrophobic interactive adsorbents derivatized with PEG [6,13]. Such approaches show promise for method scouting in the *de novo* design and assembly of partition systems to achieve particular separations.

4.2.2 CHOICE OF PHASE-FORMING SALTS

Phase separation in PEG–salt systems appears to be largely determined by the charge density and size of the salt anions present in systems. Small multivalent anions of high charge density are observed to be the most effective, and this largely correlates with their effectiveness as protein precipitants [4,5]. Systems composed of alkali metal salts such as potassium phosphate show the greatest capacities for proteins, exceeding 20 mg/mL in particular separations. Sodium sulfate may also be suitable, but work in our laboratory [9,10] indicates that the solubilities of many proteins in ammonium and magnesium sulfate systems are too low to facilitate the useful operation of large-scale ATPSs exploiting such salts.

Since phase separation is so dependent upon the presence of the divalent anion, it can readily be appreciated that phase separation is pH dependent in potassium phosphate and other systems where the phase-forming salt is part of the buffering system. Thus, biphasic systems of PEG–citrate can be operated in a lower pH range (4–8) than PEG–phosphate. In contrast, the latter may be employed at more alkaline pH values (5–9), where protein solubility (but not necessarily stability) may be improved. This operating range can be extended to pH 12 with trivalent *ortho*-phosphate.

4.2.3 CHOICE OF PEG MOLECULAR WEIGHT

Work in our laboratory has indicated that, for the fractionation of intracellular yeast proteins, the useful applicable range of PEG extends from (estimated) molecular weights of 700 daltons to a limit of about 3000 daltons [7]. In contrast, β-galactosidase, a large (approx 470 kDa) multisubunit enzyme sourced in constitutive mutants of *Escherichia coli*, is almost unique in exhibiting high partition coefficients (i.e. strong top-phase preference; see Figure 4.1) in systems exploiting PEG nominal masses up to 6000 daltons [14]. Similar or identical species, expressed as recombinant heterologous proteins expressed, intracellularly or extracellularly, in yeast hosts, might selectively and simply be fractionated from host native proteins by the employment of such high molecular weight polymers. Current capabilities in the genetic engineering of purification tags fused with target products are currently rapidly advancing [15] (See also Chapter 7). This encourages the view that protein constructs might be specifically assembled to exhibit exceptional phase preferences in ATPSs, thereby facilitating simple fractionation from bulk impurities [16,17].

However, proteins in general will be more soluble in the upper PEG-rich phases when lower molecular weight polymers are employed, since they will be least excluded by such phase-forming components. At the same time, they are likely to approach solubility limits in the salt-rich lower phase of the same systems. In systems containing higher molecular weight PEG, the binodal is displaced to lower concentrations of PEG and salt, and exclusion from the top phase dominates partition whilst the influence of salting out in the lower phase is reduced. In all systems, solute solubility will be greatest at the critical point. However, it should be emphasized that exception to such general rules are common enough — and, like all quantifiable variants in a herd, may indicate mechanisms for separation. For example, cytochrome c from yeast (and other species) exhibits a lower salt-phase preference throughout the range of PEG molecular weights (700–3000 daltons) and operational pH values (pH 7–9) which generate transient and variable responses with other bulk intracellular proteins sourced in yeast [18].

4.2.4 INCREASE OF SOLUTE EXCLUSION AND SALTING-OUT

In an ATPS composed of a defined molecular weight range of PEG (e.g. PEG 1450) and a selected phase-separating salt (e.g. potassium phosphate), further manipulation of the protein partition coefficient, yield and relative purity can be achieved in a variety of ways. In moving from a position close to the phase boundary (the binodal in Figure 4.1) towards higher concentrations of polymers and salts ('deeper into the phase diagram'), the concentrations of phase-forming solutes will rise in their respective phases. Thus both salting-out and polymer exclusion phenomena would be expected to increase, whilst the balance between the two might enhance or depress the partition coefficient ($K\hat{p}$) of a target protein product or unwanted solute impurities. For the majority of proteins, either such manipulations increase $K\hat{p}$ through the predominance of salting-out effects or material is lost as a precipitate at the interface when exclusion dominates. Under favourable conditions, as demonstrated with yeast intracellular proteins, both effects can be harnessed to advance product purity or the concentration and yield of product in PEG-rich top phases [4–9].

4.2.5 ELECTROSTATIC EFFECTS

It has generally been found that a reduction in the nominal pH of a particular ATPS will reduce the partition coefficient ($K\hat{p}$) of contained proteins [5,7]. Under favourable circumstances, this can lead to the separation and potential recovery of proteins of lower isoelectric point (pI) but similar hydrophobicity contained in a complex mixture.

However, it should be noted that proteins of similar pI may partition very differently in a given system. Thus cytochrome *c* shows strong bottom-phase preference in PEG–phospate ATPSs until the system pH approaches the isoelectric point of the protein (>pH 10). In contrast, lysozyme has a stronger preference for the PEG-rich top phase despite a similar pI and molecular mass [6,10]. In addition, it is difficult to uncouple pH effects from those of salt in such systems. Manipulation of pH will necessarily alter the balance of buffering salts, as for instance between mono- and dibasic species, with consequent effects on the phase diagram (see Figure 4.1). PEG–sulfate systems buffered by dilute phospate salts would successfully uncouple such effects, but (as noted earlier) are compromised by excessive protein precipitation [9].

As well as manipulating protein charge through the system pH, similar effects can also be achieved by addition of so-called neutral salts. For example, the charge shielding effect of these salts appears to delicately modify the interaction of negatively charged proteins with PEG polymers in the top phase. As a result, under selected conditions protein partition in defined systems can be directed from the top to the bottom phase simply by salt additions. Effective salts are alkali metal halides and thiocyanates which should be used at concentrations of 1–500 mmoles/kg of phase system. At higher concentrations, thiocyanates tend to disrupt protein tertiary structures whilst halides promote dispersion forces. Both treatments counter the original electrostatic effect and serve to reverse the original depression of partition coefficients [4–8].

A balance of forces may be exploited in such systems. By variation of the polymer excluding and salting-out effects (changing PEG molecular weights or concentrations of phase-forming salts), subtle influence may be brought to bear on the susceptibility of different protein species (product or impurity) to the influence of neutral salt or pH manipulation. For example, in a primary-phase separation, the selection of proteins partitioned to the top phase can be influenced by the choice of PEG and/or tie-line length (TLL related to both PEG and salt concentrations) (see Figure 4.1). Subsequently, selective removal of a fraction of these top-phase proteins (products or impurities) may be achieved by reduction of TLL, pH variation or neutral salt addition with the introduction of a second bottom phase. These features permit the introduction of a degree of subtlety unexpected in such a simple, two-step process.

4.3 Strategic Selection of Systems for the Recovery of Intracellular Proteins

In the spirit of the introduction, and the key parameters outlined above, the practical philosophy of strategic system selection in generic applications is much more important than the retailing of detailed protocols which are commonly compromised by aspects of system specificity. However, procedures adopted in our laboratory over the last five years [4–13] for screening ATPSs for effectiveness in solving individual separation problems are worth an extended examination and discussion.

At the outset, the common aim is always to undertake the minimum of partitioning steps (preferably two) to transform a complex particulate suspension into a refined feedstock. This should stand alone as the end-product or be compatible with further conventional operations of molecular fractionation (ultrafiltration, precipitation, chromatography, etc.) for additional concentration, purification and formulation. In a standard procedure, two sequential steps of primary forward partition and secondary back extraction should yield the product in the final, salt-rich, bottom phase. Such a solution is ideally suited to subsequent processes of precipitation, ultrafiltration or hydrophobic chromatography should further purification be necessary [8,19]. By contrast, the polymeric content of the top phase is far less tractable. In optimal circumstances with disrupted waste brewers' or bakers' yeasts, such a process eliminates cell debris, nucleic acids, charged carbohydrates and pigments (flavins, tannins, polyphenols and products of time-dependent 'browning'). Our experience with other source species indicates that such behaviour could be expected of all microbial extracts generated from fermentations as whole broths or cell disruptates.

General process development, with accompanying investigative experimentation, is outlined point by point below. A summary is offered in Table 4.1.

Point 1   Yeast cells suspended at 40% wet v/v are disrupted by wet milling in a bead mill (type KDL 0.6 1 Dynomill) in a suitable, low ionic strength buffer of the same salt type and pH chosen for the primary-phase system. This enables accurate quantitation of the additional salt required for phase formation. Other buffers and additives (inhibitors, antioxidants, chelators, detergents, etc.) may be added at or post disruption, but their effects upon subsequent two-stage phase partitioning are likely to be unknown.

Table 4.1   Summary of practical strategies in two-stage (S1, S2) partition utilizing PEG–salt aqueous two-phase systems

---

Objective selective recovery of protein fractions from microbial cells
  *Establish phase diagrams for two-phase systems
  *Experiment initially with clarified feedstock
  *Determine sensitivity to PEG MW and phase composition
  *Determine sensitivity to pH and salt content
  *Use highest PEG MW for product-rich top phase
  *Adjust S1 volume ratio to maximize yield
  *Examine salt and pH effects in S2 for product-rich bottom phase
  *Exploit any variation of K with V_r to downscale bottom S2 phase

---

Point 2 In early trials, cell disruptates are partitioned in 20 g ATPS at solute concentrations that are substantial enough to reflect the final process but at values less than those finally required in order to eliminate the complication of any concentration effects [7]. Thus, in the yeast systems, trials are routinely run with a double dilution of the disruptate generated by wet-milling 40% wet w/v cell suspensions. Such disruptates are commonly difficult to clarify completely by processes of high-speed centrifugation or refined microfiltration [8,19]. Initial experimental conditions may be established with partially clarified material, but should be confirmed with whole suspension feeds. High-speed bench centrifugation of 20 g ATPS at forces ($>12\,000\,g$) greater than those anticipated in process separations maximizes the yield of liquid phases in development work. This facilitates the extensive characterization of the distribution of both product and contaminants essential for the full description of a working system.

Point 3 The overall practical strategy should be established at the outset. In the majority of instances, two sequential partitioning steps are sufficient to eliminate different groups of common contaminating impurities at each stage. In the primary, forward extraction (S1; see Table 4.1), a protein product is partitioned to the PEG-rich top phase to the exclusion of cell debris, nucleic acid and carbohydrate. Debris commonly occupies both the salt-rich bottom phase and the interface between the top and bottom phases.

Increases of volume ratio (see Figure 4.1) may be manipulated to maximize protein recovery in this first extraction [5,8]. In the secondary or back extraction (S2; see Table 4.1), the first top phase is contacted and mixed with a fresh, modified salt phase to re-partition the target produce away from protein and other contaminants (notably pigments) remaining from the forward extraction. However, in the light of the individual behaviour of proteins such as cytochrome *c* discussed above, it should be apparent that particular molecular characters may recommend modification of such a partitioning strategy.

Point 4 Potential practitioners should be aware that the components of ATPS, deriving both from the phase-forming chemicals and the biological feestock, may have profound effects on the biological activities of species and the assays exploited for their quantitation. For example, the activities of acidic proteases may be permanently lost in alkaline systems ($>$ pH 8). In addition, both phases of a PEG–phosphate ATPS are characterized by a significant salt concentration and potential buffering capacity which may affect assay systems, whilst the high working concentrations of both PEG and phosphate may influence measurable activities and binding kinetics with assay reagents, even after extensive dilution (or dialysis) of samples. The high osmotic pressures of sampled phases will also affect sample preparation, particularly for electrophoretic studies, and controlled desalting and lyophilization are advised for the most accurate quantitation.

Point 5 It is essential that developmental ATPSs are accurately located on tie-lines within a working-phase diagram which can be assembled by simple cloud-point procedures [1,7] (see Figure 4.1). Systems characterized by average molecular weights which lie between the nominal values attributed to the 'discrete' fractions available from commercial sources can be assembled by the appropriate weighted mixing of selected materials. Such a blending strategy is most accurate where the nominal molecular weights of polymers differ by only small amounts. In our work, the following molecular weights have been found to be most useful in manipulating the recovery of intracellular proteins from disrupted yeasts: PEG 400, 700, 1000, 1450, 3350 and 6000 daltons.

It should be strongly emphasized that commercial materials are not in themselves monodisperse with respect to molecular mass, and those mass distributions may well vary between commercial suppliers. System development should therefore be conducted with materials whose long-term supply and quality can be guaranteed.

Point 6 When preparing the primary, forward extraction system (S1), it is essential to partition a protein product to the top phase using as high a molecular weight of PEG as possible without incurring constraints of polymer exclusion. As will be clear from earlier discussion of β-galactosidase and potential heterologous constructs, the highest possible molecular masses of PEG applicable at this stage will yield the best overall product purities.

Systems should be chosen close to the binodal curve (Figure 4.1) to avoid or minimize product precipitation, but subsequent practical systems may be developed by working into the phase diagram on tie-lines characterized by higher concentrations of phase-forming components.

When establishing system conditions with dilute, clarified feedstock as discussed under point 1, it should be remembered that the presence of cells, cell debris and macromolecular solutes will critically affect the phase-forming characteristics of systems (e.g. biphasic systems resulting from apparently monophasic concentrations of PEG and salt; see Figure 4.1) and also limit phase recoveries, except at relatively high volume ratios.

Once prospective systems are identified, they should also be examined in respect of overall system pH and (in exceptional circumstances) addition of neutral salt (the latter is more applicable to the back extraction). In all cases, favourable systems should be critically judged in terms of overall product yield, concentration and purity.

Point 7 Having successfully established a potential forward extraction, a second system (S2) must be identified to efficiently back-extract the target product into a fresh, salt-rich bottom phase. Favourable systems should be examined for the effect of reduced tie-line length (TLL) (Figure 4.1) and system pH upon product partition. This may be best achieved by either the admixture of the acid and base salts or by addition of the appropriate acid, e.g.

orthophosphoric acid. The former reduces the phase-forming propensities of systems rather more than the latter. In addition, the impact of the addition of neutral salts up to concentrations of 0.5 moles/kg system mass should be examined in conjunction with system pH reduction, with the objective of reducing the partition coefficient of target products in the back extraction with benefit to overall yield and purity.

Point 8   It should now be possible to specify two separate systems differing in pH and/or salt content in which partition is directed to different phases. Such systems should be operated sequentially by taking the primary, product-rich top phase (S1) and reconstituting a new bottom phase by calculating the necessary amounts of phase-forming salt and water to be added from the phase diagram (see Figure 4.1). Appropriate adjustment to system TLL, pH and neutral salt content should be made to transfer the product across the interface to the new bottom phase.

Point 9   Further system optimization is now required. Product yield to the first top phase should be examined as the concentration of cell debris and solutes in the original feedstock is increased to maximize throughput. In the same context, if not already established, the addition of solid PEG and phosphate (as opposed to the use of concentrated stock solutions or other possible approaches to system assembly) should also be validated in terms of maintenance of quantitative phase partition and intermediate product purity. Adjustment of the volume ratio will be necessary to maximize the intermediate product yields. In the second stage, particularly in systems removed from the critical point, some dependence of the partition coefficient upon volume ratio may be observed [8]. However, in such circumstances, a useful reduction of the volume of the bottom phase may be achieved without profound effect on the yield of product.

Certain multivariate statistical techniques are conventionally applicable to these types of experimental development of a working strategy [20]. However, whilst the exact weighting of particular parameters remains obscured by an incomplete understanding of the mechanics of partitioning, the practical insight gained from the systematic examinations outlined above remain invaluable. Applications of statistical methods at later stages of system refinement may nevertheless prove to be rewarding.

4.4 Recovery of Protein Products from Phase Systems

A variety of conventional options are available for subsequent processing of the final, product-rich phase which for reasons outlined above is commonly the second, salt-rich phase. After appropriate adjustment of the salt content (increased or decreased concentration), the phase may be directly applied to hydrophobic chromatographic adsorbents derivatized with short alkyl, phenyl

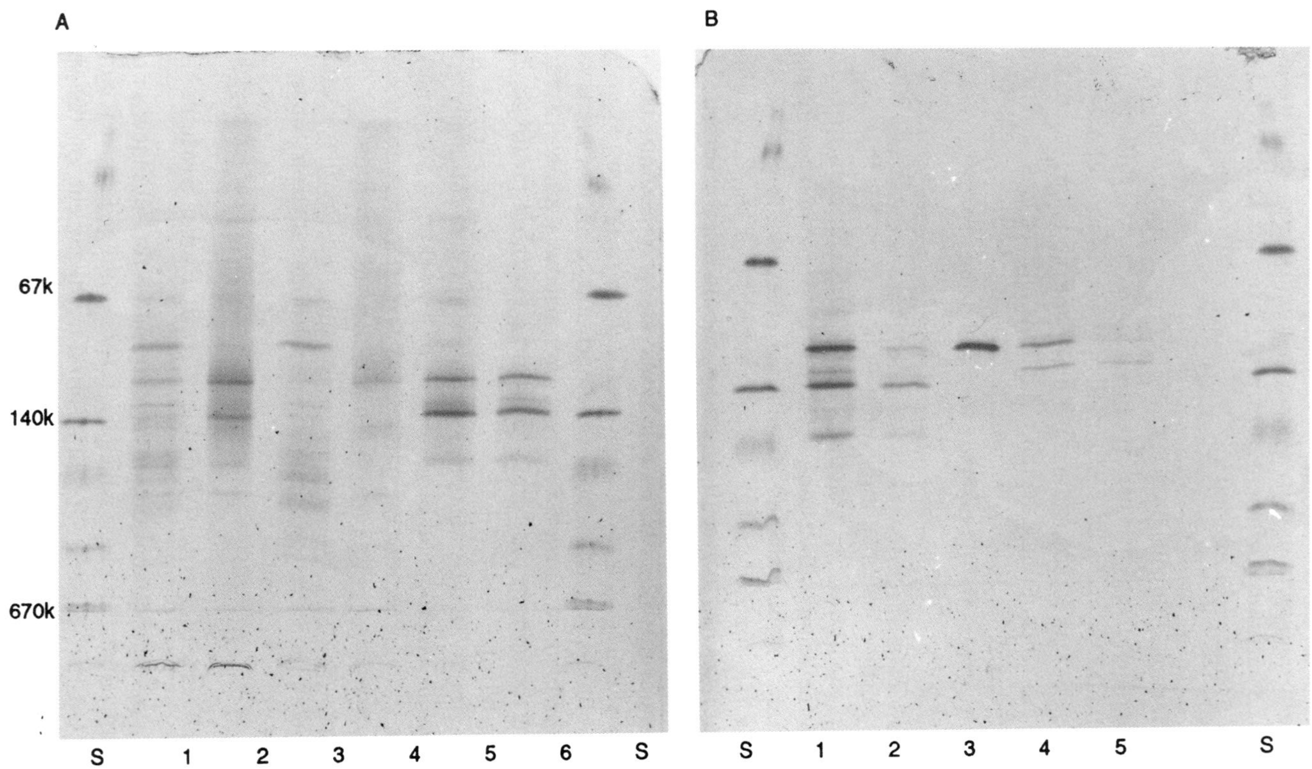
A
67k
140k
670k
S 1 2 3 4 5 6 S
B
S 1 2 3 4 5 S

or even polyether ligands [6,21]. It is not so widely recognized that bioaffinity interactions are also remarkably robust in high salt concentrations, particularly those possessed of a hydrophobic component, as demonstrated with the application of immunoadsorbents (Huddleston and Lyddiatt in Ref. [2]). It has been demonstrated that suitable adsorbents of this type may be directly employed in the two-phase system [22], and thus by inference and practical observation [2] could cope well with direct feeds of the second bottom phase in conventional fixed-bed contactors. Direct adsorption of product by both methods has the added advantage of facilitating buffer exchange (elimination of phase-forming components) since evolution will generally be achieved in a very different solvent system. In the case of hydrophobic interaction chromatography (HIC), the eluant is commonly of low ionic strength and thus desalting of the product is additionally achieved. Ion exchange adsorption would be possible in place of HIC, but only if instigated after successful diafiltration of the second bottom phase (S2).

Desalting by dia/ultrafiltration poses a number of practical problems. The high osmotic pressure of all system phases means that low permeate fluxes [19] and protein productivities are initially attained, even with the expensive utilization of large membrane areas. Permeate fluxes are further diminished by the low concentrations (1–3%) of residual PEG present in secondary bottom phases where the linear polymeric mature of even low molecular weight PEG species ($<$ 3000 daltons) constrains transport across commonly used ultra-filtration membranes (10 000 or 30 000 molecular mass rejection). Ultra- and diafiltration of secondary, PEG-rich top phases has proved to be near impossible without excessive, initial dilution and unacceptable time-scales in our laboratory [19].

An example of the integration of ultrafiltration and subsequent ion exchange adsorption/desorption with a two-stage ATPS applied to the recovery to near purity of a discrete, intracellular protein from waste brewers' yeast is illustrated in Figure 4.2.

Figure 4.2 Purification to near homogeneity of an intracellular enzyme from waste brewers' yeast by integrated processes of an ATPS and ion exchange adsorption. Native polyacrylamide gel electrophoresis was undertaken using clarified process samples and standard molecular weight markers (tracks S in Gels A and B; molecular weight range 67–670 kDa). Gel A illustrates crude yeast extract (track 1) which was partitioned in a low molecular weight PEG–phosphate system (S1) to yield top and bottom phases (tracks 2 and 3 respectively). Top phase from S1 was back-extracted into a second system (S2) by the introduction of a fresh bottom phase, distinguished by fresh phosphate, and adjustments to pH and salt content to yield new top and bottom phases (tracks 4 and 5 respectively). The latter was concentrated as a retentate (track 6) in ultrafiltration (10 kDa molecular rejection membrane). Gel B shows desalted material (track 1) fractionated by fixed-bed chromatography on DE-52 cellulose to yield unbound material (track 2) and successive peak fractions (tracks 3 to 5) collected from linear gradient elution (0–0.4 M NaCl)

4.5 Recovery of Phase-forming Components

The significant costs (albeit low compared to a polymer–polymer ATPS) entailed in formulating phases having high concentrations of polymers and salts (see Figure 4.1) has promoted a strong academic interest in the implications for process economy. One route, which is currently actively researched, concerns the adoption of novel polymers to reduce the amounts of solute required for phase formation. Increased product selectivity is an additional objective here. An alternative approach would involve the utilization of low cloud point temperatures to effectively eliminate a salt-phase recovery step from a polymer–polymer ATPS [23].

A second approach has concerned the investigation of ways of closing the process through recycling of phase-forming components to reduce or eliminate waste streams requiring costly treatment before disposal to the environment. In the simplest approach, the direct recycling of PEG-rich phases without any treatment or adjustment has been reported [24]. This has stimulated many developmental studies which assume the simple achievement of phase recycling [25,26]. However, in such schemes the presence of added neutral salts and accumulation of other solutes (particularly proteins and pigments resistant to back extraction) in the secondary top phase may be expected to promote cumulative, adverse effects upon the phase diagram, as well as product capacities, yields and purities. Activated carbon adsorbents may be effective in scavenging low molecular weight contaminants (particularly pigments) if suitable contractors can be devised to work with a rather viscous feedstock. Selected chlorinated hydrocarbons have been shown to be effective in extracting PEG from aqueous solutions. However, adoption of such a route to closed processing would incur high additional capital and operational costs in respect of solvent recycling.

Recent re-examination [27,28] of direct recycling of PEG-rich secondary top phases [24] has illuminated the value of carefully assembling two-phase systems specifically optimized for both system productivity *and* constituent recycling [27,28] (see Figure 4.3). In this work, repetitive recycling of a significant proportion of PEG (plus some naturally associated phosphate) has been achieved whilst maintaining constant productive recoveries of bulk proteins (42%) and discrete enzyme activities (fumarase 52% and pyruvate kinase 80%) over five complete cycles of a two-stage extraction as outlined above. Economic savings estimated on chemical inventories alone account for 40% of the cost of a process having no phase recycle.

Careful selection of operating tie-lines and pH for the back extraction is crucial here in avoiding (for example) accumulating concentrations of neutral salt, although it is perhaps surprising (yet encouraging) how robust the system is in the face of visibly accumulating pigment contaminants. The longer term prognosis for such a strategy of phase recycling is currently under test [29]. Recycling of phosphate-rich bottom phases from primary (S1) after product

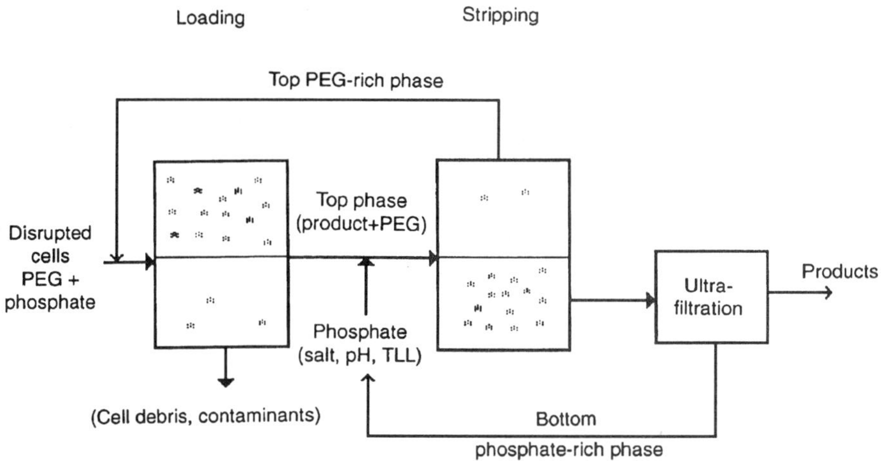

Figure 4.3 Schematic representation of a two-stage ATPS process with integrated phase recycling. The target protein product (*) is contained in a suspension feedstock derived from microbial cell disruptates comprising cell debris, other proteins, nucleic acids, carbohydrates and disparate macro- and micromolecular contaminants. An initial system (S1) preferentially recovers product in a PEG-rich top phase which, in contact with a fresh bottom phase (S2), is back-extracted to a salt-rich phase for ultra/ diafiltration and further processing (see Figure 4.2). Polymer and salt recovered as indicated from S2 can be recycled to lower the chemical inventory for this repetitive batch process (discussed further in Refs. [27] and [28])

depletion may prove more difficult in respect of contamination with cell debris. In addition, the PEG content (1–3%; see Figure 4.1) of secondary bottom phases (S2; depleted of product by ultrafiltration or other means) may compromise direct-phase recycle, but not necessarily the specific recovery and re-use of phase-forming solutes.

Countercurrent ethanol extraction has been proposed for the recycling of phosphate phases [30]. Such an approach has been judged to be cost effective in the face of the significant waste treatment charges associated with high phosphate concentrations. In principle, evaporation could be usefully adapted to effect dewatering in the recovery of both phases. This would be particularly true of the phosphate-rich phase where crystallization procedures could return material at high purity. However, the cost of evaporative recovery of phase components from an ATPS will be high due to the low vapour pressures involved. Such an approach might only prove economic where waste heat was available from other sectors of a process or adjacent process plants.

At the present time, the authors are unaware of any wholly closed, pilot or production plant designs for the production of biomolecules in an ATPS cited or outlined in the current literature. However, informal exchanges with practitioners in the field suggest that some such operations are planned or are under way. Except for the highest value products (for which an inefficient, conventional downstream process may be tolerated or even mandated by early implementation of clinical trials), it is essential that plant and operational design take account of any cost savings (real or relative) associated with successful phase recovery.

4.6 Conclusion

Summarized work with yeast intracellular products indicates that the technique of phase partitioning in PEG–phosphate aqueous two-phase systems (ATPSs) has a remarkable facility for achieving the recovery and partial purification of both intracellular and extracellular proteins sourced in microbial fermentations. In particular, the technique copes exceptionally well with obdurate debris, nucleic acid and carbohydrate contaminants in low force fields (low-speed centrifugation) which contrasts with the energy demands and technical sophistication required of refined methods of centrifugation and microfiltration.

In the absence of a mechanistic description of the partition process, attention to the detailed understanding of the impact of critical parameters will enable the assembly of a working system of protein recovery. This can readily be achieved by attention to the practical points detailed above. Successful systems may then be further optimized by empirical iteration or experimentation based upon statistical methodologies. Practical experience indicates that the key controlling variables include choice of polymer molecular weight, polymer and phosphate concentrations (indicated by tie-line length (TLL) in Figure 4.1), volume ratio, working pH and neutral salt concentrations.

Such methods are particularly suited to protein recovery from complex biological suspensions, and may be adaptable for the isolation of relatively discrete particles. This category would include refractile bodies, protein aggregates, membrane preparations, etc. [10], which are currently dependent upon scale- and throughput-limited processes of ultracentrifugation.

Acknowledgements

The authors are grateful for unpublished data and valuable practical commentaries offered by Julia A. Flanagan, Marco Ritɔ-Palomares and past and present members of the Biochemical Recovery Group.

4.7 References

1. H. Walter, D. E. Brooks and D. Fisher (Eds.) (1985) *Partitioning in Aqueous Two-Phase System*, Academic Press, New York.
2. D. Fisher and I. A. Sutherland (Eds.) (1989) *Separations Using Aqueous Phase Systems*, Plenum Press, New York.
3. M.-R. Kula, K. H. Kroner and H. Hustedt (1982) in A. Fiechter (Ed.), *Advances in Biochemical Engineering*, Vol. 24, Spinger-Verlag, Berlin, p. 73.

4. J. G. Huddleston and A. Lyddiatt (1991) *Appl. Biochem. Biotechnol.*, **26**, 249.
5. J. G. Huddleston, A. Viede, K. Kohler, J. A. Flanagan, S.-O. Enfors and A. Lyddiatt (1991) *Trends in Biotechnol.*, **9**, 381.
6. J. G. Huddleston, R. Wang and A. Lyddiatt (1994) *Biotechnol. Bioengng.*, **44**, in press.
7. J. G. Huddleston, K. W. Ottomar, D. M. Ngonyani and A. Lyddiatt (1991) *Enzyme and Microbiol. Technol.*, **13**, 24.
8. J. A. Flanagan, J. G. Huddleston and A. Lyddiatt (1991) *Bioseparation*, **2**, 43.
9. J. G. Huddleston, S. O'Brien, H. Picard and A. Lyddiatt (1993) in D. H. Logsdail and M. J. Slater (Eds.), *Solvent Extraction in the Process Industries*, Elsevier Applied Science, Oxford, p. 1048.
10. J. G. Huddleston (1994) Unpublished experiments.
11. C. J. Dale, S. G. Walker and A. Lyddiatt (1994) in D. L. Pyle (Ed.), *Separations for Biotechnology*, Vol. 3, Royal Society of Chemistry, Cambridge, p. 141.
12. K. P. Anathapademanaban and E. D. Goddard (1987) *Colloids and Surfaces*, **25**, 393.
13. J. G. Huddleston, R. Wang, J. A. Flanagan, S. O'Brien and A. Lyddiatt (1994) *J. Chromatogr.*, **668**, 1.
14. A. Veide, T. Lindback and S.-O. Enfors (1984) *Enzyme and Microbiol. Technol.*, **6**, 325.
15. H. M. Sassenfeld (1990) *Trends in Biotechnol.*, **8**, 88.
16. C. Hassinen, K. Kohler and A. Veide (1994) *J. Chromatogr.*, **668**, HM.
17. S. G. Walker, J. Huddleston and A. Lyddiatt (1994) in *The 1994 IChemE Research Event*, Vol. 1, Institution of Chemical Engineers, Rugby (ISBN 0 85295 320 8), p. 156.
18. J. G. Huddleston, K. W. Ottomar, D. Ngonyani, J. A. Flanagan and A. Lyddiatt (1990) in D. L. Pyle (Ed.), *Separations for Biotechnology*, Vol. 2, Elsevier, Oxford, p. 181.
19. J. A. Flanagan. Thesis submitted for the degree of PhD, University of Birmingham.
20. M. W. Routh, P. A. Swartz and M. B. Denton (1977) *Anal. Chem.*, **49**, 1422.
21. J. M. Duval, C. Delestre, M. C. Carre, P. Hubert and E. Dellacherie (1991) *Carbohydrate Polymers*, **115**, 233.
22. J. G. Huddleston, A. Lyddiatt, R. Kuboi, R. Suzaka, W.-H. Wang and I. Komasawa (1992) in T. Sekine (Ed.), *Process Metallurgy*, Vol. 7B, *Solvent Extraction. Proc. ISEC'90 Part B*, Elsevier, Oxford, p. 1767.
23. P. A. Harris, G. Karlstrom and F. Tjeneld (1991) *Bioseparation*, **2**, 237.
24. H. Hustedt (1986) *Biotechnol. Lett.*, **8**, 791.
25. N. Papamicael, B. Borner and H. Hustedt (1992) *J. Chem. Tech. Biotechnol.*, **54**, 47.
26. S. L. Mistry, J. A. Asenjo and C. A. Zaror (1993) *Bioseparation*, **3**, 343.
27. M. Rito-Palomares, J. G. Huddleston and A. Lyddiatt (1994) *Trans. IChemE Part C*, **72**, 11–13.
28. M. A. Rito-Palomares, J. G. Huddleston and A. Lyddiatt (1994) in D. L. Pyle (Ed.), *Separations for Biotechnology*, Vol. 3, Royal Society of Chemistry, London, p. 413.
29. M. A. Rito-Palomares (1994) Unpublished experiments.
30. A. Greve and M.-R. Kula (1990) *J. Chem. Tech. Biotechnol.*, **50**, 27.

5 SOLVENT EXTRACTION OF FERMENTATION BROTH

Laurence R. Weatherley

5.1 Introduction

The direct extraction of valuable products such as pharmaceuticals from untreated fermentation broths is of growing importance as the manufacture of both low-volume and bulk products via biotechnological routes increases. The separation problem presented by production of soluble material via such routes has a number of aspects which influence the nature of the extraction technique and the feasibility of using liquid–liquid extraction advantageously in comparison with other potential separation techniques such as ion exchange and membrane filtration. Generally the liquors exiting a fermenter are dilute solutions of the product in a large volume of water; many of the secondary components are poorly defined and are present in concentrations that can vary according to the precise conditions under which the fermentation was carried out. Table 5.1 shows the primary concentrations typically encountered for a

Table 5.1 Examples of products from biochemical reactions

High volume	$> 200\,m^3$ per batch	Reactor exit concentration (wt%)
Single-cell proteins		3.5
Ethanol		7–12
Lactic acid		5
Citric acid		10
Medium volume	50–$200\,m^3$ per batch	Reactor exit concentration (wt%)
Penicillin G		3.5
Cephalosporin		3
Streptomycin		1
Extracellular enzymes		0.5–1.0
Small scale	$< 50\,m^3$ per batch	Reactor exit concentration (wt%)
Vitamin B		0.005
Riboflavin		0.1

Downstream Processing of Natural Products. Edited by Michael S. Verrall
©1996 John Wiley & Sons Ltd

selection of bioproducts. Another important factor is the stability of the product. In the case of unstable products a rapid separation process is necessary in order to avoid significant loss of product.

There are also a number of positive features of biological routes for product synthesis. In certain cases the biotechnological route is the sole method of manufacturing the product. In general, processes are conducted at near ambient conditions of temperature and pressure and therefore there may be significant savings with respect to equipment capital costs and energy costs compared with conventional technology operating at high temperature and pressure. The selective production of stereo-specific compounds is also possible via microbiological routes, an option that may not be readily available using conventional chemical synthesis.

Liquid–liquid extraction is a versatile separation process which has been developed and applied extensively in a number of important process industries including nuclear fuel recycling, the hydrometallurgical industry, the petrochemicals industry and in the pharmaceutical industry. The principle of the technique (see Figure 5.1) involves efficient contact of two liquid phases, a feed phase and an extracting solvent. The phases are contacted by dispersing one liquid as a drop dispersion into the second liquid which remains as a continuous phase. The chemical properties of the solvent are chosen to facilitate selective uptake of the desired product from the feed phase. In this stage of the process, mass transfer of solute from the feed phase into the solvent extract phase takes place. The rate of extraction during contact exerts a fundamental influence upon the size and throughput of equipment. High mass transfer rates are required to achieve short residence times and thus small equipment size.

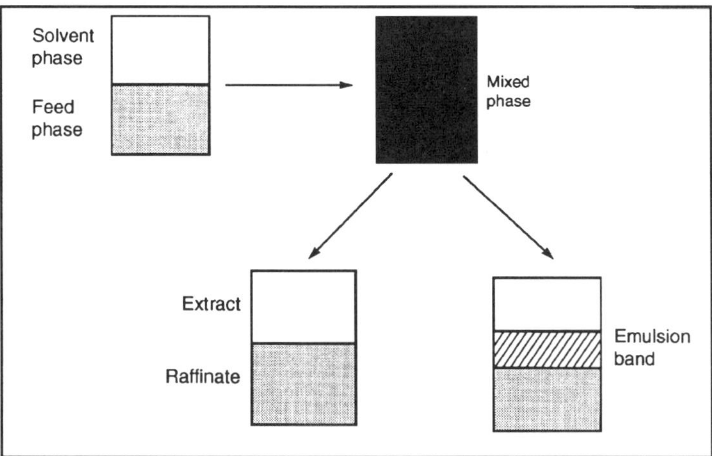

Figure 5.1 Principle of liquid–liquid extraction

There is an extensive literature dealing with the principal extraction mechanisms which included physical extraction, ion-pair extraction and liquid ion exchange (e.g. see Refs. [1] and [2]). The separation conditions during liquid–liquid extraction are controlled according to the chemical environment reflected by, for example pH, salt concentration and control of valency state.

The physical properties of the solvent are chosen to facilitate efficient mass transfer and to ensure rapid disengagement of the two phases after contact. This is also of fundamental importance to the efficiency of the extraction process. Significant entrainment of either phase in the other can represent appreciable loss of process efficiency and should be minimized. The chemical basis of solvent extraction means that a wide degree of control of the process is possible and makes solvent extraction well suited to very effective separation of complex mixtures into their individual constituent components in a sequential process, conducted entirely in the liquid phase. This flexibility has been used to great effect in the metals separation industry and in the nuclear industry and has led to the adoption of simpler process flowsheets which avoid significant solid–liquid handling and separation operations and has reduced materials inventory [3].

Further potential improvements in the overall efficiency of separation processes may be achieved by reducing the number of process units required. In the context of biological product recovery by extraction this potential may be realized by reducing or removing pretreatment processes such as filtration and extracting the desired product directly from whole broth. There have been demonstrable advantages in adopting direct extraction in terms of a simpler process and in higher overall product recoveries. For example, in the case of penicillin recovery it is estimated that losses of up to 15% of the penicillin can be incurred during the filtration of the fermentation broth. The use of prefilters can also give rise to problems of infection and blockage. There is also evidence that, despite filtration, solids contamination of the rich feed liquor can still occur, leading to maloperation of extractors which are not designed to operate in the presence of solid material. Therefore there is a strong incentive for the design of whole broth extraction processes and appropriate equipment.

Implementation of the concept of 'whole broth' extraction is not without potential difficulties which arise due to the presence of solid biomass during extraction and due to the presence of highly surface active impurities such as proteins, polysaccharides and lipids. These materials can seriously inhibit interfacial mass transfer kinetics and can also significantly prejudice phase separation.

The reliable operation of commercial-scale liquid–liquid extraction processes is heavily dependent upon the ability to maintain consistent rates of mass transfer and efficient phase disengagement kinetics. These two basic criteria apply to all classes of extraction operation including batch processes and continuous contact processes, such as in column contactors, and to stagewise operations such as in mixer settlers.

The design of solvent extraction processes for biological products is significantly influenced by mass transfer and by phase separation considerations. The complex nature of process streams exiting from fermentation or other bioreactor systems gives rise to potential difficulties with respect to slow mass transfer and with respect to complex interfacial behaviour arising from the highly surface active nature of the mixture.

Therefore the feasible operation of direct extraction processes requires that the effects of these adverse phenomena be addressed in the design of contacting equipment. In order to achieve this, certain aspects of the fundamental behaviour of whole broth extraction systems require careful consideration. The principle considerations include modifications to drop behaviour, interfacial phenomena and contacting, and these are now considered.

5.2 Drop Behaviour

The physical behaviour of liquid–liquid mixtures in the presence of gross impurities and biomass solids is of fundamental importance in many industrial extraction processes. In direct extraction processes where there is less pretreatment of rich feed liquors prior to contacting, modifications in drop behaviour can influence interfacial mass transfer in a number of ways. Intradroplet mixing and circulation processes may be inhibited by the presence of mycelial solids and by modifications in fluid rheology due to the presence of soluble biopolymers. These effects exert an important influence upon the overall rate of mass transfer and also strongly influence physical contacting processes including coalescence, drop motion dynamics and drop size distribution.

Early fundamental work by Garner and Skelland [4] proposed that the unhindered motion of fluid droplets dispersing from a nozzle in the non-jetting region into a second immiscible fluid differed from the free fall motion of solid spheres in three respects:

(a) Frictional drag of the surrounding fluid may induce circulation of the interface and the interior of the drop.
(b) Gravitational forces acting on the drop may cause its shape to depart from exact sphericity and assume a degree of oblateness.
(c) Prolate–oblate type oscillations may occur during the motion of the drop.

It is generally accepted that modelling of single-drop motion in a second immiscible liquid medium can conveniently use the motion of rigid spheres as a reference point for determining the deviation of terminal velocity behaviour of non-rigid drops due to the above effects.

Recent research [5] into the drop behaviour of mycelial broth liquid–liquid systems attributes substantial modification in drop mechanics to the presence of:

(a) surface active agents,
(b) solid biomass.

The behaviour of untreated and filtered fermentation liquors may be compared with the corresponding behaviour of drops of pure water dispersing under the same conditions. The experimental relationship between drop size and terminal velocity in unhindered motion may be compared with the predictions of correlations validated for 'solid-free' liquid–liquid systems. Research shows that surface active agents and the presence of undissolved biomass could both play an important role in determining the mechanics of drop motion and that there are important differences between the behaviour of truly pure systems and that of contaminated systems such as those involving fermentation liquors.

Predicted terminal velocity behaviour of pure water deviates significantly from that predicted using correlations by Hu and Kintner [6] and by Klee and Treybal [7] which were apparently based upon experiments in slightly contaminated systems. This was corroborated by data presented by Thorsen *et al.* [8] proposing that the presence of very small concentrations of surface active agents modify drop rigidity, internal circulation and thus terminal velocity. This view was further confirmed by Skelland *et al.* [9] in which drops which were artificially doped with surface active molecules showed significant modification with respect to drop size and terminal velocity behaviour.

One of the difficulties encountered during the experimental characterization of whole broth behaviour is that of variability, which is reflected in the scatter of experimental data. Despite this, research has shown a tendency of fermentation broth derived liquors to behave as rigid spheres, which is in contrast to the behaviour of a pure liquid phase such as water. The

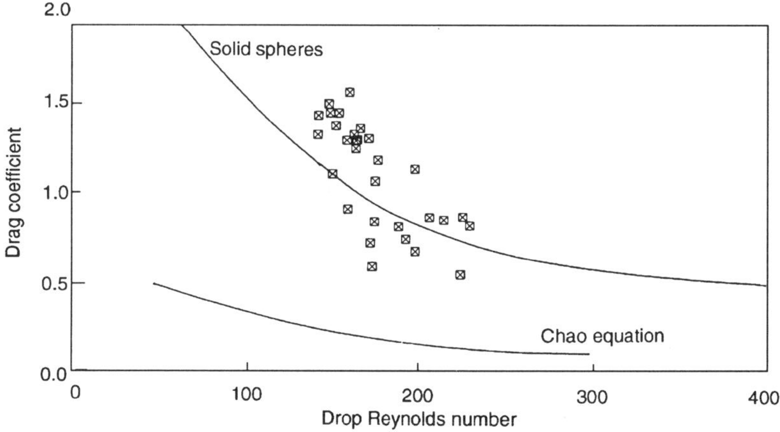

Figure 5.2 Drag coefficient–droplet Reynolds number relationship for single whole broth drops: comparison with the Winnikow and Chao predictions. (Drop phase: *Penicillium chrysogenum* mycelial culture broth. Continuous phase: 30% v/v tri-n-butyl phosphate in heavy distillate)

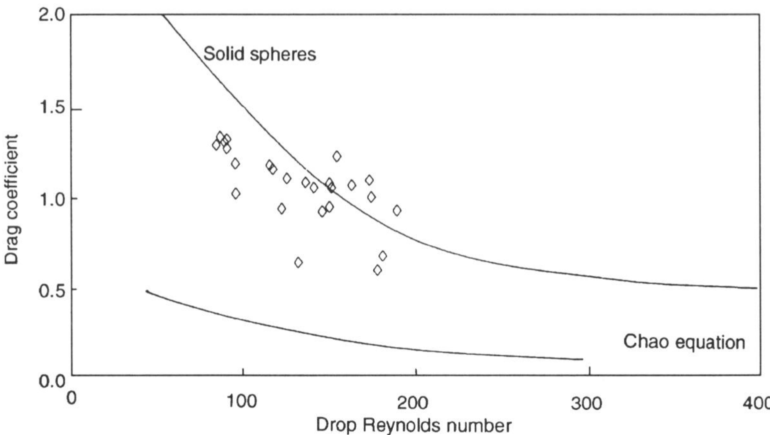

Figure 5.3 Drag coefficient–droplet Reynolds number relationship for single droplets of filtered broth: comparison with the Winnikow and Chao predictions. (Drop phase: *Penicillium chrysogenum* mycelial culture broth. Continuous phase: 30% v/v tri-n-butyl phosphate in heavy distillate)

experimental relationship between the drag coefficient and the Reynolds number based on the terminal velocity for a mycelial culture broth showed that it is comparable with that for rigid spheres (see Figure 5.2), as predicted using the correlation of Winnikow and Chao [10]. Drops of filtered broth exhibit an intermediate behaviour (see Figure 5.3) in which a lesser increase in drop rigidity could be attributed to the presence of surface active agents, but not to the presence of mycelia. It was concluded that the presence of mycelia and the presence of soluble contaminants affect the terminal velocity behaviour.

Further evidence of the effects of impurities upon drop behaviour in whole broth systems is also forthcoming from comparisons of experimental terminal velocity data with other correlations. Experimental terminal velocity measurements for drops of mycelial culture broth may be compared with each of the predictions of the Hu and Kintner model and the Klee and Treybal model, in addition to the model of Chao.

Figure 5.4 shows the comparison with the Hu and Kintner model. The calculated parameters of X and Y match the experimental values determined using the droplets of an untreated broth and using drops of a filtered broth with reasonable accuracy. The points for pure water, however, are well away from the line of the model and confirm that Hu and Kintner's development appears to be more accurate for liquid–liquid systems in which the dispersed phase is contaminated. It is reasonable to assume that the suspensions of biomass contain traces of surface active material which would explain the closeness of fit of this model.

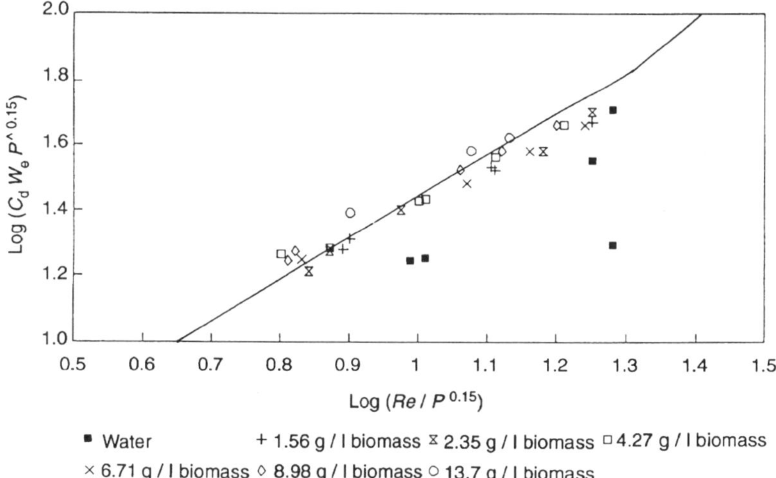

Figure 5.4 Terminal velocity behaviour of droplets: comparison of experimental data with the Hu and Kintner model for whole broth drops, filtered broth drops and pure water drops. (Drop phase: *Penicillium chrysogenum* mycelial culture broth. Continuous phase: 30% v/v tri-n-butyl phosphate in heavy distillate)

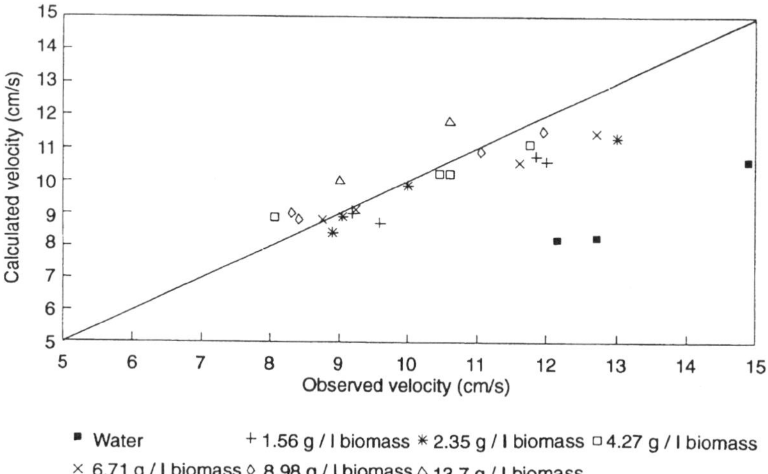

Figure 5.5 Terminal velocity behaviour of droplets: comparison of experimental data with the Klee and Treybal model for whole broth drops, filtered broth drops and pure water drops. (Drop phase: *Penicillium chrysogenum* mycelial culture broth. Continuous phase: 30% v/v tri-n-butyl phosphate in heavy distillate)

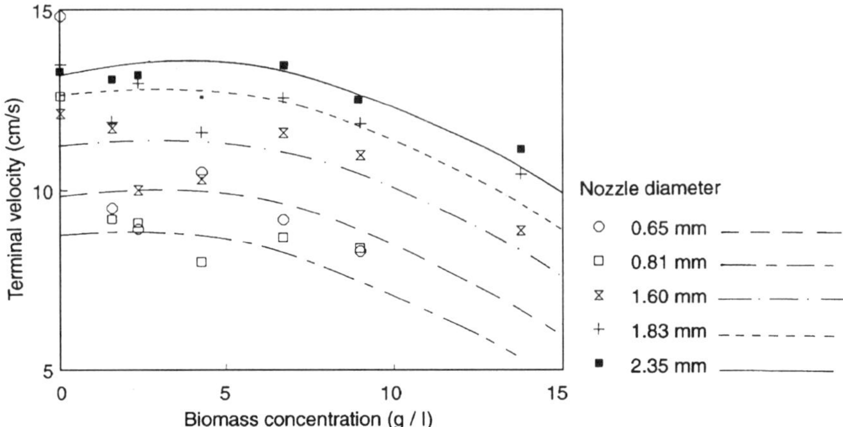

Figure 5.6   The effect of biomass concentration upon terminal velocity of single drops in unhindered motion. (Drop phase: *Penicillium chrysogenum* mycelial culture broth. Continuous phase: 30% v/v tri-n-butyl phosphate in heavy distillate)

A similar observation may be made in Figure 5.5 where the experimental values of droplet terminal velocity are compared with the predictions of the Klee and Treybal model. Again the data for the untreated broth and for the filtered broth demonstrate a better fit compared with that for the droplets of pure water. This is consistent with the fact that Klee and Treybal's correlation was formulated from experimental data obtained using commercial-grade reagents.

Figure 5.6 shows the effect of biomass concentration upon the terminal velocity [11] and there is clear evidence of a maximum value of terminal velocity with respect to biomass concentration. Even if the effects of soluble surface active agents are ignored, the effects of biomass upon droplet rheology can to some extent explain the observed behaviour. There are three possible contributing effects which could influence the final velocity attained by the descending droplet: density, rigidity and internal circulation and surface topology.

Density is not thought to have a significant effect here since the changes with respect to biomass concentration are small and would not influence terminal velocity to the extent observed.

Droplet rigidity increases as the viscosity of the phase increases with respect to biomass, and this would be consistent with the observed tendency towards rigid sphere behaviour (see Figure 5.2). On this basis alone the terminal velocity would be expected to increase with respect to biomass content. Internal circulation has an effect upon terminal velocity which is well known [4], and in general terms decrease in drop internal circulation is accompanied by a decrease in terminal velocity. The droplets containing biomass have a structure resulting from the presence of mycelia.

Drop rigidity increases with respect to biomass concentration and thus the degree of internal circulation would be expected to decrease with an increase in biomass concentration. Consideration of the drop rigidity and internal circulation together thus presents two opposing effects with respect to terminal velocity. On this basis it could be that the positive slopes of the curves shown in Figure 5.6 are explained by predominance of the positive effect of drop rigidity upon velocity. As the biomass content increases, however, the negative effect of biomass content upon internal circulation becomes the dominant effect and results in the observed maximum value in terminal velocity and the subsequent negative slope portions of the curves. An increase in frictional drag due to a reduction in the smoothness of the drop surface is a further possibility as the biomass concentration within the drops increases. It is possible that micelle type entities deriving from the biomass orientate across the interface, thus changing the topology of the drop surface, although microscopic evidence of such an effect has yet to be determined.

Results of this recent work clearly indicate possible mechanistic reasons for the reduction in the mass transfer rate. The following propositions were made:

(a) Droplet rigidity increases with respect to biomass concentration, leading to reduced internal droplet circulation, implying a reduction in eddy diffusion.
(b) The presence of surface active agents stabilizes the interface, an effect that manifested itself in lower droplet terminal velocities behaviour of both *filtered* broths and unfiltered broths. There are substantial difficulties in obtaining reliable terminal velocity data for dispersions of untreated, unhomogenized mycelial fermentation broths. The role of the biomass solids is clearly important in terms of their effect on droplet rheology and thus internal circulation and terminal velocity behaviour.

The tendency towards rigid droplet behaviour exhibited by both untreated and filtered broth droplets indicates an important role of surface active agent species in the inhibition of drop internal circulation. The observed modifications in terminal velocity behaviour indicate a strong influence of the presence of insoluble biomass and surface active agents upon droplet dynamics.

There are very clear implications for mass transfer behaviour. The reduction in internal circulation and droplet oscillation evident from the terminal velocity behaviour suggest that significant reductions in internal drop mass transfer would be expected. The reduction in terminal velocity compared to pure solution behaviour has negative implications with respect to contactor capacity in the context of continuous processing, and in addition would suggest reductions in continuous phase mass transfer for droplets undergoing either hindered or unhindered settling in a continuous contactor.

 LAURENCE R. WEATHERLEY

5.3 Mass Transfer

Direct extraction of untreated broths involves fewer process steps and in certain cases improves overall yield. However, in addition to difficulties of phase separation there is also evidence of slow mass transfer during whole broth extraction. A clear understanding of the factors that influence the retardation of the mass transfer rate and an insight into the relative importance of the various mechanisms that are potentially rate limiting are vital for future process and equipment design.

There are three major factors that would prejudice mass transfer kinetics in the context of whole broth extraction:

(a) complex rheology,
(b) the presence of surface active agents,
(c) the presence of solid biomass.

Published data on these possible effects within the context of whole broth extraction are scarce. In principle, complex rheology in either the dispersed or continuous phase can reduce diffusional fluxes. For example, within a single drop the eddy diffusion component of mass transfer flux will be reduced as the effective viscosity increases and the dominant mechanism for transport of

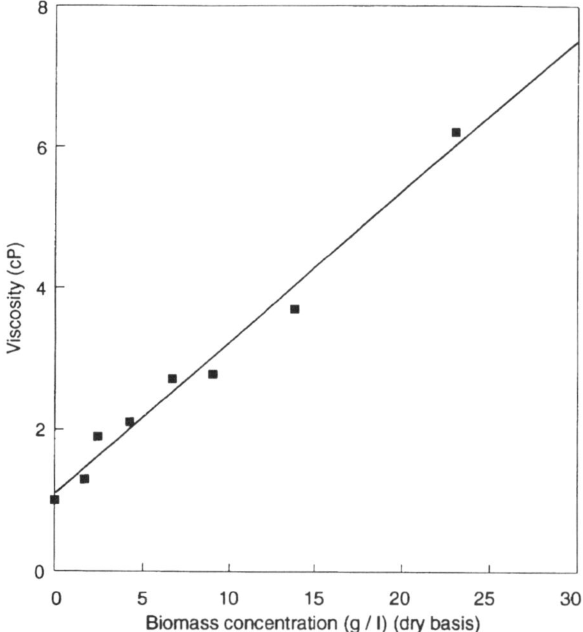

Figure 5.7 Variation of mean viscosity with biomass concentration for a *Penicillium chrysogenum* mycelial culture broth

solute through the drop will tend to molecular diffusion, thus reflected in an overall reduction in the mass transfer rate. Figure 5.7 shows how the mean viscosity of a mycelial culture broth varies with respect to the concentration of biomass [12]. Translated into a change in the dispersed-phase mass transfer coefficient, using, for example, the correlation of Handlos and Baron [13], the increase in viscosity would translate into a reduction in the drop-phase mass transfer coefficient by approximately a factor of three (see Figure 5.8).

An increase in viscosity would not be the only factor with a negative effect upon mass transfer. Another well-known factor is the effect that the presence of surface active agents has upon interfacial mass transfer and there are many data for systems involving non-biological surfactants that support this. Skelland and Caenpeel [14] summarize the major mechanisms that can inhibit mass transfer due to the presence of surface active agents and these are listed as follows:

(a) formation of an interfacial barrier which may be of a mechanical, physical or chemical nature;
(b) modification of the hydrodynamics by reduction of internal circulation in drops, and damping of interfacial turbulence;

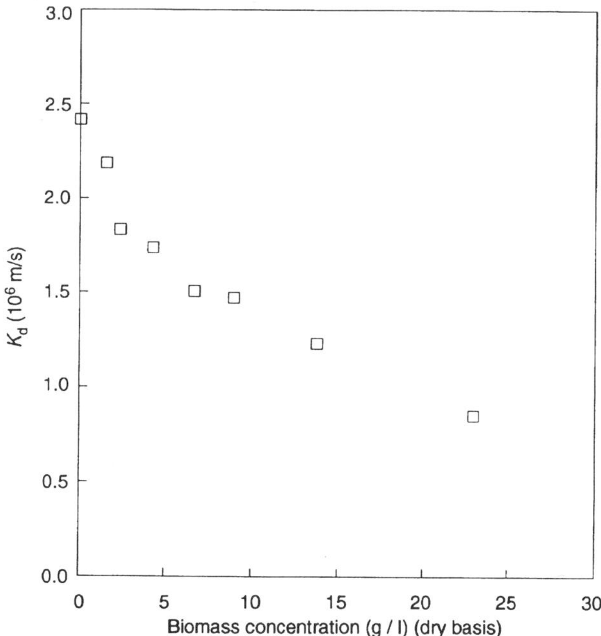

Figure 5.8 Variation of predicted drop side mass transfer coefficient with respect to biomass concentration due to changes in mean viscosity using the Handlos and Baron equation

(c) decrease in the terminal velocity of drops, which is referred to in an earlier section;

(d) reduction in interfacial tension, which lessens transmission of turbulence across the interface;

(e) retardation of the rate of coalescence, which is a particularly important mechanism for increasing drop side mass transfer coefficients in swarming drop systems.

The presence of biomass solids is the third important factor in determining the kinetics of whole broth extraction and can significantly modify rate processes in the vicinity of the liquid–liquid interface. Crabbe *et al.* [15,16] present data for solvent extraction of ethanol from aqueous yeast suspensions (*Saccharomyces cerevisiae*) into decan-1-ol. These showed substantial reductions in the rate of mass transfer. A sample of their data is replotted in Figure 5.9 and illustrates the stepwise reduction in the mass transfer coefficient as the yeast concentration in the feed phase is increased across the range 0.001–20 gL/L (dry basis). One of the most significant aspects of these observations is the very steep reduction in the

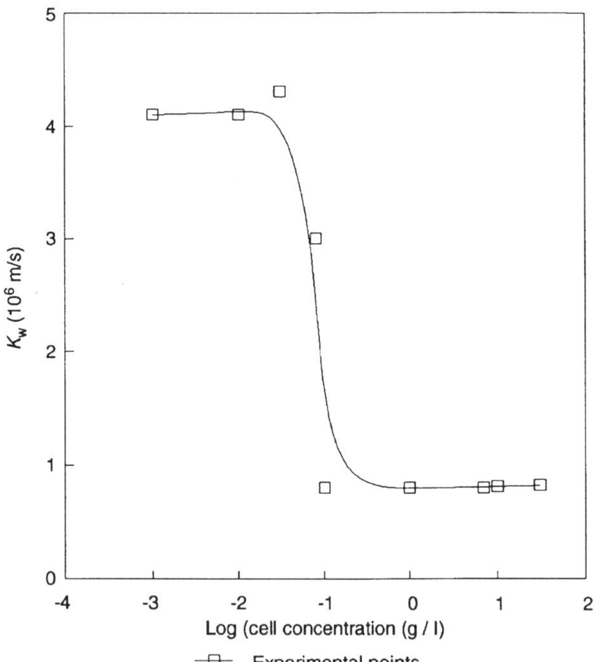

Figure 5.9 Experimental relationship between observed mass transfer coefficient and biomass concentration for ethanol extraction from suspensions of *Saccharomyces cerevisiae* [15] into decan-1-ol

mass transfer coefficient — in this case an 83% reduction was observed — which occurs across a very narrow range of biomass concentrations — in this case between 0.02 and 0.1 g/L. Microscopic examination of the liquid–liquid interface suggested that the reduction could be explained by adsorption of yeast cells at the interface. the effect of adsorbed biomass at the interface could significantly dampen turbulence at the interface and thus explain the reduction in the mass transfer coefficient. At concentrations beyond the 0.1 g/L level there is very little further reduction in the mass transfer coefficient observed and this suggests that the interface 'fills', with biomass rapidly producing a large and sudden reduction in mass transfer. At this stage the interface is relatively insensitive to the presence of further biomass which may be present in the bulk liquid phase at higher concentrations. Micrographic data presented by Doherty *et al.* [17] support this view and show that yeast cells tended to line up along a liquid–liquid interface.

These data were for suspensions of yeast cells in water and did not involve actual fermentation broths. Measurements for the extraction of ethanol from yeast based fermentation broths and from dosed mycelial culture broths in spray columns confirm reductions in the observed mass transfer rate (see Refs. [18] and [19]).

The general observations relevant to mass transfer in whole broth extraction systems are as follows. Firstly, the mass transfer coefficient declines rapidly as the biomass concentration increases for each system. Secondly, it is clear that the mass transfer reduction is likely to be greater in the case of the mycelial broth systems. Available data from observations of extraction of filtered broth systems also suggest that the presence of soluble species such as natural surfactants exerts a significant influence upon the rate of mass transfer, probably due to a combination of rheological effects and the inhibition of interfacial turbulence. This observation is consistent with other observations pertinent to drop dynamic behaviour alone referred to earlier in the chapter.

5.4 Contacting

Literature reference to the use of continuous contact equipment for either whole broth extraction or indeed any type of bioproduct recovery is largely confined to pilot-scale column contactors and to centrifugal contactors.

The reciprocating Karr column has been used with some success for a range of pilot-scale biological product extractions. The Karr column (see Figure 5.10), has been used widely for a range of products recovered by liquid–liquid extraction and include antibiotics from whole broth, penicillin, copper, ethanoic acid and other products. Muller *et al.* [20] describe the use of a reciprocating Karr column for the reactive extraction of penicillin from concentrated buffer solutions at high pH with encouraging results. The success of this application depended to some degree upon the absence of any stringent requirement for a short residence time contact.

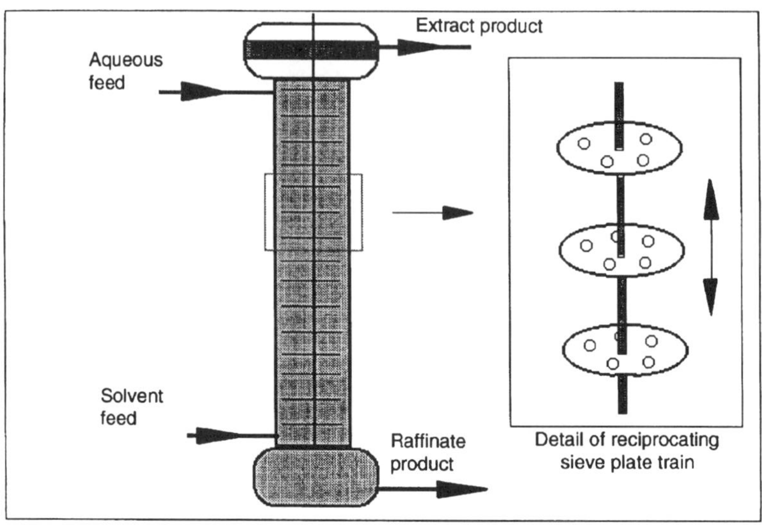

Figure 5.10   The reciprocating Karr column. Outline arrangement of flow

Karr *et al.* [21] discuss applications of the Karr column for whole broth separation. They compared the performance of the Karr reciprocating column with the Podbielnak centrifugal extractor for whole broth extraction. In the study it was found that the broth required three passes through the Podbielnak extractor compared with just one for the Karr column in order to receive the same recovery. Lower overall operating costs are claimed in the case of the Karr column with a reduction of 55% in solvent costs. The Karr column has also been used successfully in the demonstration of extractive fermentation processes in which the ability to achieve successful direct extraction is highly desirable [22].

The Karr column is well suited to whole broth extraction as the shear rate is relatively uniform throughout the contactor and can be controlled according to the rate of reciprocation. The drop size within the column can be controlled and therefore the formation of stable emulsions is less likely. Another advantage of the Karr column is that scale-up on the basis of pilot-scale data is relatively straightforward.

Other work on continuous contact for biochemical recoveries also includes the work of Minier and Gomer [23], demonstrating the potential of such an approach for simultaneous bioconversion and recovery of ethanol via immobilized yeast fermentation.

De Filipi and Moses [24] describe the solvent extraction of ethanol from concentrated broths using near-critical carbon dioxide as solvent in continuous sieve plate extraction columns. There are circumstances in which the hydraulic

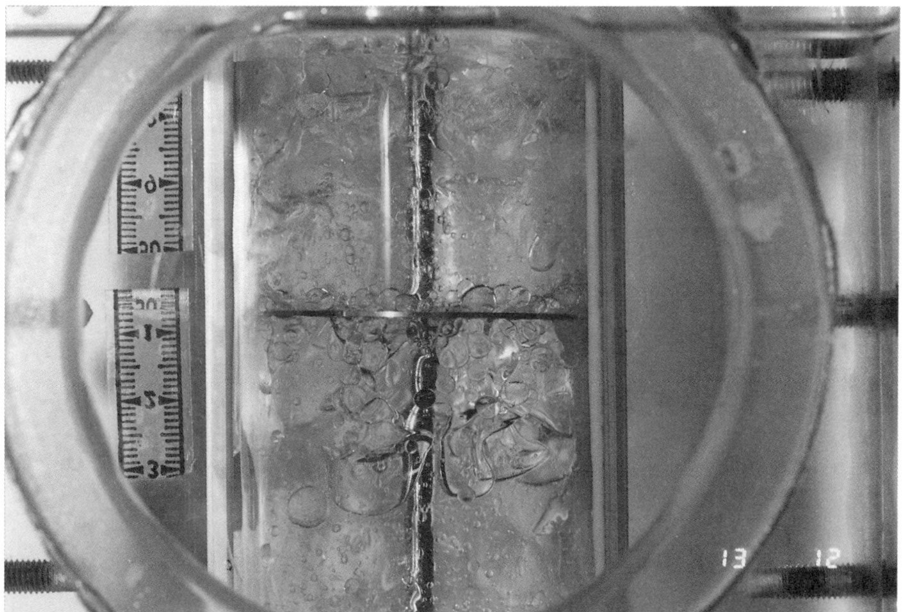

Figure 5.11 Water lamella foam formation observed during the filtered broth extraction of fermented malt extract liquor into 30% v/v tri-n-butyl phosphate/heavy distillate solvent in a pulsed plate column

behaviour in columns in the presence of fermentation debris can cause major problems.

Grintzos [25] reported on the contact of fermented malt extract solutions in the presence of *Saccharomyces cerevisiae* with solvent solutions of tributyl phosphate in the absence of mass transfer. The experiments showed that it was not possible to achieve any significant aqueous flowrate without flooding the column. However, the flooding was not of a conventional nature and appeared to be caused by the creation of a biliquid foam in the immediate vicinity of the dispersed phase inlet. The observed results of the phenomena are shown in Figure 5.11 and depict the development of the water lamella foam as the fermentation broth was introduced into the top of the active section of the column. The droplets shown are in fact droplets of organic phase surrounded by a thin film of aqueous material suspended in a continuous organic phase. The nature of this new phase appeared to be that of a water–lamella biliquid foam defined by Sebba [26].

The photograph in Figure 5.12 shows a top view of the biliquid foam formed by intimate mixing of the broth and organic phases. The region near the outer edge of the foam shows that certain parts of the biliquid phase have a filamentous morphology. Further work showed that addition of propan-2-ol can destabilize the biliquid foam and significantly reduce its collapse time. It is

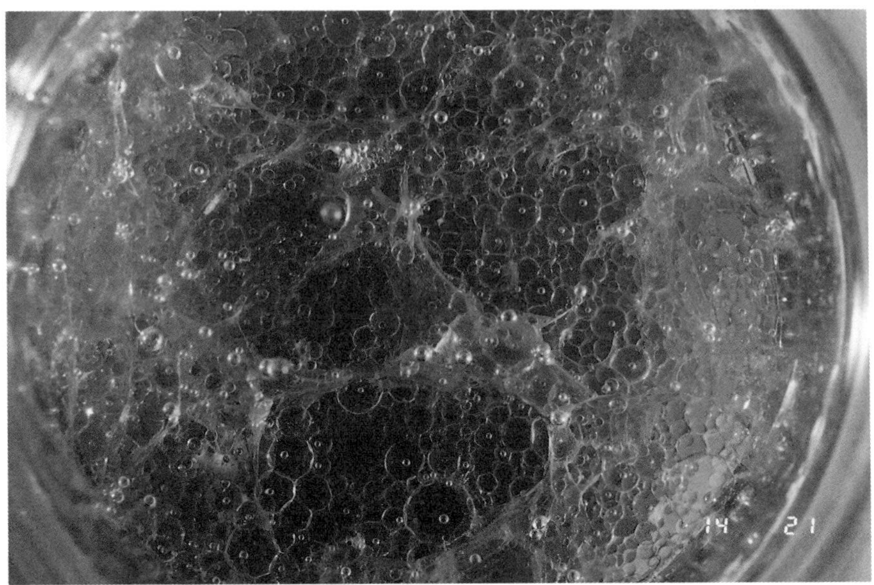

Figure 5.12   Biliquid foam formation observed during the mixing of fermented malt
extract liquor with 30% v/v tri-n-butyl phosphate/heavy distillate solvent

probable that the presence of soluble proteins play an important role in the
formation of the additional interfaces observed in the biliquid system. The role
of such molecules could possibly be altered by either modification of the ionic
strength in the broth prior to contact or by altering the ambient temperature
within the column.

Centrifugal contactors have been used with great success for many years for
the recovery of antibiotics from fermentation broths. The centrifugal contactor
offers the facility of rapid short residence time contact which is necessary for
efficient recovery; this will be discussed here briefly. One of the earliest
centrifugal contactors used for broth extraction is the Podbielnak contactor
(APV Baker) which is still in extensive use. A cutaway of the contactor is
shown in Figure 5.13 (APV Chemical Machinery Inc.). The unit is made up of
a stainless steel rotor mounted on a central rotating shaft. The rotor comprises
a number of perforated concentric metal strips which act to disperse and mix
the immiscible liquid phases in a highly efficient counter-contact cycle. The
rotor arrangement is also equipped with ports for the introduction and
removal of the contacting phases at the centre of the unit and at the periphery.

Early work with the Podbielnak contactor by Anderson and Lau [27]
highlighted the potential for improving fermentation broth recoveries by direct
extraction of the untreated liquor. Experience of whole broth extraction of
chloramphenicol broth and penicillin broth showed that considerable cost

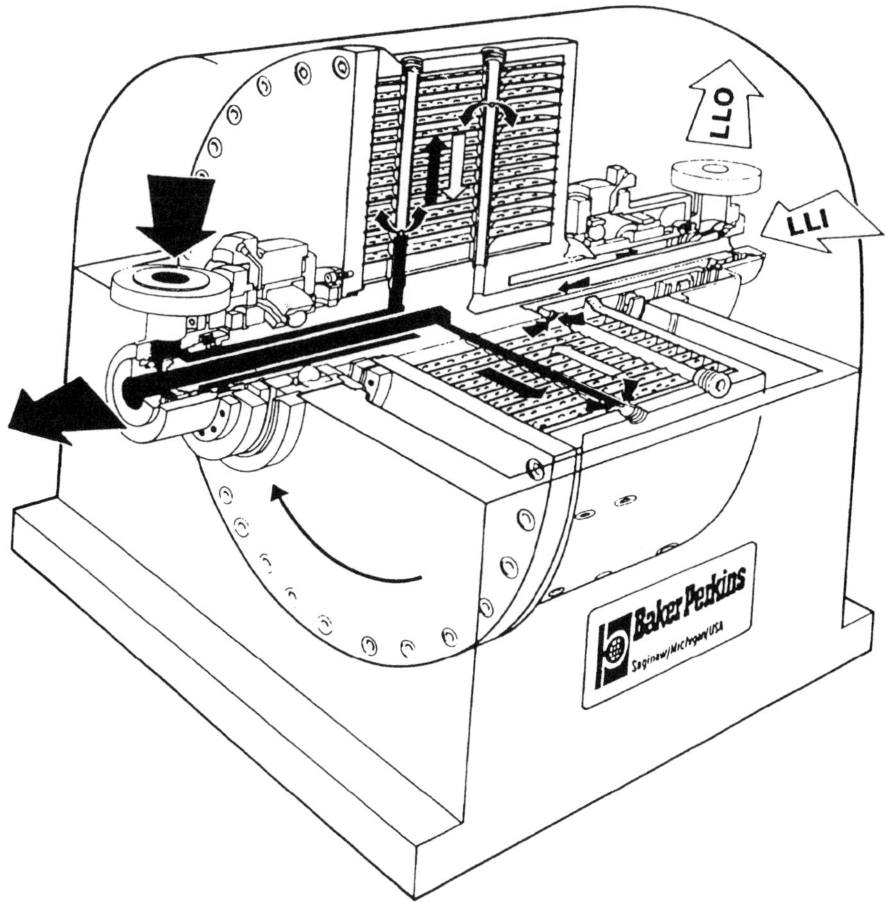

Figure 5.13 The Podbielnak contactor. (APV Chemical Machinery Inc.)

advantages could be gained by employing direct extraction. Cost comparisons based upon the experience with chloramphenicol which were presented then showed that reductions in processing costs of approximately 30% could be achieved. The major savings were achieved by reduced material costs associated with broth filtration. A significant reduction in labour costs was also achieved.

The actual performance of the contactor during whole broth penicillin extraction was reduced by slow mass transfer, necessitating a significant recycle of raffinate in order to maintain good recovery. An additional problem was the need to recover solvent from the final raffinate by stripping.

Another reported problem associated with whole broth extraction in the centrifugal contactor is the accumulation and buildup of solids, which can

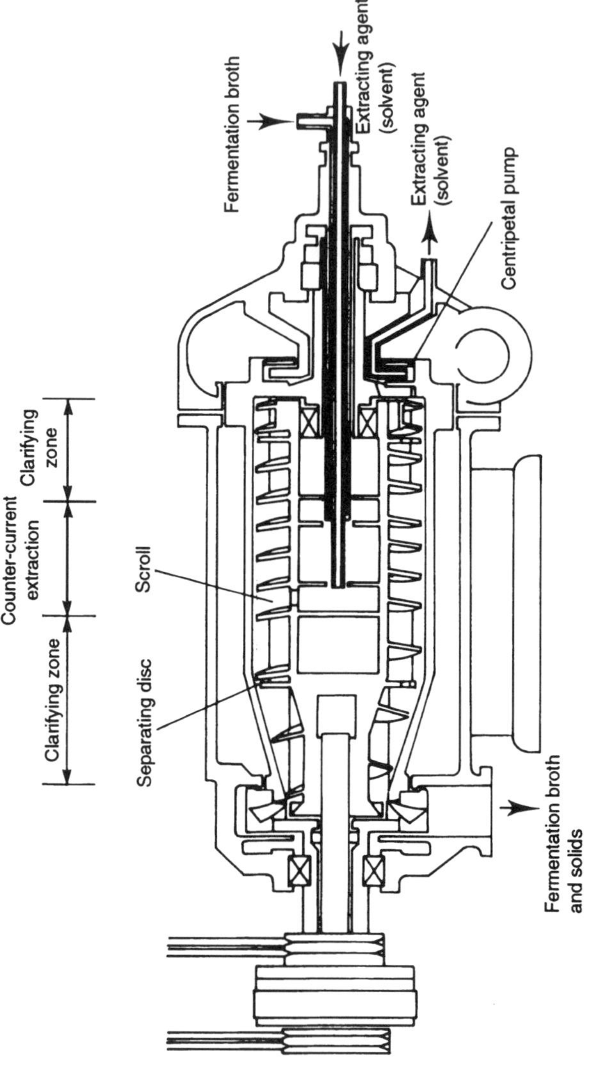

Figure 5.14 The extraction decanter. (Westfalia Separator)

under certain circumstances lead to rotor imbalance and increase in levels of vibration.

A more recent development in centrifugal contact equipment is the extraction decanter [28], introduced by Westfalia Separator (see Figure 5.14). The two contacting liquid phases are introduced into the rotating bowl at separate points and contact occurs both in the inlet area and inside the flights of the scroll. Differences in phase densities may be accommodated in the decanter by means of a ring dam at the cylindrical end of the bowl. Following contact in the countercurrent extraction zone, the extract phase is discharged under pressure at the cylindrical bowl end of the device. The raffinate phase discharges at the opposite end of the bowl by gravity. The extraction decanter is specifically designed for whole broth extraction and the presence of the scroll ensures that separated solids are collected and transported to the raffinate discharge end of the contactor.

5.5 Future Developments

It is very likely that the search for new and improved techniques for whole broth extraction will continue because of the potential economic and environmental benefits. The move to simpler processes involving fewer operations will continue and in the case of biological product separation this implies a continuing demand for innovation in methods of primary separation. In the context of whole broth extraction, further refinements in the design and performance of centrifugal contactors will be sought and there is scope for innovation in the design and arrangement of contactor internals. An excellent example already under evaluation is the HIGEE contactor [29], which is a high-speed rotary device involving the use of novel synthetic packing materials specifically tailored to the interfacial properties of the process mixture for optimum performance. The primary development of this concept has been in the context of gas–liquid contacting and is currently being extended to liquid–liquid extraction.

There is also a potential role for other high-gradient techniques, as yet not demonstrated at commercial scale, such as electrostatically enhanced extraction [19]. This technique involves electrostatically spraying the untreated broth into the solvent, thus producing a very fine dispersion of charged drops in intimate contact with the extractant. So far, laboratory-scale studies have demonstrated that large increases in extraction rates are possible during the extraction of untreated yeast based whole broths and mycelial culture broths.

This technique has two advantageous features which are directly relevant to whole broth extraction. Firstly, a large increase in mass transfer is achieved at very low additional energy input to the system. Secondly, the disengagement of dispersed and continuous phases after contact is facilitated by the presence of electrostatic charge on the dispersed phase. Significant engineering development is required for this technique to be exploited commercially. Environmental demands will encourage improvements in phase separation performance of all types of contactor and there may be an important role for electrostatic coalescence systems in whole broth extraction processes.

5.6 References

1. J. D. Thornton (Ed.) (1992) *Science and Practice of Liquid–Liquid Extraction*, Oxford University Press, Oxford.
2. M. H. B. Baird, A. Lo and C. Hanson (1992) *Handbook of Solvent Extraction*, 2nd edn, Wiley Interscience, New York.
3. J. E. Barnes and J. D. Edwards (1982) 'Solvent extraction at Inco's Acton precious metal refinery', *Chemistry and Industry*, March, 151–5.
4. F. H. Garner and A. H. P. Skelland (1955) 'Some factors affecting droplet behaviour in liquid–liquid systems', *Chem. Engng. Sci.*, **4**(4), 149–57.
5. L. R. Weatherley and C. Turmel (1992) 'Terminal velocity studies of whole broth single drops in a liquid–liquid system', *Ind. Engng. Chem. Res.*, **31**, 1739–44.
6. S. Hu and R. C. Kintner (1955) 'The fall of single drops through water', *AIChEJ*, **1**(1), 42–8.
7. A. J. Klee and R. E. Treybal (1956) 'Rate of rise or fall of liquid droplets', *AIChEJ*, **2**(4), 444–7.
8. G. Thorsen, R. M. Stordalen and S. G. Terjesen (1968) 'On the terminal velocity of circulating and oscillating liquid drops', *Chem. Engng, Sci.*, **23**, 413–26.
9. A. H. P. Skelland, S. Woo and G. G. Ramsay (1987) 'Effects of surface active agents on drop-size, terminal velocity and drop oscillation in liquid–liquid systems', *Ind. Engng. Chem. Res.*, **26**, 907–11.
10. S. Winnikow and B. T. Chao (1966) 'Droplet motion in purified systems', *Physics Fluids*, **9**, 50–61.
11. L. R. Weatherley and C. Turmel (1993) 'Terminal velocity studies of whole broth single drops in a liquid–liquid system — the effect of biomass concentration', *Proc. 1st Jordanian Chemical Engineering Congress*, University of Amman, 18–21 October 1993.
12. C. Turmel (1989) 'Droplet velocity studies in fermentation broth liquid–liquid extraction systems', MSc Thesis, Heriot-Watt University, Edinburgh.
13. A. E. Handlos and T. Baron (1957) 'Mass and heat transfer from drops in liquid–liquid extraction', *Ass. Ind. Chem. Engrs. J.*, **3**(1), 127–36.
14. A. H. P. Skelland and C. L. Caenpeel (1972) 'Effects of surface active agents on mass transfer during droplet formation, fall, and coalescence', *AIChEJ*, **18**(65), 1154–63.
15. P. G. Crabbe (1983) 'Recovery of fermentation products by liquid extraction: effect of microbial organisms on the rate of extraction', *Proc. CHEMECA 83. The 11th Australian Conference on Chemical Engineering*, Brisbane, Australia, 4–7 September 1983, pp. 253–9.
16. P. G. Crabbe, C. W. Tse and P. A. Munro (1986) 'Effect of microorganisms on rate of liquid extraction of ethanol from fermentation broths', *Biotechnol. Bioengng.*, **28**(7), 939–43.
17. E. A. Doherty, J. A. S. Goodwin, J. C. Slaughter and L. R. Weatherley (1994) 'A novel method of studying cell behaviour at the solvent/broth interface', *Bioseparation (International Journal of Separation Science in Biotechnology)*, **4**, 85–8.
18. L. R. Weatherley, I. Campbell, J. C. Slaughter and K. S. Sutherland (1987) 'Electrically enhanced solvent extraction of biochemicals', in M. S. Verrall and

M. J. Hudson (Eds.), *Separations for Biotechnology*, Ch. 25, Ellis Horwood, Chichester, pp. 353–9.

19. G. J. Laughland, M. K. Millar and L. R. Weatherley (1987) 'Electrostatically enhanced recovery of ethanol from fermentation liquor by solvent extraction', IChemE Symposium Series 103, *Extraction '87*, EFCE Publication Series 347, pp. 263–78.

20. B. Muller, E. Schlichting, L. Bischoff and K. Schugerl (1987) 'Reactive extraction of penicillin G in a pilot plant Karr-column', *Appl. Microbiol. Biotechnol.*, **26**, 36–41.

21. A. E. Karr, W. Gebert and M. Wang (1980) 'Extraction of whole fermentation broth with a Karr reciprocating extraction column', *Can. J. Chem. Engng.*, **58**(2), 249–52.

22. S. R. Roffler, H. W. Blanch and C. R. Wilke (1988) 'In-situ extractive fermentation of acetone and butanol', *Biotechnol. Bioengng.*, **31**, 135–43.

23. M. Minier and G. Gomer (1980) 'Ethanol production by extractive fermentation', *Proc. EEC Conf. On Energy from Biomass*, Brighton, November 1980, pp. 298–305.

24. R. P. De Filip and J. M. Moses (1982) 'Extraction of organics from aqueous solutions using critical fluid carbon dioxide', *Biotechnol. Bioengng. Symp.*, Vol. 12, pp. 205–19.

25. G. A. Grintzos (1989) 'Hydraulics of a pulsed plate column during whole broth extraction', MPhis Thesis, Heriot-Watt University, Edinburgh, October 1989.

26. F. J. Sebba (1972) 'Biliquid foams — a preliminary report', *J. Colloid and Interface Sci.*, **40**(3), 468–74.

27. D. W. Anderson and E. F. Lau (1955) 'Commercial extraction of unfiltered fermentation broths in the Podbielnak Contactor', *Chem. Engng. Prog.*, **51**(11), 507–10.

28. H. Katinger, F. Wibbelt and H. Scherfler (1981) 'Kontinuierliche direkte Extraktion von Antibiotika mit Separatoren und Dekantern', *vt-Verfahrenstechnik*, **15**, 179–82.

29. C. R. Howarth, J. G. M. Lee and C. Ramshaw (1993) 'The hydrodynamic and mass transfer characteristics of a Hi-Gee solvent extractor', in D. H. Logsdail and M. J. Slater (Eds.), *Solvent Extraction in the Process Industries*, Vol. 1, Elsevier Applied Science, London and New York, pp. 25–32.

6 CHEMICALLY ASSISTED SOLVENT EXTRACTION

C. Judson King

6.1 Introduction

Products of bioprocessing typically contain oxygen and/or nitrogen, which means that they bear chemically active functional groups that can be used to advantage for recovery and fractionation of the products. The term 'chemically assisted solvent extraction' is meant there to connote extraction processes wherein chemical interactions govern the separation obtained. An example is the use of an organic base extractant, e.g. trioctylamine, to recover a carboxylic acid from a fermentation process.

There is not a sharp demarcation between chemical interactions and general solvation. The usual criteria are whether or not a stoichiometric relationship appears to be involved and whether chemical equilibrium expressions are capable of describing the equilibria involved.

6.1.1. INCENTIVES

Features of chemically assisted extraction that are useful for recovery of products from bioprocessing include selectivity, capacity and low-energy consumption [1].

6.1.1.1 Selectivity

Chemical interactions are specific to solutes in a particular chemical class. They are therefore useful for removing a particular substance, or class of substances, selectively from a complex solution, such as a fermentation medium.

6.1.1.2 Capacity

Chemical interactions can also enhance solvent capacity and thereby reduce the solvent circulation rates that are needed in an extraction process. Because of

Downstream Processing of Natural Products. Edited by Michael S. Verrall
©1996 John Wiley & Sons Ltd

the nature of chemical equilibria, equilibrium distribution ratios increase at lower solute concentrations, where less of the extractant is consumed. Hence capacity is particularly enhanced for extraction of components having low concentrations in the source stream. Most products formed in bioprocessing operations are dilute.

6.1.1.3 Low-energy Consumption

There are several reasons why chemically assisted extraction can consume less energy than alternative processes.

(a) Extraction, and for that matter adsorption and some membrane processes, remove the solute rather than the solvent (i.e. water) from solution. Such processes generally consume less energy than those in which the solution is concentrated by removing water, such as by evaporation.
(b) Chemically interacting extractants take up less water along with the solute than do non-interacting solvents, thereby lessening needs for subsequent dewatering of the product.
(c) The higher solute-uptake capacities afforded by chemically interacting extractants lead to energy consumptions lower than for extraction processes utilizing agents having lower capacities, because of lower solvent circulation rates.
(d) Extraction can be accomplished using countercurrent contactors, leading to more efficient use of the separating agent than occurs with fixed beds and elution and displacement chromatography, which bring about more dilution.

6.1.2 LIMITATIONS

There are, of course, constraints and limitations associated with chemically assisted extraction. These include the need for facile regeneration of the extractant and avoidance of contamination and irreversible product degradation.

6.1.2.1 Regeneration

For a variety of reasons, it is virtually always necessary to regenerate the extractant and prepare it for re-use. The desired product(s) will need to be isolated from the extractant; the expense of make-up chemical extractant would usually become prohibitive without regeneration and re-use; even if costs were not controlling, there would be the problem of disposing of used extractant. Regeneration is usually a major component of process cost.

6.1.2.2 Contamination and Toxicity

Contamination issues arise in at least two ways:

(a) If it comes into contact with the organism, the extractant and/or the diluent with which it is mixed in the solvent phase may retard the biochemical reaction or, in extreme cases, kill the organism.
(b) The extractant and/or the diluent may be soluble and/or may emulsify in the aqueous raffinate, creating a need for processing the aqueous reject before it is released.

6.1.2.3 Product Degradation

Because of the very same chemical effects that assist the extraction, the extractant may alter the product in a deleterious way. Prime examples are denaturing of proteins and loss of biological activity of hormones, enzymes, etc.

6.1.3 LIKELY USES OF CHEMICALLY ASSISTED EXTRACTION

Product degradation considerations tend to discourage the use of chemically assisted extraction for recovery of biologically active macromolecular substances. An exception occurs for affinity interactions, described below. Because of the product degradation problem and the potential energy efficiency, the most likely uses for chemically assisted extraction are for recovery and fractionation of the larger volume products which have lower unit costs. This category includes commodity chemicals, many pharmaceuticals, amino acids and other speciality chemical products.

6.2 Types of Chemical Interactions

6.2.1 ACID–BASE INTERACTIONS

Regeneration needs provide strong incentives for the use of readily reversible chemical interactions, i.e. those with lower free energies of reaction. Interactions suitable for regenerable, chemically assisted extraction include hydrogen bonding, acid–base interactions, ion exchange, ion pairing, affinity interactions and complexation in general.

Nearly all interactions useful for separations of biological products involve acid–base interactions in the most general (Lewis) sense. There are gradations in the types of acid–base interactions, ranging from 'soft' to 'hard' and from hydrogen bonding to ion-pair formation.

6.2.1.1 Amine Extraction of Carboxylic Acids

A prime example of chemically assisted extraction of bioproducts has been the extraction of carboxylic acids by amine extractants [2]. Depending upon the pK_a values and polarities of the amine and carboxylic acid, as well as the nature of the solvent medium, the acidic hydrogen of the carboxylic acid either hydrogen bonds to the amine or transfers partially or totally to it [3,4]. In the case of a total shift of the proton to the amine, an ammonium carboxylate ion pair is formed.

Long-chain tertiary amines (i.e. trioctyl or tridecyl) are used commercially for recovery of citric acid from fermentation broths [5]. Phase equilibria for extraction of a number of carboxylic acids by a tertiary amine extractant in a variety of different organic diluents have been interpreted quantitatively [6] in terms of chemical complexation models reflecting:

(a) the acid strength of the carboxylic acid;
(b) the ability of the diluent to solvate or complex with the acid–amine complex;
(c) the tendency of the acid and the diluent towards stoichiometric overloading, i.e. more than one acid taken up per amine.

Overloading occurs as a result of hydrogen bonding between a second carboxylic acid and the available second oxygen of the carboxylate group in the acid–amine ion pair (see Figure 6.1) [3,7–9]. It is a weaker interaction than the primary acid–amine interaction.

Amine extractants can also be used for extraction of more complex acids, such as aromatic carboxylic acids, amino acids and penicillins, in addition to aliphatic carboxylic acids [10].

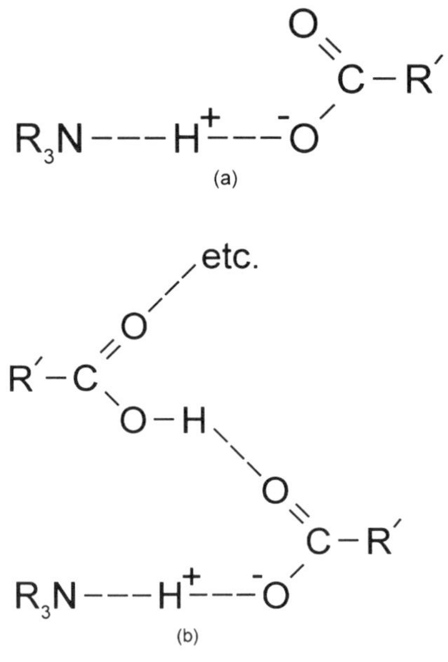

Figure 6.1 Extraction of carboxylic acids with amine extractants: (a) 1:1 complexation; (b) addition of one or more carboxylic acid molecules to the acid–amine complex

6.2.1.2 *Extraction of Carboxylic Acids and Phenols by Phosphoryl Compounds*

Carboxylic acids can also be extracted with any of various phosphoryl compounds, of which phosphine oxides have been most commonly considered [11]. Extraction of a carboxylic acid with a phosphine oxide takes place through strong hydrogen bonding, without a shift of the proton. Such phosphoryl compounds are effective extractants for phenols as well [12], again interacting through a strong hydrogen bond.

6.2.1.3 *Extraction with Quaternary Ammonium Extractants*

Quaternary ammonium extractants can also be used to extract carboxylic acids, in which case an ion pair is necessarily formed. This is a case of liquid ion exchange (LIX), a term that can also be attached to the stronger interactions of carboxylic acids with amines. Regeneration is much more difficult for quaternary ammonium carboxylates. Extraction with quaternary ammonium compounds has been investigated as a means of recovery and fractionation of amino acids [10,13,14].

6.2.2 ION-PAIR EXTRACTION

The use of a high molecular weight amine to extract a carboxylic acid as an ion pair could also be called 'ion-pair extraction', a term that is applied to situations where an organic cation (or anion) pairs with an aqueous anion (or cation), and thereby extracts the anion (or cation) from the aqueous phase into the organic phase. This technique has been developed extensively for use in chemical analyses [15].

A related approach is to pair an interfacially active cation (or anion) with a solute anion (or cation), and then to draw the interfacially active ion pair into a reverse micelle phase [16]. In principle, the interfacially active ion pair could be removed by foam or bubble fractionation or solvent sublation as well, with due regard to product degradation issues.

A different use of ion-pair extraction is the situation where the complexing agent itself must be in the ionized form, i.e. cationic or anionic, in order to be effective. In that case formation of an ion pair in an organic phase can make the cation or anion available in a reactive extraction system. As one example, ion-pair extractants composed of a quaternary ammonium cation and an organoboranate anion in a suitable diluent [17,18] have been used for

$$[\text{Quaternary ammonium}]^+ \, [\text{boronate}]^- + \text{diol}$$

$$\longrightarrow [\text{quaternary ammonium}]^+ \, [\text{boronate:diol}]^-$$

Figure 6.2 Extraction of a diol with a quaternary ammonium–organoboronate ion pair

extraction of glycols and sugars through formation of cyclic complexes between the anionic organobornate and *cis*-vicinal diols (see Figure 6.2).

6.2.3 AFFINITY INTERACTIONS

Enzymes and other substances for which molecular conformation is important can be recovered by affinity partitioning, wherein a specific affinity ligand is coupled to one of the polymer components present in a two-aqueous-phase extraction system. This approach provides an extractive separation based upon the chemical and steric specificity of the affinity ligand, while preserving the product against degradation through use of the aqueous environment [19,20]. The technique is analogous to affinity chromatography.

6.3 Regeneration Alternatives

Regeneration of a chemically assisted, or reactive, extraction requires reversing the chemical reaction. There are, in general, three ways of doing this:

(a) Change the equilibrium constant of the reaction.
(b) Force the reaction in the reverse direction by continual removal of one or more products, e.g. by vaporization.
(c) Carry out additional reactions and/or separations to recover the desired product and prepare the extractant for recycle.

The use of additional reactions is generally difficult, complex and expensive.

In acid–base extractions, the most straightforward way to alter the equilibrium constant is to change the pH, and that is often done. However, if we take the example of an organic base extracting a carboxylic acid, the change of pH would require adding a strong acid, e.g. H_2SO_4, to displace the carboxylic acid, which could then be back-extracted into water. To prepare the extractant for re-use, one would then have to add a strong base, e.g. NaOH, forming Na_2SO_4, in turn could be leached out of the solvent with water. The result, unfortunately, is consumption of both H_2SO_4 and NaOH, and formation of a waste salt stream.

Another way to change the reaction equilibrium is through a change in temperature, in cases where the enthalpy of reaction is sufficiently large. This is the basis of the successful process for recovery of citric acid from fermentation media by extraction with a tertiary amine, regenerated by back-extraction into water at a higher temperature [5]. Extraction equilibria can also be altered considerably by changing the diluent that is mixed with the extractant. Thus, a carboxylic acid can be extracted by an amine in a diluent that strongly solvates or complexes with the acid–amine complex, after which the diluent is changed to a weakly solvating one, allowing back-extraction into water. The change of diluent can be accomplished by distillation [21]. Temperature swing and diluent change can be used together.

If the chemical interaction used for extraction is weak enough, it may be possible to regenerate without either changing the equilibrium relationship or forcing the back-reaction. In the case of extraction of carboxylic acids with tertiary amines this situation can be achieved if the extraction is carried out largely in the realm of stoichiometric overloading, i.e. if the ratio of acid to amine in the extract never drops much below 1.0 [22]. As already noted, the overloading interaction is weaker than the primary acid–amine interaction. Of course, in such a case back-extraction into an aqueous phase must lead to a product that is more dilute than the aqueous feed, unless there are special factors at play, such as a high ionic content in the feed. However, the extraction process does still serve to free the product of impurities that were present in the original aqueous feed, and the dilution effect may not be overly disadvantageous if sufficient evaporation capacity is available [23].

Distillation can be used to force the back-reaction if the product is sufficiently volatile. Thus, a tertiary amine extractant laden with ethanoic acid can be regenerated by distillation of ethanoic acid from the extract [24]. Less volatile acids can sometimes be regenerated in the same way if the extracted acid is esterified during regeneration.

6.4 Extraction at $pH > pK_a$

Most fermentations to produce carboxylic acids operate at aqueous solution pH above pK_{a1} of the carboxylic acid, in which case the acid is present largely as the carboxylate. A sufficiently basic extractant can nonetheless take the carboxylic acid up with satisfactory capacity [25]. Tertiary amine extractants with diluents (e.g. alcohols) that solvate the acid–amine complex retain capacity up to and somewhat above pH 6. Pyridyl extractants are weak enough bases so that they do not sustain capacity much above the pK_a of the acid.

What is needed, then, is a regeneration method powerful enough to remove the carboxylic acid from the basic extractant. Figure 6.3 displays this need in terms of pH driving forces. A solution of a chemical base, such as NaOH or NH_4OH, could back-extract the acid from the extractant, but places the acid in

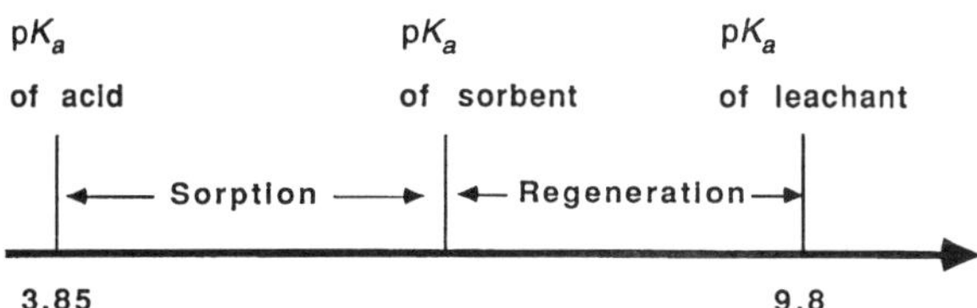

Figure 6.3 Driving forces for regenerable sorption or extraction of a carboxylic acid from a solution at $pH > pK_a$. The arrow is in the direction of increasing basicity. $pK_a = 3.85$ and 9.8 for lactic acid and trimethylamine respectively

the salt form. The ammonium carboxylate resulting from back-extraction with aqueous NH_4OH could be decomposed thermally, but such processes are subject to irreversible formation of amides. As noted in Figure 6.3 trimethylamine (TMA) has $pK_a = 9.8$, and therefore an aqueous solution of TMA can back-extract a carboxylic acid (e.g. lactic acid) from a tertiary amine extract. (The aqueous tertiary amine ionizes and is therefore a much stronger base than an organic tertiary amine.) A trialkylammonium carboxylate cannot form an amide upon heating. Thermal decomposition of the carboxylate provides product acid and recycle TMA [26].

6.5 Toxicity and Contamination

Studies of toxicity of extractants, including reactive extractants, should be carefully scrutinized, since toxic effects can come from impurities which could be washed out of the solvent before use. Often solvents and extractants used in toxicity studies have not been adequately prepurified.

For chemically assisted extraction, as well as in general, there are three approaches to coupling separation and product recovery steps with a bioreactor. These are *in situ* separation, use of an external recycle loop and sequential processing.

6.5.1 *IN SITU* SEPARATION

Since products often serve to inhibit the reaction, it can be attractive to integrate separation into the reactor itself. With chemically active extractants this may cause problems of contamination, i.e. toxicity to the organism and consequent reduction or cessation of production. A reactive extractant can affect a micro-organism through the solubility of the extractant in the aqueous cell matter and through coating of cells with immiscible organic phase, thereby affecting transport to and from the organism and interfacial tension [27,28]. Approaches to the toxicity issue for *in situ* extraction include separation of the extractant from the aqueous phase by a macroporous membrane [29], immobilization of the cells in a substance such as carrageenan and scavenging of residual extractant with a harmless oil, such as soybean oil [30].

6.5.2 EXTERNAL RECYCLE

In a recycle processing loop external to the bioreactor, additional steps (e.g. carbon adsorption, stripping and/or membrane processes) can follow the separation and can serve to remove any undesirable components that enter during the extraction. An external recycle processing loop with a sufficiently large flowrate can fulfil the same function as *in situ* separation.

6.5.3 SEQUENTIAL PROCESSING

Strictly sequential processing without recycle removes the issue of toxicity to the organism but brings a need for continual make-up of the bioreaction medium, as well as disposal of the remainder of the effluent.

6.6 Reactive Solid Sorbents

Reactive sorbents parallel the functions of reactive extractants, except that they are not readily used in countercurrent flow. Solid sorbents can function as adsorbents or ion exchangers, with the dividing line again not being distinct. Properties of basic polymeric sorbents, including those with tertiary amine, pyridyl and imidazole functionalities, for uptake of carboxylic acids and phenols have been reported [31–35]. The equilibria can be interpreted in terms of one-to-one complexation, with the complexation constant relating to acid and basic strengths [35]. Whereas amine extractants exhibit stoichiometric overloading, solid amine sorbents do not [25], presumably because competition from water in the aqueous environment precludes hydrogen bonding of additional acid molecules of the sort shown in Figure 6.1(b).

Solid sorbents can be regenerated by the various methods described above for extractants, except for changing the composition of the diluent. An additional method of regeneration for solid sorbents is leaching with a solvent, which works well, for example, for recovery of sorbed carboxylic acids from poly(vinylpyridine).

As for extractants, the ability of sorbents to recover acids from aqueous solutions at $pH < pK_a$ of the acid relates to basic strength [25]. Pyridyl sorbents lose capacity rapidly as pH increases above pK_a. Tertiary amine sorbents sustain capacity up to pH 6 and somewhat above and are regenerable, e.g. by back-extraction with aqueous TMA. Quaternary ammonium sorbents maintain capacity to very high values of pH, but are difficult to regenerate.

Reactive sorbents can be less subject to toxicity effects when used for *in situ* separation within a fermentation vessel. Provided that residual monomer has been leached out of the sorbent by pretreatment, there should be no solubility of the sorbent in the aqueous phase. Adsorption of cellular material on the sorbent could still be problematic. Holst and Matiasson [36] have reviewed various means of using solid sorbents in fermentation and controlling toxicity. Srivastava *et al.* [37] used Amberlite IRA-400 (Rohm and Haas Co.), a quaternary ammonium anion exchanger in the –OH form, successfully, in an external recycle loop to remove lactate continuously and thereby greatly increase the rate of production of lactate. They report no detrimental effect of the anion exchanger on the growth of *Lactobacillus delbrueckii* in the fermentation vessel.

Acknowledgement

Our own research in this area has been supported by the Biochemical and Chemical Technology Research Program, Advanced Industrial Concepts Division, Office of Industrial Technologies of the US Department of Energy under Contract DE-AC03-76SF00098.

6.7 References

1. C. J. King (1987) 'Separation processes based upon reversible chemical complexation', in R. W. Rousseau (Ed.), *Handbook of Separation Process Technology*, John Wiley, Chichester, pp. 760–74.
2. C. J. King (1992) 'Amine-based system for carboxylic acid recovery; tertiary amines and the proper choice of diluent allow extraction and recovery from water', *CHEMTECH*, **22**, 285–91.
3. G. M. Barrow (1956) 'The nature of hydrogen bonded ion pairs: the reaction of pyridine and carboxylic acids in chloroform', *J. Am. Chem. Soc.*, **78**, 5802–6.
4. V. S. Shmidt and E. A. Mezhov (1965) 'The structure and extractive power of amines and their salts', *Russian Chem. Rev.*, **34**, 585–99.
5. A. M. Baniel, R. Blumberg and K. Hajdu (1981) 'Recovery of acids from aqueous solutions', US Patent 4,275,234, 23 June 1981.
6. J. A. Tamada, A. S. Kertes and C. J. King (1990) 'Extraction of carboxylic acids with amine extractants. 1. Equilibria and law-of mass action modeling', *Ind. Engng. Chem. Res.*, **29**, 1319–26.
7. G. M. Barrow and E. A. Yerger (1954) 'Acid–base reactions in non-dissociating solvents', *J. Am. Chem. Soc.*, **76**, 5211–16.
8. A. A. Chaikorshii, B. P., Nikol'skii and B. A. Mikhailov (1966) 'Complex formation in nonaqueous solutions X. Interaction of tridecylamine with acetic acid', *Sov. Radiochem. (Engl. Trans.)*, **8**, 163–71.
9. J. A. Tamada and C. J. King (1990) 'Extraction of carboxylic acids with amine extractants. 2. Chemical interactions and interpretation of data', *Ind. Engng. Chem. Res.*, **29**, 1327–33.
10. K. Schügerl and W. Degener (1989) 'Gewinnung niedermolekularer organischer Verbindungen aus komplexen wässerigen Gemischen durch Extraktion', *Chem.-Ing. Tech.*, **61**, 796–804.
11. R. W. Helsel (1977) 'Removing carboxylic acids from aqueous wastes', *Chem. Engng Prog.*, **73**(5), 55–9.
12. J. D. MacGlashan, J. L. Bixby and C. J. King (1985) 'Separation of phenols from dilute aqueous solution by use of tri-n-octyl phosphine oxide as extractant', *Solvent Ext. Ion Exch.*, **3**, 1–25.
13. E. Schlicting, W. Halwachs and K. Schügerl (1985) 'Reactive extraction of salicyclic acid and D,L-phenylanine in a bench-scale pulsed sieve column', *Chem. Engng. Process.*, **19**, 317–28.
14. T. Hano, T. Ohtake, M. Matsumoto, D. Kitayama, F. Hori and F. Nakashio (1991) 'Extraction equilibria of amino acids with quaternary ammonium salt', *J. Chem. Engng. Japan*, **24**, 20–4.
15. G. Schill (1974) 'Isolation of drugs and related organic compounds by ion-pair extraction', in J. A. Marinskey and Y. Marcus (Eds.), *Ion Exchange and Solvent Extraction*, Dekker, New York, pp. 1–57.
16. E. B. Leodidis and T. A. Hatton (1989) 'Interphase transfer for selective solubilization of ions, amino acids and proteins in reversed micelles', in M. P. Pileni (Ed.), *Structure and Reactivity in Reversed Micelles*, Elsevier, Oxford.

17. T. Shinbo, K. Nishimura, T. Yamaguchi and M. Sugiura (1986) 'Uphill transport of monosaccharides across an organic liquid membrane', *J. Chem. Soc. Chem. Commun.*, 1986, 349–51.
18. L. A. Randel, T. K.-F. Chow and C. J. King (1994) 'Ion-pair extraction of multi-OH compounds by complexation with organoboronate', *Solvent Extr. Ion Exch.* **12**(4), 765–78.
19. S. D. Flanagan and S. H. Barondes (1975) 'Affinity partitioning', *J. Biol. Chem.*, **250**, 1484–9.
20. G. Johansson (1985) 'Partitioning of proteins', in H. Walter, G. E. Brooks and D. Fisher (Eds.), *Partitioning in Aqueous Two-Phase Systems*, Academic Press, New York.
21. J. A. Tamada and C. J. King (1990) 'Extraction of carboxylic acids with amine extractants. 3. Effect of temperature, water co-extraction and process considerations', *Ind. Engng. Chem. Res.*, **29**, 1333–8.
22. A. M. Baniel (1982) 'Extraction of organic acids from aqueous solutions', US Patent 4,334,095, 8 June 1982.
23. A. M. Baniel and D. Gonen (1991) 'Production of citric acid', US Patent 4,994,609, 19 February 1991.
24. N. L. Ricker, E. F. Pittman and C. J. King (1980) *J. Separ. Process Technol*, **1**(2), 23–30.
25. L. A. Tung and C. J. King (1994) 'Sorption and extraction of lactic and succinic acids at pH > pK_{a1}', *Ind. Engng. Chem. Res.*, **33**(12), 3217–23.
26. L. J. Poole and C. J. King (1991) 'Regeneration of carboxylic acid–amine extracts by back-extraction with an aqueous solution of a volatile amine', *Ind. Engng. Chem. Res.*, **30**, 103–25.
27. V. P. Lewis and S. Yang (1992) 'A novel extractive fermentation process for propionic acid production from whey lactose', *Biotechnol. Prog.*, **8**, 104–10.
28. V. M. Yabannavar and D. I.-C. Wang (1987) 'Bioreactor system with solvent extraction for organic acid production', *Ann. New York Acad. Sci.*, **506**, 523–35.
29. T. Cho and M. L. Shuler (1986) 'Multimembrane bioreactor for extractive fermentation', *Biotechnol. Prog.*, **2**, 53–60.
30. V. M. Yabannavar and D. I.-C. Wang (1991) 'Strategies for reducing solvent toxicity in extractive fermentations', *Biotechnol. Bioengng.*, **37**, 716–22.
31. N. Kawabata and K. Ohira (1979) 'Removal and recovery of organic polutants from aquatic environment. 1. Vinylpyridine–divinylbenzene copolymer as a polymeric adsorbent for removal and recovery of phenol from aqueous solution', *Environ. Sci. Technol.*, **13**, 1396–402.
32. N. Kawabata, J. Yoshida and Y. Tanigaway (1981) 'Removal and recovery of organic pollutants from aquatic environmrnt. 4. Separation of carboxylic acids from aqueous solution using cross-linked poly(4-vinylpyridine)', *Ind. Engng. Chem. Prod. Res. Dev.*, **20**, 386–90.
33. M. Chanda. K. F. O'Driscoll and G. L. Rempel (1983) 'Sorption of phenolics onto cross-linked poly(4-vinylpyridine)', *React. Polymers*, **1**, 281–93.
34. M. Chanda, K. F. O'Driscoll and G. L. Rempel (1985) 'Sorption of phenolics and carboxylic acids on polybenzimidazole', *React. Polymers*, **4**, 39–48.
35. A. A. Garcia and C. J. King (1989) 'The use of basic polymeric sorbents for the recovery of acetic acid from dilute aqueous solution', *Ind. Engng. Chem. Res.*, **28**, 204–12.
36. O. Holst and B. Matiasson (1991) 'Solid sorbents used in extractive bioconversion processes', in B. Matiasson and O. Holst (Eds.), *Extractive Bioconversions*, Marcel Dekker, New York, pp. 189–205.
37. A. Srivastava, P. K. Roychoudhury and V. Sahai (1992) 'Extractive lactic acid fermentation using ion-exchange resin', *Biotechnol. Bioengng.*, **9**, 607–13.

7 PRODUCT MODIFICATION TO ASSIST ISOLATION (FACILITATED PROCESSING)

Alan R. Thomson

7.1 Introduction

Downstream processing (DSP) of protein products can account for up to 80% of total production costs. With present technology, each product requires development of a different DSP scheme starting *ab initio*, which is usually carried out *after* optimization of the product system. Recombinant DNA technology can completely revolutionize this situation since it enables:

(a) heterologous genes to be expressed at high levels in virtually all types of cell,
(b) product distribution to be designed into the system (as inclusions, in the periplasm),
(c) properties to aid recovery and purification to be incorporated into the product expressed,
(d) the downstream process to be designed *ab initio* and hence to be integrated more fully with the production stage.

Increasing use is being made of this new concept of *facilitated* processing, not only in research but also in commercial production, and there is increasing emphasis on technology suitable for larger-scale use.

This chapter is focused almost exclusively on product modification *in vivo*, it being assumed that appropriately high-level expression in the producing cell has been achieved. Attention will be concentrated on proteins and peptides, but given the pace of developments in molecular biology, it is likely in the longer term that engineering of specifically designed multicomponent systems and even metabolic pathways will be possible in the future.

Downstream Processing of Natural Products. Edited by Michael S. Verrall
©1996 John Wiley & Sons Ltd

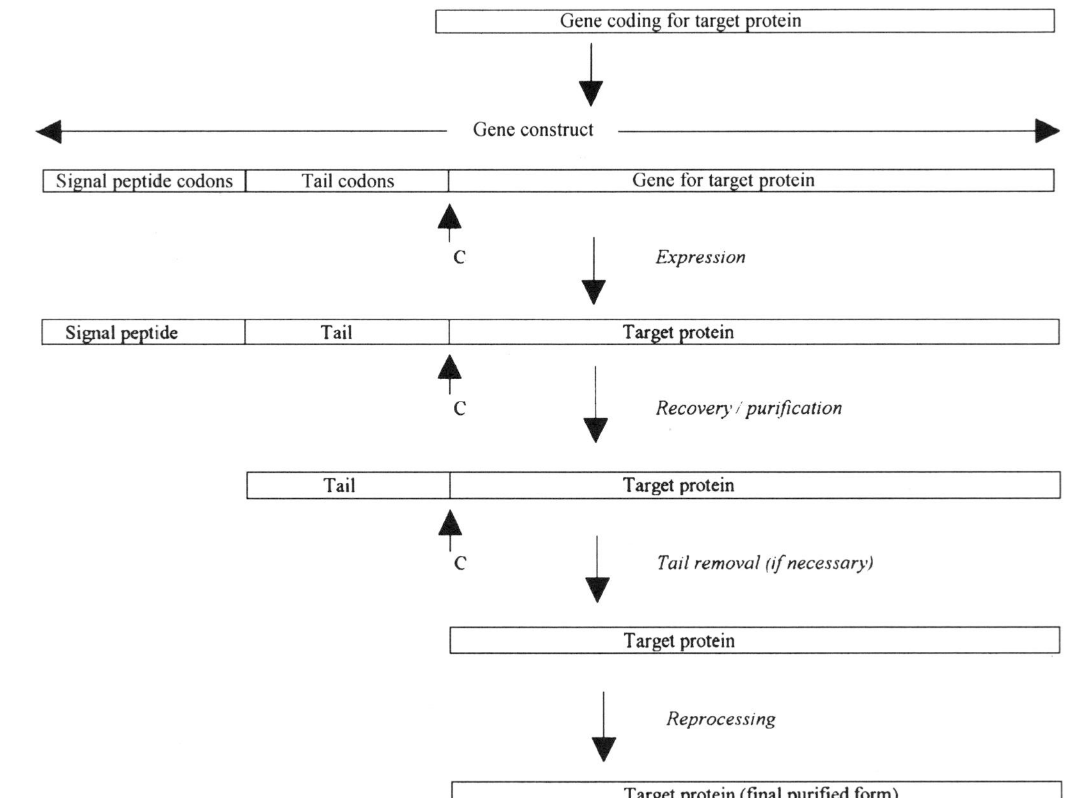

Figure 7.1 Principle of facilitated processing. Note that the tail can be C- and/or N-terminal (N-terminal shown); C = cleavage site; signal peptide is incorporated in this example

The concept of facilitated processing is straightforward, (Figure 7.1) (see also Refs. [1] to [3]). In essence, DNA is constructed incorporating the gene for the desired product (target), together with DNA coding for additional amino acid sequences ranging from single amino acids to complete proteins. These are fused at the N- and/or C-terminals of the target gene. The fusions can also include protease-sensitive cleavage sites, signal peptides for secretion, 'caps' to minimize intracellular proteolysis and additional 'tails' (e.g. enzymes) for specific applications. Thus, the hybrid protein expressed can have a wide range of predetermined, additional properties such as specific binding, differences in solubility and/or hydrophobicity, enzyme activity, etc., which can be used to facilitate recovery and purification, in a largely predictable way, and which can be removed specifically to yield authentic target. On removal of the tail, the properties of the fusion-protein/polypeptide are altered radically and independently of those of contaminants; hence reprocessing gives further purification. Contrast this with conventional biotechnological processes for producing proteins and peptides. The cell system used is that producing the target in nature, and the purification process used depends entirely on the properties of the target and the producing system. Facilitated processing can completely change this, since by incorporating selected additional proteinaceous moieties in the target *new and radically different properties* can be conferred on the fusion-protein and can be utilized *predictably* to design DSP. Furthermore, the *same tail* can be used in general for a wide range of targets, further simplifying DSP for new products. It is clear that this approach will enable bioprocesses to be designed *ab initio* so that the total process of production and downstream processing can be integrated to an extent hitherto impossible.

Best practice in DSP of bioproducts is now well established and utilizes a number of basic principles including maximizing production levels, minimizing the number of process steps, early volume reduction, mild processing conditions, use of well-tried technology wherever possible, etc. Effective *facilitated recovery* depends on three main additional factors:

(a) product modification (which defines the principal separation method),
(b) efficient and accurate tail removal,
(c) modification of the fusion product location in the system (if advantageous).

Attention will be focused mainly on factor (a). Firstly, a brief word on factor (c). In many prokaryotes (e.g. *Escherichia coli*), high-level expression of heterologous genetic material, and sometimes of homologous genes, results in the formation of inclusion bodies (insoluble product aggregates). Often these can only be processed using chaotropic agents, and have to be refolded to regain biological activity. Any advantages in having discrete particles of product may be offset by the difficulty of recovering native, functionally active product. Signal peptides can direct fusion products into the periplasm (inclusions less likely) or the culture medium [2,4]. Product recovery from the latter is clearly advantageous since neither cell separation nor cell disruption is required. However, some organisms naturally secreting proteins also secrete proteases (e.g. *Bacillus subtilis*), in which case protease-deficient strains can be used [5,6]. Inclusion body formation is much less common in eukaryotes, many of which secrete products into the culture medium [3,7,9,10].

7.2 Tail Types

Table 7.1 lists tail types reported (in this chapter, the term 'tail' includes all sequences additional to the target product). They have been attached at the N- and/or C-terminals of the target and can have a wide range of properties. Many allow separations based on specific interactions (e.g. enzyme for substrate/analogue). More attention is now being devoted to using less specific properties (e.g. charge) for larger-scale use, to improve cost effectiveness. Only the major types used will be discussed here; for details of others reported see Table 7.1 and Refs. [1] to [3].

7.2.1 ENZYMES

7.2.1.1 *β-Galactosidase (β-gal)*

This enzyme is attractive since efficient expression systems are available for *E. coli*, also an effective affinity chromatographic process with substrate analogue as ligand. C- and N-terminal fusions expressible at high levels (up to 100 mg/L), equivalent to ~40% of cell protein are available, for a range of cell types including bacteria, (e.g. *E. coli*, *B. subtilis*), yeast and human and animal cell lines [1,3,7,9]. The *β*-gal tail can protect small peptides from intracellular degradation, an additional advantage [2]. Collagenase and blood factor Xa have been used to specifically cleave the *β*-gal fusion products. An early example of a *β*-gal tail is the isolation in high purity of the very labile replication initiator of plasmid R6K [11]. The R6K fusion including the cleavage site –pro–X–gly–pro–Y– (for collagenase) was produced in *E. coli* MC1000, and the fusion product was isolated from cell lysate by one-step affinity chromatography on a thiogalactosidyl-Sepharose. Following digestion with purified collagenase, R6K was collected in the *breakthrough peak* on the same column; the contaminants, not being modified, were still adsorbed. *β*-Gal fusions tend to be expressed as inclusions in bacteria, overcome by incorporating signal peptides resulting in periplasmic accumulation in *E. coli* and secretion into the medium in many other systems [12]. The large size of *β*-gal (monomer = 116 kDa) is also a drawback.

7.2.1.2 *Glutathione-S-Transferase (GST)*

GST from *Schistosoma japonicum* (26 kDa) is being used increasingly as a purification tail, and regulated, high expression vectors have been developed.

Table 7.1 Facilitated processing; fusion tails

Fusion tail	Size	N- or C-terminal	Processing technique	References
7.2.1 Enzymes				
β-Galactosidase	116 kDa	N,C	IS,IA	1–3,8,9,11,12
Glutathione-S-transferase	26 kDa	N	IS(GSH)	1–3,13–16
Human carbonic anhydrase II	29 kDa	N	IS(sulfonamide)	17
Chloramphenicol acetyl transferase	24 kDa	N	IS	1–3,18
trpE	27 kDa	N	RPC(HIC)	1–3,19
Autolysin	16 kDa	N	IS(DEAE exchanger)	1–3,20
7.2.2 Polypeptide-binding proteins				
Staphylococcal protein A (SPA)	14–31 kDa	N	IA(on IgG)	1–3,21–23
Light meromoysin	200–263 aa	N	Precipitation (salt)	24
7.2.3 Antigenic epitopes				
FLAG Peptide	8 aa	N	IA(Ca-dependent binding)	1–3,25
HPC4-binding peptide	12 aa	N	IA(Ca-dependent binding)	26
E.coli recA polypeptide	144 aa	C	IA	1–3,27
7.2.4 Carbohydrate-binding domains				
Maltose-binding protein	40 kDa	N	IS(maltose)	1–3,29,31
Starch-binding domain	119 aa	C	IS(starch)	1–3,30,32
Mannose-binding domain	?	N	IS(mannose)	33
Glucanase-binding domains	111, 128 aa	N,C	IS(cellulose)	2
Lectin-binding domain	?	?	IS(galactose)	34
7.2.5 Poly(amino-acid) tails				
Poly(arg)	5–15 aa	C	I-Ex	1–3,35,36
Poly(asp)	5–16 aa	N,C	I-Ex;pptn;2-PAE; IExM;RME	1–13,38,39
Poly(his)	1–10 aa	N,C	IMAC;2-PAE; 2-PAAE	1–13,40–44
Poly(cys)	4 aa	N	Thiopropyl-Sepharose	2
Poly(phe)	11 aa	N	Phenyl-Superose	2
7.2.5.4 Amino acids and peptides				
Glu-X dipeptide	2 aa	N	I-Ex	2,45
(–ala–trp–trp–pro–) peptide	4 aa	C	2-PAE	46
7.2.6 Biotin-binding domain	8 kDa	N	IStreptavidin (monomer)	1,2,47,48

Abbreviations: IS, immobilized substrate; IA, immobilized antibody; GSH, reduced glutathione; RPC, reversed phase chromatography; HIC, hydrophobic interaction chromatography; I-Ex, ion exchange; pptn, precipitation; 2-PAE, two-phase aqueous extraction; IExM, ion exchange membrane; RME, reversed micelle extraction; IMAC, immobilized metal affinity chromatography; 2-PAAE, two-phase aqueous affinity chromatography; IStreptavidin, immobilized streptavidin.

GST fusions in *E. coli* tend to be relatively soluble, though inclusion bodies are not unknown. Such fusions have been expressed at up to 30 mg/L in *E. coli* [13] and in the baculovirus/insect cell system at up to 10 mg/L, ~10% of cell protein [14,15]. GST fusions can be purified in one step from cell extracts/culture medium on reduced glutathione (GSH) columns. With a suitable cleavage site (for Xa, thrombin), authentic target protein can be isolated by digestion and reprocessing as before on GSH, contaminants being retained on the column. Yields ten times those obtained by conventional methods have been reported [16]. In the baculovirus/insect cell system, fusions often show good solubility [15]. Examples include isolation of plant FD-NADP+FNR oxidoreductase [16], calrectulin [13], proteins from HSV and HIV-1 (in *E. coli*), a tyrosine kinase [14], hLDL receptor and cystic fibrosis proteins [15] (baculovirus/insect system).

7.2.1.3 Human Carbonic Anhydrase II (hCAII)

Very recently, the merits of hCAII (29 kDa) have been demonstrated. High-level expression vectors have been developed for the enzyme linked N-terminal to the target via a cleavage site [17]. Expression levels of up to >50% as soluble protein have been reported in *E. coli*, and fusions were stable in the cell for ~6 h. The availability of an inexpensive, efficient affinity purification system using sulfonamide resins is a major advantage [17]. This has been used batchwise to recover hCAII hybrids, >99% pure in one step from crude cell lysates. After digestion with enterokinase and reprocessing, pure authentic target was recovered and further purified in high yield in the flow-through fractions from sulfonamide–resin columns. Sites for factor Xa and thrombin can be incorporated, and CNBr (for –met–) and hydroxylamine (for –asn–gly–) have also been used. The resin is very stable, is re-usable, has a reasonable capacity (25 mg protein/mL) and processing conditions are mild, yet the adsorbent is selective ($K_d \sim 10^{-8}$M). The hCAII system has been used to produce a wide range of targets containing from 8 to 160 amino acids, including angiotensin and *E. coli* F1 subunits. Although with hCAII, fusions generally accumulate within the cell, they tend to be soluble; presumably a leader sequence could be incorporated to encourage secretion. It has two additional advantages: viz. the enzyme is easily detected and the fusions can be used directly to raise antibodies. The hCAII system seems to have many of the desired properties for facilitated DSP, and may even be suitable for low-value proteins. It looks very promising, and many of the disadvantages of other systems have been deliberately avoided by its *design*.

7.2.1.4 Other Enzymes

Chloramphenicol acetyl transferase [18], autolysin [19] and the trpE gene [20] of *E. coli* have also been used. For more details see Table 7.1 and Refs. [1] to [3].

7.2.2 POLYPEPTIDE-BINDING PROTEINS

7.2.2.1 Staphylococcal Protein A (SPA, 14–31 kDa)

Specific binding of SPA to mammalian IgG ($K_d \sim 10^{-8}$M) has been widely utilized [2,21,22]. Suitable vectors are available for expressing a range of IgG binding sites intracellularly in *E. coli* (e.g. ZZ [21]), or with the SPA signal sequence, in the periplasm of *E. coli* or extracellularly by *Staphylococcus aureus*. Expression levels can be very high (up to 550 mg/L), and the fusions can be recovered and purified very effectively on immobilized IgG. Enzyme cleavage sites have been incorporated and are generally preferred to chemical cleavage. SPA tails can protect small peptides against degradation [2]. The high affinity of the IgG ($K_d \sim 10^{-8}$M) may make elution of sensitive target proteins difficult, but coupled with its specificity is a great advantage for recovery of fusions from dilute extracts. For larger-scale use, the cost and relative instability of IgG present problems. Nevertheless, SPA fusion has been used on the 1000 L scale in the production of insulin-like growth factor (IGF-1) [2]. Following recovery from cell extracts by IgG chromatography, cleavage of the included –asn–gly– site with hydroxylamine and rechromatography, the unadsorbed fractions contained IGF-1 with an authentic N-terminal. An important application is the isolation of antibodies where tail removal may not be necessary.

Dual tails have proved useful for the production of peptides very unstable in *E. coli* [23]. Here the target peptide was sandwiched between N-terminal IgG binding sites and a C-terminal albumin binding domain (from streptococcal Protein G), thus protecting it against *E. coli* peptidases. The fusion product was purified by sequential affinity chromatography on immobilized IgG and immobilized albumin. The large tails also reduced the formation of inclusion bodies.

The power of fusion technology has been dramatically demonstrated recently [21]. Here the fusion protein consisted of two synthetic IgG binding domains (ZZ) derived from SPA, and a repeat structure (M5) from a major malaria antigen. The ZZ–M5 construct is a possible component of a malarial vaccine. The fusion protein was *designed* to have a low pI (~ 4.5), enabling it to be adsorbed on DEAE exchangers at pH 5.5 where most *E. coli* proteins were not adsorbed. ZZ–M5 was expressed at > 550 mg/L in *E. coli* and $> 65\%$ was exported into the medium by incorporating the SPA signal sequence. The crude fermentation broth was passed upwards through an expanded bed of DEAE adsorbent; the cells, debris and most of the proteins were not adsorbed. The ZZ–M5 was then eluted downwards with 0.5 M NaCl. In this way, a sixteen–fold volume reduction with $> 90\%$ recovery of ZZ–M5 was achieved in one step. ZZ–M5 was further purified using immobilized IgG with virtually 100% recovery.

7.2.2.2 Light Meromyosin (LMM)

C-terminal polypeptides derived from rabbit fast muscle meromyosin retain the ability to form coiled-coil polymers in low-salt solutions. These will resolubilize in high-salt conditions [24]. This property is the basis of a highly novel new technique. A series of high-level expression vectors have been constructed for *E. coli* in which target proteins are fused to the N-terminal of LMM. The fusions can be recovered from crude cell extracts by low/high-salt cycling. Xa-sensitive sites can be incorporated, enabling native target protein to be readily recovered. Many of the LMM fusions occur as inclusion bodies, possibly overcome using signal peptides. LMM fusions have been used in heterologous production of the *tat* and protease proteins of HIV-1, a neurofibromatosis type 1 protein and several others. LMM fusions could also be useful for production of proteins toxic to the cell and for peptide antibody production since the peptide is likely to be exposed. This approach highlights alternatives to affinity, which might be more suitable for larger-scale use, although the resolution achievable is likely to be lower.

7.2.3 ANTIGENIC EPITOPES

The octapeptide FLAG (–asp–tyr–lys–asp–asp–asp–asp–lys–) was specifically designed to be highly hydrophilic and immunogenic with a built-in protease sensitive site [2,25]. It is usually placed at the N-terminal of the target. Being very hydrophilic, it should be highly exposed in solution, and hence is unlikely to interfere with protein folding yet should be available for specific binding. A monoclonal antibody to FLAG has been isolated requiring Ca^{2+} for binding, hence affinity binding and release can be controlled by varying Ca^{2+} concentration. The FLAG epitope is about the minimum length for strong immunological binding. Cleavage is achieved with purified enterokinase (specific for $-(asp)_3-lys-$), releasing the target with an authentic N-terminal. The ompA signal from *E. coli* can be used to manipulate cellular location. In spite of its elegance and major advantages, the cost of immobilized antibody is a drawback, and although used successfully with interleukins and other proteins, it has been less useful for peptides. Low cleavage yields have been reported. A similar though not identical epitope, with Ca^{2+}-dependent binding to a specific antibody (HPC4) has also been developed [26]. The epitope can be attached to the C- or N-terminal, and a factor Xa site can be incorporated.

Additional examples of the use of epitopes as tails include production of somatostatin in *E. coli* via the *rec*A polypeptide [27] and of adenyl cyclase in yeast with a short antigenic peptide at the N-terminal, purified 700-fold in a single step by immunoaffinity chromatography with anti-*rec*A monoclonal antibody [28].

7.2.4 CARBOHYDRATE-BINDING DOMAINS

The substrate-binding domains of carbohydrases and lectins have also been used as purification tails [1–3,29–34]. The main rationale is to utilize relatively inexpensive and widely available carbohydrates such as starch and cellulose as adsorbents. The tail size varies from maltose-binding protein (MBP, ~40 kDa) to an endocellulose-binding site (111 amino acids). MBP is the best characterized and has been used to purify a range of proteins expressed intracellularly in *E. coli* [1,29,31]. A recent example is the production of an *E. coli* heat-stable enterotoxin (STb), expressed in *E. coli* and accumulating in the periplasm (via signal sequence), accounting for ~33% of the cell extract protein [31]. The fusion protein was adsorbed on crosslinked starch, and was eluted in high yield with maltose. After cleavage with Xa, the authentic enterotoxin which was now *not* adsorbed on the starch was recovered in high yield. In another example, starch-binding domains (SBDs) of differing lengths have been produced via MBP fusions, cleaved by Xa [32]. This route avoided the very tedious and difficult exercise of purifying SBDs of different sizes from proteolytic digests of glucoamylase-1. Binding domains for starch [30,32], cellulose [2] and mannose [33] have been used in a similar way (Table 7.1).

Earlier reports of low yields appear to have been overcome, at least in part [30]. Lectins have also been utilized [34]. For example, a construct of insulin signal sequence–rat hepatic lectin–human placental alkaline phosphatase could be expressed in rat fibroblasts and in the baculovirus/insect cell system. The fusion was exported into the medium and could be recovered and purified on immobilized galactose in the presence of Ca^{2+}, using reduced Ca^{2+} for elution. Given the range of lectins available, the method has considerable potential.

7.2.5 POLY(AMINO ACID) TAILS

7.2.5.1 *Poly(arg)*

(Arg)$_5$ tails impart very basic properties to proteins (pI > 12), and were one of the first product modifications investigated [35,36]. The short peptide β-urogastrone with C-terminal –(arg)$_5$– expressed in *E. coli* was adsorbed efficiently from cell lysates, batchwise on a sulfated ion exchanger. Most of the *E. coli* proteins, being neutral or acidic, did not adsorb. On elution and treatment with carboxypeptidase B, followed by rechromatography on the same exchanger, the now 'neutral' authentic β-urogastrone was recovered in good yield ~95% pure. This approach highlights two important features specifically designed into the tail:

(a) the introduction of very basic characteristics into the target, thus distinguishing it *specifically* from host proteins, and enabling inexpensive exchangers to be used;
(b) the *dramatic change in pI* on removal of the poly(arg) tail (cf. unmodified contaminants).

A C-terminal –arg– or –lys– in the target can be retained by incorporating a non-basic amino acid N-terminal to the poly(arg) tail. Treatment with carboxypeptidase B can then be followed with carboxypeptidase A (blocked by arg or lys). Poly(arg) fusions can also be processed by polyelectrolyte precipitation (e.g. polyacrylic acids [37]), prior removal of nucleic acids being unnecessary, and possibly with ion exchange membranes and two-phase aqueous systems. Further advantages of poly(arg) tails are small size and the ease of detection (cf. bio- or immunoassay). Problems include the need in *E. coli* to use strains lacking the *ompA* protease to minimize degradation, while plasmid instability and reversal have been reported [2].

7.2.5.2 Poly(asp)

Considerable work has been done with poly(asp) tails [38,39]. The negatively charged fusions can be precipitated with polyelectrolytes (e.g. polyethyleneimine), and can be purified by partition in two-phase aqueous systems, by reversed micelle extraction and by ion exchange [2]. With β-gal fusions the optimal tail length for precipitation was $-(asp)_{11}-$. The modest separation factors reported with β-gal ($5\times$) contrasts with much higher purification factors achieved with other poly(asp) fusions [2]. Yields were higher with C-terminal tails, but prior removal of nucleic acids was essential. Poly(asp) tails are expressed stably and are in general stable to proteases [2]. A new approach using ion exchange membranes looks promising for larger-scale use [39]. In this, fusions of β-gal with a tail of the form:

$$-gly-asp-pro-met-ala-(asp)_n-tyr-$$

have been processed in a DEAE membrane module. In a single pass up to 40% of the fusion protein was adsorbed with a maximum at $-(asp)_{11}-$ and could be eluted with M NaCl. Prior nuclease treatment and diafiltration were required. It is likely that this system could be further optimized.

7.2.5.3 Poly(his)

Transition metals such as Cu^{2+}, Ni^{2+}, Zn^{2+} and Co^{2+} form stable complexes with proteins via $-his-$ and $-cys-$ residues. IMAC (immobilized metal affinity chromatography) is based on this effect through the use of chelating resins [40] and depends on the relatively few $-his-$ residues in proteins [41]. Fusions with poly(his) tails show stronger binding than normal proteins, and are being used increasingly for purification. The most widely used tail is $(his)_6$, but a variety of his-containing sequences have been reported [2]. Expression levels of constructs are generally maximal with $-(his)_{5,6}-$, up to 40 mg/L in the baculovirus/insect cell system [42].

Poly(his) fusions are selectively adsorbed on IMAC resins, and a number of one-step purifications have been reported [42,43] using Ni^{2+}, Zn^{2+} and Co^{+2}. Strength of binding is approximately proportional to chain length/number of $-his-$ residues. Similarly, ease of elution varies with tail length and composition, e.g. decreasing pH gradient for $-(his)_2-$ to only 10% of $-(his)_6-$ being eluted under the same conditions. For insoluble fusions, where 6M guanidine $-HCl$ was required to keep the fusion protein in solution, only $-(his)_6-$ fusions could be processed with IMAC. The various other –his– containing tails vary in their avidity for IMAC adsorbents, and hence offer useful flexibility in processing [2].

In general his-containing tails do not affect folding and functional properties of the target. C-terminal poly(his) tails can be removed with carboxypeptidase A. Retention of a specific C-terminal amino acid removed by carboxypeptidase A can be achieved by incorporating a –lys– or –arg– residue at the C-terminal of the poly(his) tail, blocking further digestion by carboxypeptidase A. The basic amino acid can then be removed with carboxypeptidase B. In many poly(his) constructs, an additional protease site has been incorporated, e.g. for thrombin, Xa or rennin. In a recent system for expression of heterologous proteins in *E. coli*, using the N-terminal –(his)$_6$– tail, a polylinker was incorporated to allow cloning in three different reading frames [42]. With this system, the HIV-1 protein produced accounted for ~30% of total cell protein, yielding ~40 mg/L culture of highly purified product. Recently, poly(his) tails have been shown to alter dramatically partition in two-phase aqueous systems of PEG/K phosphate (incorporating PEG–chelate/metal ion complexes) [44] (See also Chapter 4). This could be a very important development. Also, polymer–chelated-metal ion reagents have been used to precipitate poly(his) fusions fractionally, further extending the range of usable separation techniques [41].

7.2.5.4 Other Poly(Amino) Acid Tails

Poly(cys) and poly(phe) fusions have been developed for *E. coli*, the products being recovered in good yield with columns of thiopropyl-Sepharose and phenyl-Superose respectively [2].

7.2.5.5 Amino Acids and Peptides

A dipeptide with one charged amino acid (glutamate) has been useful in the production of human growth hormone (HGH), the single additional charge being sufficient to facilitate recovery by ion exchange [11]. The glutamate (N-terminal), could be removed specifically with an exodipeptidase (DAP I) since the authentic N-terminal of HGH is proline (blocks DAP I). For a target protein without N-terminal proline, –pro– can be included in the tail and after digestion with DAP I, the –pro– can be removed specifically with DAP IV. A peptide tail recently reported is –(ala–trp–trp–pro–)$_x$– ($x = 1$ or 3) [45]. This enhances dramatically the partition of fusions into the upper phase of a PEG–4000/K phosphate system (by 7× and 60× respectively). Good expression levels were reported (7–30 mg/g cell weight). This offers perhaps the first systematic predictable approach to using two-phase aqueous partition, notoriously unpredictable for protein processing, and could represent a very significant breakthrough, given the advanced technology available for liquid–liquid extraction (see also Chapter 4).

7.2.6 BIOTIN-BINDING DOMAINS

The remarkable affinity and specificity of biotin for certain proteins (avidin, streptavidin; $K_d \sim 10^{-15}$M) has been utilized in a highly ingenious manner. All cells contain a small number of proteins with covalently attached biotin (only one in *E. coli*). Biotin is coupled *in vivo* by biotin ligase to a unique –lys– residue in a specific, proline-rich, 75 amino acid region of the protein. The DNA sequence for the biotin-binding domain (BBD) has been isolated from *Pseudomonas shemanii*, and vectors for its expression in *E. coli* have been developed [46]. The BBD has been incorporated as an N-terminal construct, using the indigenous *E. coli* biotin ligase to attach intracellular biotin to the fusion protein *in vivo*. The fusions could be recovered and purified very effectively from crude cell extracts on columns of immobilized monomeric streptavidin ($K_d \sim 10^{-7}$M) elution, being achieved with dilute buffers containing biotin in low concentration. The effectiveness of this purification process is due in part to the presence of very few biotinylated proteins in the cell. Protease-sensitive sites (e.g. for Xa, not present in streptavidin) can be incorporated. Growth inhibition due to biotin limitation can be overcome by biotin supplementation and ligase limitation at high expression levels by introducing multicopy plasmids carrying the ligase gene. With this system high levels of biotinylated fusion proteins can be expressed in both *E. coli* and yeast (up to 3×10^6 molecules/cell). Proteins produced in this way include β-gal, neomycin phosphoryltransferases, chloramphenicol acetyl transferase, yeast HIS3 protein and somatostatin release inhibitory factor [46,47]. A recent report indicates that recombinant proteins biotinylated *in vivo* in some *E. coli* strains (deficient in the DegP periplasmic protease) are exported into the culture medium (cf. β-lactamase, alkaline phosphatase, TEM [47]).

7.3 Tail Cleavage

Efficient and accurate tail removal will frequently be necessary, particularly for therapeutic materials. Both chemical and enzymatic methods are available [1–3]. Chemical cleavage can be applied only to specific amino acids (e.g. CNBr for –met– residues) or dipeptides (hydroxylamine for –asn–gly–), which are, however, relatively common in polypeptides and proteins. It is therefore more useful for oligopeptides. Enzymes are favoured for proteins and polypeptides, although the efficiency varies greatly, from near 100 to < 30% [1]. Recently, 'fusion proteases' have been developed which have greater specificity and are easier to remove after use for tail removal [48].

7.4 Present Status

It is now clear that a radically new approach to the development of bioprocesses for protein/peptide production is available. By creating gene constructs incorporating additional DNA, coding for moities ranging from short peptides to complete proteins, it is demonstrably possible to:

(a) greatly simplify recovery and purification,
(b) predetermine the location of the product in the producing system,
(c) design *ab initio*, much more rational, predictable and integrated processes,
(d) generate multifunctional products (e.g. antibody–enzyme hybrids).

It should also be possible to develop more standardized production processes for proteinaceous materials, which could speed up regulatory approval.

At this stage no particular system is pre-eminent, each has some advantages. Those that are proven or look promising for production-scale use include:

(a) ZZ epitopes of SPA: very high expression levels in *E. coli*, immunoaffinity separation, demonstrated on a 1000 L scale, widely applicable, tends to give 'soluble' fusions;
(b) hCAII: useful expression levels, inexpensive affinity purification method, widely applicable, tends to give 'soluble' fusions;
(c) poly(amino acids), especially:
   - poly(his): purification by inexpensive ion exchangers, one-step purifications reported, small size, ease of tail removal, range of separation methods possible;
   - poly(arg): small size, inexpensive ion exchangers, ease of removal, highly basic, less suitable for small peptides;
   - poly(asp): wide range of separation methods possible, small size, reduces pI;
(d) GST: high expression levels, one-step affinity separations reported, 'soluble' fusions;
(e) biotinylation: avidity and specificity of adsorbent, tends to give 'soluble' fusions;
(f) LMM: simple and inexpensive processing (precipitation), possibly useful for toxic proteins;
(g) CBDs (carbohydrate binding domains): very inexpensive/biodegradable adsorbents, 'soluble' fusions;
(h) FLAG peptide: very small, built in protease site, very hydrophilic, ease of elution from antibody;
(i) β-galactosidase: well studied, high expression, affinity separation (via saccharide ligand), 'protects' small peptides.

Much further work is still needed to realize the full potential of the facilitated processing approach. Areas requiring attention include:

(a) Extending the range of hosts available and of high-level expression systems (*E. coli* is by far the preponderant host used at present). This is essential to ensure that accurate glycosylation can be carried out, since most therapeutic proteins are glycosylated [5,7,9,10,49].

(b) Widening the range of separation technologies used, particularly for the larger scale. At present affinity predominates, the future lies possibly in combining specific interactions with techniques other than chromatography (e.g. ion exchange membranes, expanded bed adsorption, two-phase acqueous extraction, precipitation).

(c) Improving the understanding of and ability to manipulate signal peptides.

(d) Improving the range and efficiency of cleavage systems (at present very variable).

(e) Improving intracellular stability of fusions (e.g. with dual tails, by incorporating ubiquitin [50]).

7.5 Process Design Strategy

When considering the use of facilitated DSP, factors to be considered include:

(a) End use of product: purity/yield required?, removal of tail necessary?, scale envisaged?, sterility for pharmaceutical use?

(b) State of development of conventional DSP alternatives.

(c) Availability of target gene and of high expression systems.

(d) Availability of suitable host cell. Is an alternative to *E. coli* desirable (for glycosylation?); if so consider yeast (not only *Saccharomyces cerevisiae* which 'over' glycosylates), slime moulds [51] or animal systems (baculovirus/insect cells)?

(e) Is there a processing window/affinity technique available (e.g. few biotinylated proteins in cells; most *E. coli* proteins have neutral or acidic pIs)?

(f) The substantial advantage of achieving secretion (especially if into the culture medium)?

(g) Expression levels attainable may determine preliminary processing required (e.g. concentration, dialysis) and potential advantages of affinity (high resolution).

(h) Scale of DSP required (e.g. gel filtration—less suitable for proteins on a large scale)?

(i) Is stabilization of the target fusion needed (e.g. dual tails, large protein tail, fusion with ubiquitin [50])?

(j) Tail size to maximize productivity (1 kDa *target* fused to 100 kDa *tail* = 1% yield)?

(k) Tail value for assay/monitoring?

(l) How can the total production process be best integrated?

Armed with clearly defined objectives for the process envisaged, and with answers to at least some of the above questions, an outline DSP design can readily be generated. Thus for small-scale and research purposes, affinity tails are attractive, and many one-step purifications have been reported. They are also very effective for recovering a specific product from low concentrations, thus reducing the need to maximize expression levels (see also Ref. [2]). For larger-scale use, obtaining a high expression level is vital, and overall cost considerations will be very important even for high-value products (e.g. pharmaceuticals). Thus, for example, a poly(amino acid) tail having a low molecular weight and processable with conventional ion exchangers, by bulk adsorption, by precipitation with membranes or by liquid–liquid extraction may be more cost effective, at least for the initial stages [41,44]. The generally high resolving power of affinity methods must be borne in mind, however, even for the larger scale [2,21]. In addition, the purification protocols for *different* products with the *same tail* will be very similar, and after tail cleavage reprocessing by the same separation technique provides substantial further purification, often to near homogeneity, since contaminants will be unaffected. Overall, the *predictability* and flexibility of the facilitated processing approach, together with the possibility of logical process design, confer great advantages.

7.6 Conclusions

The genetic engineering aspects of *facilitated processing* are now reasonably well developed and generation of high-level expression systems appears generally feasible though by no means a trivial task. Neither is the development by conventional means of a new process for each new protein/peptide target. Furthermore, since each tail can be used with different targets, the development of recovery and purification processes for new products should be greatly simplified. In addition, it should also be possible to develop more standardized production processes which could significantly speed up regulatory approval.

By combining the concept and technology of *facilitated* downstream processing with basic knowledge now being generated on the properties of host systems [52,53], it should be possible to design *ab initio*, highly integrated and more cost effective DSP for producing proteins and peptides, much more predictable and much less affected by technical and biological variation.

7.7 References

References [1] to [3] are reviews of facilitated processing covering the literature to early 1992, Refs. [7] to [8] are specific Journal issues devoted to expression systems and Refs. [9] to [10] are reviews on the baculovirus/insect cell system. The remaining references are papers on specific aspects of facilitated processing of particular importance or those published in 1992–September 1994. Journals containing relevant articles include *Protein Expression and Purification, Bio Technology, Biotechnology and Bioengineering, Enzyme and Microbial Technology* and *Methods in Enzymology* (1990), Vol. 185.

1. M. Uhlen, G. Forsberg, T. Moks, M. Hartmanis and B. Nilsson (1992) *Curr. Opin. Biotechnol.* **3**, 363.

2. C. F. Ford, I. Souminen and C. E. Glatz (1991) *Prot. Exp. Pur*, **2**, 95.

3. M. Uhlen and T. Moks (1990) *Meth. Enzymol.*, **185**, 129.

4. K. Johnson, C. K. Murphy and J. Beckwith (1992) *Curr. Opin. Biotechnol.*, **3**, 481.

5. X.-C. Wu, S.-C. Ng, R. I. Near and S.-L. Wong (1993) *Bio/Technol.*, **11**, 71.

6. B. K. Dalmia and Z. L. Nikolov (1994) *Enzyme and Microbial. Technol.*, **16**, 18.

7. Various articles on all species (1993) *Curr. Opin. Biotechnol.*, **4**(5).

8. Various articles on all species (1992) *Curr. Opin. Biotechnol.*, **3**(5).

9. V. A. Luckow (1993) in M. L. Shuler, H. A. Wood, R. R. Granados and D. H. Hammen (Eds), *Insect Cell Cultures: Production of Improved Biopesticides and Proteins from Recombinant DNA*, John Wiley, New York.

10. L. A. King and R. D. Possee (1992) *The Baculovirus Expression System: A Laboratory Guide*, Vol. 1, Chapman & Hall, London.

11. G. Germino and D. Bastia (1984) *Proc. Natl Acad. Sci. USA*, **81**, 4692.

12. K-P. Koller, G. Reiss, K. Sauber, E. Uhlmann and H. Wallmeier (1989) *Bio/ Technol.*, **7**, 1055.

13. S. Baksh, K. Burns, J. Busaan and M. Michalek (1992) *Prot. Exp. Pur.*, **3**, 322.

14. C. Spana, E. C. O'Rourke, J. B. Bolen and J. Fargnoli (1993) *Prot. Exp. Pur.*, **4**, 390.

15. S. Peng, M. Sommerfelt, J. Logan, Z. Huang, T. Jillig, K. Kirk, E. Hunter and A. Sorscher (1993) *Prot. Exp. Pur.*, **4**, 95.

16. E. C. Serra, N. Carillo, A. R. Krapp and E. A. Ceccarelli (1993) *Prot. Exp. Pur.*, **4**, 539.

17. G.van. Heeke, R. Shaw, R. Schnitzer, J. M. Couton, S. M. Schuster and F. W. Wagner (1993) *Prot. Exp. Pur.*, **4**, 265.

18. J. Robben, G. Massie, E. Bosmans, B. Wellens and G. Volkaert (1993) *Gene*, **126**, 109.

19. B. Ortega, J. L. Garcia, M. Zazo, J. Varela, I. Munoz-Willery, P. Cuevas and G. Giminez-Gallego (1992) *Bio/Technol.*, **10**, 795.

20. D. G. Yansura (1990) *Meth. Enzymol.*, **185**, 129.

21. M. Hanson, S. Stahl, R. Hjorth, M. Uhlen and T. Moks (1994) *Bio/Technol.*, **12**, 285.

22. B. Nilsson and L. Abrahamson (1990) *Meth. Enzymol.*, **185**, 144.

23. B. Hammarberg, P.-A. Nygren, E. Holmgren, A. Elmblad, M. Tally, U. Hellman, T. Moks and M. Uhlen (1989) *Proc. Natl. Acad. Sci. USA*, **86**, 4367.

24. V. Wolker, K. Medea, R. Schumann, B. Brandmeier, L. Weismuller and A. Wittighofer (1992) *Bio/Technol.*, **10**, 900.

25. T. P. Hopp, K. S. Prickett, V. L. Price, R. T. Libby, C. J. March, D. P. Ceretti, D. L. Urdal and P. J. Conlon (1988) *Bio/Technol.*, **6**, 1204.

26. A. R. Rezaie, M. M. Fiore, P. F. Neuenschwander, C. T. Esmon and J. H. Morrissey (1992) *Prot. Exp. Pur.*, **3**, 453.

27. G. G. Krivi, M. L. Bittner, E. Rowold, Jr, E. Y. Wong, K. C. Glenn, K. S. Rose and D. C. Tiemeier (1985) *J. Biol. Chem.*, **260**, 10263.

28. J. Field, J.-I. Nikawa, D. Broek, B. MacDonald, L. Rodgers, I. A. Wilson, R. A. Lerner and M. Wigler (1988) *Mol. Cell. Biol.*, **8**, 2159.

29. F. Bregere and H. Bedonelle (1992) *C. R. Acad. Sci III*, **314**, 527.

30. L. Chen, C. Ford, A. Kusnadi and Z. Nikolov (1991) *Biotechnol. Prog.*, **7**, 225.

31. C. D. Handl, J. Harel, J.-I. Flock and J. D. Dubreuil (1993) *Prot. Exp. Pur.*, **4**, 275.

32. A. R. Kusnadi, C. Ford, Z. Nikolov and L. Zivko (1993) *Gene*, **127**, 193.

33. M. E. Taylor and K. Drickamer (1992) *Prot. Exp. Pur.*, **3**, 1308.

34. M. E. Taylor and K. Drickamer (1991) *Biochem. J.*, **274**, 575.

35. H. M. Sassenfeld and S. J. Brewer (1984) *Bio/Technol.*, **2**, 76.

36. S. J. Brewer and H. M. Sassenfeld (1985) *Trends in Biotechnol.*, **3**, 119.

37. I. Suominen, C. Forney, J. Luther, M. Niederauer and C. Glatz (1990) 'Enhanced selectivity in bulk separations using charged-peptide fusions', paper presented at the Engineering Foundation Conference on *Recovery of Bioproducts*, St Petersburg Beach, Florida, May 1990.
38. I. Suominen, C. Ford, D. Stachon, H. Heimo, M. Niederauer, H. Nurmila and C. Glatz (1993) *Enzyme and Microbiol. Technol.*, **15**, 593.
39. M. H. Heng and C. Glatz (1993) *Biotechnol. Bioengng.*, **42**, 333.
40. J. Porath (1992) *Prot. Exp. Pur.*, **3**, 263.
41. M. E. Van Dam, G. E. Waenschell and F. H. Arnold (1989) *Biotechnol. Appl. Biochem.*, **7**, 492.
42. V. Cheynet, B. Verrier and F. Millet (1993) *Prot. Exp. Pur.*, **4**, 367.
43. C. A. Franke and D. E. Hruby (1993) *Prot. Exp. Pur.*, **4**, 101.
44. F. H. Arnold (1991) *Bio/Technol.*, **9**, 150.
45. K. Kohler, C. Ljungquist, A. Kondo, A. Veide and B. Nilsson (1991) *Bio/Technol.*, **9**, 642.
46. J. E. Cronan Jr (1990) *J. Biol. Chem.*, **265**, 10327.
47. K. E. Reed and J. E. Cronan Jr (1991) *J. Biol. Chem.*, **266**, 11425.
48. P. A. Walker, L. E.-C. Leong, P. W. P. Ng, S. H. Tan, S. Waller, D. Murphy and A. G. Porter (1994) *Bio/Technol.*, **12**, 601.
49. C. E. Warren (1993) *Curr. Opin. Biotechnol.*, **4**, 596.
50. T. R. Butt, S. Jonnalagadda, B. P. Monia, E.J. Sternberg, J. A. Marsh, J. M. Stadel, D. J. Ecker and S. T. Crooke (1989) *Proc. Natl Acad. Aci. USA*, **86**, 254.
51. W. Dittrich, K. L. Williams and M. Slade (1994) *Bio/Technol.*, **12**, 614.
52. A. R. Thomson (1985) in M. S. Verrall (Ed.), *Discovery and Isolation of Microbial Products*, Ch. 3, Ellis Horwood, Chichester, pp. 52–64.
53. R. E. Turner, B. S. Baines and J. A. Asenjo (1994) in L. J. Pyle (Ed.), *Separations for Biotechnology*, Vol. III, Royal Society of Chemistry, London, p. 518.

8 PROCESS INTEGRATION IN BIOTECHNOLOGY

Juan A. Asenjo and E. W. Leser

8.1 Introduction

Process integration in the production of biological molecules in modern biotechnology encompasses a broad range of scientific and engineering disciplines. Modern biotechnology began to be developed during the second half of this century, particularly when the production of antibiotics and amino acids became important. The outcome of the progress in genetic engineering, which is known today as the 'new biotechnology', has brought a completely different approach to this field. This change can be perceived in a new perspective in what was known as 'fermentation technology'. Instead of giving priority to the biological reactor, this field is now clearly divided into three parts, and what happens inside the culture system shares importance with the upstream and downstream set of operations and activities. It is important to understand what caused this change of focus.

Before the advent of genetic manipulation of micro-organisms, the achievement of high productivities within a process was only possible with the improvement of the production organism by classical mutation. This was done using traditional strain selection methods, or the optimization of the fermentation process, by manipulation of operating parameters. Also, in the former systems, based on biological processes, the concentration of products in the fermented broth was high (e.g. ethanol, citric acid) [1], the separation and purification of products was based on well-defined unit operations (e.g. distillation, extraction) and extensive equilibrium data, and, finally, the products, which had low molecular weight, were secreted into the broth.

The change of perspective following the new biotechnology suggests a different trend. The products usually have an important commercial value, due not only to the intrinsic relation associated with 'high value–low volume' products but also because most of them are an important and rapidly growing

Downstream Processing of Natural Products. Edited by Michael S. Verrall

class of drugs. Since 1981, fourteen biopharmaceutical products, including twelve therapeutics, one vaccine and one *in vivo* diagnostic, have been approved by the US Food and Drug Administration [2] and 143 other biotechnology medicines and vaccines are currently in the regulatory pipeline. The market forecast for these products for the beginning of the next decade indicates a volume of sales bigger than US$10 billion. Hence, the increase of productivity is a very important goal and new solutions must be found to achieve it. As the impact of choices made in the initial stages of a bioprocess (upstream processing) is perceived in later stages (bioreactor and downstream processing), the development of efficient bioprocesses relies strongly on the interaction of rather different disciplines from the engineering sciences and the biosciences. Effective integration of the different parts of production will provide tools for increasing the productivity and/or product yield and quality.

Process integration can be adopted either involving the whole process, specific parts of it, or acting on particular unit operations as shown in Figure 8.1. In this figure the overlapping surfaces represent areas of integration between parts of the whole process, and, actually, a global integration of the three parts is still much smaller than the possibilities of integrating them two by two. The most common methods for process integration are mentioned in Table 8.1.

A general scheme for production based on biotechnological operations is presented in Figure 8.2. It is clear that some paths are simpler than others and one should take into account the fact that some of the unit operations present in the processing can be very cumbersome, as, for example, the separation of cell debris after cell disruption. If a vigorous process is necessary to break the cell envelope and release the product, it results in cell debris that needs to be separated. These residues are normally very small and their density differs very little from the density of other cell components. Consequently, the usual methods for solid liquid separation (centrifugation and filtration) are inadequate to perform this task. Although they are used, because of the lack of a better alternative, their performance is poor. If one can use an efficient enzymatic system that promotes permeability of the cell membrane with little or no cell lysis, even if there is not a complete release of the product, it could be possible to use milder conditions of mechanical disruption of the cells, consequently producing bigger debris and facilitating their separation. Thus, the concept of process integration is clearly related to the following considerations:

(a) Reduction of the number of unit operations. Rational reduction of the number of unit operations improves the process in the sense of economics and compactness. This may be performed by cutting out an operation, by moving an important operation, by the use of combined (integrated) unit operations and by a combination of the former possibilities.

(b) Reduction of process streams. Full conversion within a single, integrated unit operation instead of having voluminous recycle flows reduces process

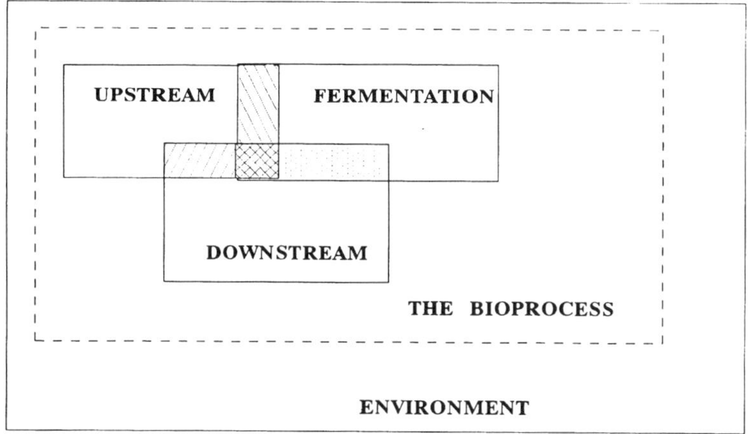

Figure 8.1 The possible areas and levels of process integration

streams in general and thereby energy consumption and waste production. Selective removal of products also decreases the water usage in general and the effluent of wastewater in particular.

(c) Control aspect. In integrated unit operations, the rate of product formation and the rate of product withdrawal from the reactor are decoupled, adding an extra degree of freedom. This improves, in principle, the control over the process. In the case of biomass retention in the fermenter by means of membrane or centrifugal techniques, the

Table 8.1 The levels of integration and the actions for integrating parts of the process

Level of integration	Type of action
Upstream	Induction of product secretion
	Induction of cell flocculation
	Induction of cell lysis
	Design of fermentation media
	Attachment of affinity tags to the product
	Induction of cell resistance to metabolites
Fermentation	Optimization of culture conditions
	Culture intensification
	in situ removal of metabolites during fermentation
	Cell recycling
	Cell immobilization
Downstream	Cell permeability
	Two-phase separation systems
	Integrated cell separation and product purification
	Design of chromatographic matrices

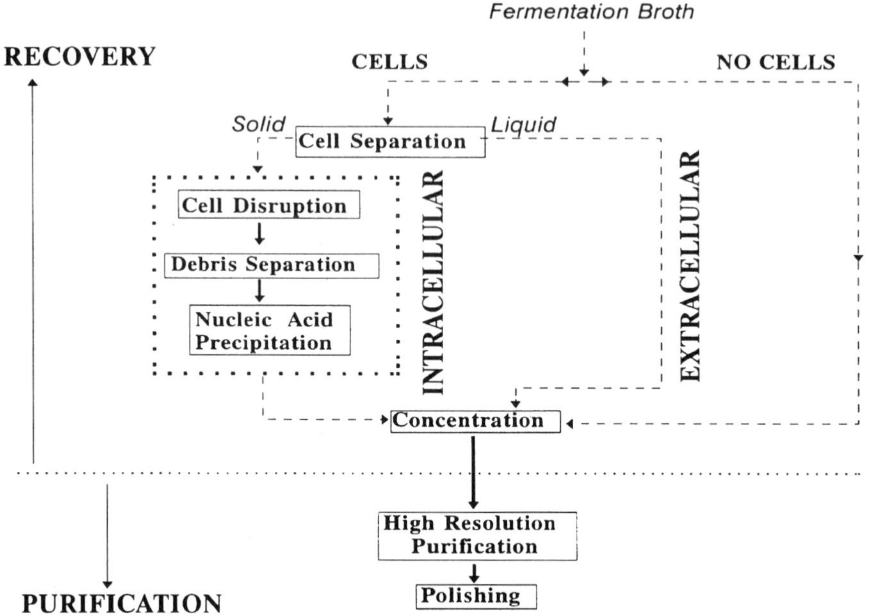

Figure 8.2   General scheme for the processing steps in downstream protein purification

productivity of the bioreactor will improve substantially because high concentrations of the biocatalyst are achieved.

8.2 Integrating Upstream

The first approach to process integration and improvement is based on the genetic manipulation of the expression system. Proteins are targeted to different compartments of cells and the subcellular localization will determine post-translational modifications. Secretion is an important example of such protein targeting [3]. The clearer understanding of these mechanisms has allowed the development of systems that lead to the secretion of the product in bacteria, yeast and mammalian cells. *Escherichia coli* is the most important organism used in bioprocesses. Being a Gram-negative bacterium, it is enveloped by an inner membrane, a cell wall and an external membrane. Secretion mechanisms normally direct an accumulation of the protein in the periplasm between the two membranes. In this case, a milder treatment for release of the product will be necessary, hence facilitating downstream processing. The use of protease-weak *Bacillus subtilis* as the production expression vector is increasing, as a consequence of the advantage present in this system: the secretion of proteins directly to the culture medium [4,5].

In yeast and other fungi, protein release is hindered by a limited cell wall porosity. The use of a lytic cell wall permeable glucanase has been recently demonstrated for selectively releasing recombinant particles, leaving the cells virtually intact [6]. Furthermore, a large fraction of intracellular proteins remained retained by the cell wall. This glucanase has recently been cloned to demonstrate the large-scale feasibility of this approach [5]. This would consist of a major improvement in simplifying the downstream processing of intracellular yeast proteins and is a clear case of process integration.

By means of genetic manipulation it is possible to induce the flocculation of the cells that will facilitate the separation of biomass after the culture [7]. The attachment of affinity tags of the protein is an important strategy as this allows the use of very specific and selective purification matrices with a high degree of resolution and high yield of purification [8].

The design of fermentation media is very important both for increasing the productivity, linking the nutritional aspects with metabolic pathways and thus enhancing the product yield, but, more importantly, by the elimination of complex nutrients, especially serum supplements, that will unnecessarily complicate the downstream processing by having to use a large number of separation steps for eliminating the serum proteins left at the end of the fermentation.

8.3 Integrating Fermentation

The adequate design of the fermentation step in biotechnology processes can provide conditions to facilitate the downstream processing. Several strategies can be adopted. The optimization of culture conditions is one of the aspects to be examined. The manipulation can be made at the level of the vessel design, aiming the adaptation of culture parameters to different tolerances towards environmental conditions present in genetically modified organisms, when compared with the wild strains [9]. This means choosing the optimal hydrodynamic regime, based on the configuration of the fermentation system and its parts.

The bioreactor conception starts with the choice of the state of the biocatalyst: bacterial, yeast, animal or insect cells, suspended cells versus the many forms of immobilization, including membrane bioreactors and cell recycle. Each of these possibilities must be evaluated not only in the light of the time courses of the bioreaction provided by stoichiometry and kinetics but also considering mixing and gas exchange requirements and heat removal This will enable the designer to decide on the reactor type and on the operation mode [10].

The optimization of the fermentation process, aiming at the integration, must take into account a balance between possible solutions and eventual constraints arising out of these actions. For example, seeking culture intensification, based on high cell concentrations, implies that an efficient

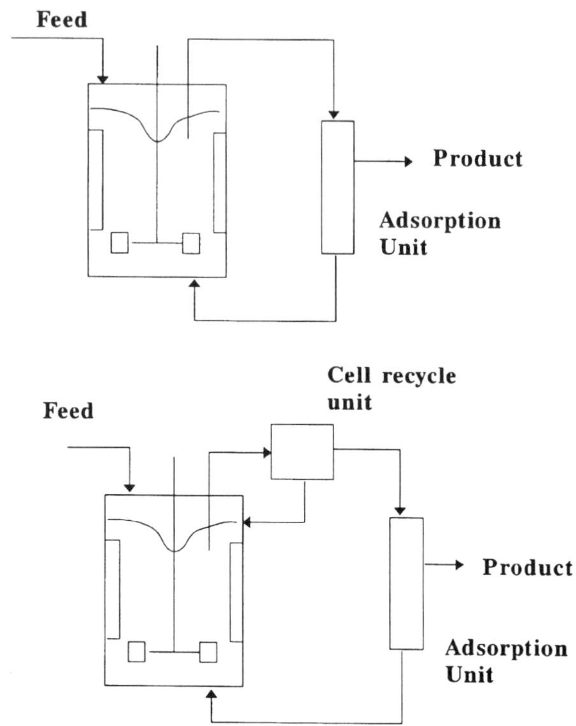

Figure 8.3   Two schemes for simultaneous fermentation and product removal

system for removing the metabolites must be adopted to prevent toxic effects of these products on the cells and possible feedback repression which reduces the levels of the enzymes required for synthesis of the product or other essential intermediates. The selective removal of metabolites can usually be achived by continuously recycling the broth through an adsorption system as shown in Figure 8.3. However, important hindrances are usually found in this mode of operation: necessity of aseptic operation, undesired retentions by the adsorbent, shear stress on the cells, fouling of the absorption system, etc. A comprehensive rcvicw of scparation integrated processes is presented by van der Wielen *et al.* [11].

Cell immobilization is a different solution for optimizing the bioreaction. In some cases it is more a necessary condition than a matter of process improvement: some animal cells lines can only grow if attached to a surface. The core of the immobilization strategy is to confine or to bind cells to a structure to keep the biological system in a particular region of the bioreactor. The most common method for immobilizing the cells is the entrapment by ionic cross linking in a layer or bead of alginate. Other methods of immobilization are entrapment in convoluted or porous structures, entrapment within the macroporous regions of asymmetric hollow fibres and adsorption of covalent attachment to the surfaces of various solid support materials such as ion exchange resins, cellulose, Sephadex, agarose, Ti^{IV} oxide and others [12]. The advantages of using immobilized cell systems are that a high cell density can be obtained, cell morphology and broth rheology can be better controlled, the washout of cells can be avoided during continuous processing and a marked enhancement of the removal of metabolites during the culture can be achieved. One example is the use of perfusion cultures in hollow fibre reactors for the production of monoclonal antibodies [13]. A controlled and homogeneous environment is achieved within the reactor by supplying nutrients and removing metabolites and, in effect, mimicking the *in vivo* capillary and lymphatic system that supports over 10^9 cells/cm^3. There is a reduction in the volume of gas headspace and, therefore, a better utilization of the culture volume and an increase in unit cell density (per volume or per surface area) by 1–2 orders of magnitude [14].

8.4 Integrating Downstream

Separation and purification of products after fermentation, the downstream processing, is probably the most critical element of modern process biotechnology. For recombinant proteins the large number of separation and purification steps required, which may include renaturation, results in the fact that downstream processing contributes 70–80% of the total processing cost. Purification strategies must be planned carefully and the techniques used at the laboratory and bench scale should only consider future scale-up and feasibility of methods for production. These strategies must include all possible solutions to facilitate the purification part of production, trying to integrate, as much as possible, upstream and downstream activities. First of all, the final utilization of the product will define most aspects of the production. It is important to characterize clearly the production expression system, and other parameters such as the composition of the stream leaving the fermenter, in terms of biological and physiochemical properties. Finally, one should have in mind possible separation procedures. Hence, any contribution to facilitate this part of the process will lead to process integration and an advantageous improvement towards productivity.

Referring again to the general scheme of downstream processing presented in Figure 8.2, it is clear that, when the product is intracellular, cells have to be disrupted, which makes the recovery and later purification stages rather cumbersome, due to the presence of microsized debris that have to be separated, to the increase of the viscosity, to the release of proteases that can degrade the product, etc. Hence, the development of appropriate expression–secretion systems is a key aspect in present industrial development as it avoids cell disintegration, as already mentioned before when discussing actions in the upstream part of the process. If an appropriate secretion system cannot be found, cell permeability, as has been recently done with yeast cells in which proteins were selectively released leaving the cell matrix almost intact [5,6], is an important approach for process integration as it will simplify the downstream processing.

Other possibilities of integration at the downstream level are concerned with the development of processes that may include two or three of the operation steps in one unit operation. The most important are liquid–liquid separation, also known as aqueous two-phase separation, and expanded-bed separation. Both methods represent paradigms of process integration in biochemical engineering. They allow the integration of the separation of whole cells or cell homogenate debris and nucleic acids from the target products as well as initial protein purification in one step.

Liquid–liquid extraction consists of exploring the use of two liquid phases formed due to the incompatibility between one polymer and an appropriate salt (e.g. polyethylene glycol (PEG) and potassium phosphate) or between aqueous solutions of two polymers (e.g. PEG and dextran). The partition coefficient is the result of various interactions of the partitioned molecules with the surrounding phases, mainly through hydrophobic, hydrogen and ionic bonding. Hydrophobic proteins usually partition to the PEG-rich upper phase of PEG–salt systems, whereas hydrophilic nucleic acids and the cell debris would partition to the lower phase. An important advantage of this technique is that it allows the continuous steady-state operation which is beneficial both for the large scale and for quality control, an essential feature of modern process biotechnology. Thus, it is clear that similarly as it happened in the production of antibiotics, as the production of recombinant proteins moves to larger scales the use of liquid–liquid extraction to separate proteins will certainly increase. (For more detail see Chapter 4.)

There is currently a great effort being made to understand the fundamentals of this operation, to find precise conditions for the characterization of operational parameters and to build mathematical models to enable scaling-up to take place [15,16]. The prediction of partition coefficients of individual proteins is a crucial aim for the design of separation and purification processes based on liquid–liquid two-phase systems. The most important properties of such a system that can be manipulated in the phases, without chemically modifying the polymers used, as is usually done for biological affinity, are hydrophobicity and charge [17]. This requires the development of suitable models and correlations to predict their responses. Similarly, it is essential to build appropriate steady-state and dynamic mathematical models to simulate and predict phase behaviour in continuous two-phase extraction and back-extraction systems.

Ion exchange and gel filtration chromatography remain the workhorses of the downstream process. They are highly selective, providing good yields of purified, biologically active product. However, process-scale columns have long cycle times because of the necessity of low-pressure operation. They also present complex scale-up engineering problems. In addition, the five or more column chromatographic steps typical of most purification schemes are cumbersome. Integration, at this point, lies in streamlining this complex multistep process. Reduction in the number of purification steps and the time necessary to perform them are major goals. Alternative separation technologies abound, with high-performance liquid chromatography (HPLC) often touted as the most promising alternative in spite of scale-up difficulties. Other chromatographic separation techniques are also being used in biomanufacturing. Chief among them are perfusion and hydrophobic interaction chromatography, which provide higher flowrates and simpler operating costs respectively [18]. Expanded-bed adsorption is a new technique that in one step accomplishes the removal of whole cells or cell debris, concentration and the initial purification of the product. Special adsorbents allow the bed to expand, keeping plug-flow profiles and with low back-mixing. The capacity is similar to those found in fixed-bed adsorption and linear velocities can be higher than 300 cm/h [19,20].

Affinity chromatography is dependent on a specific interaction between the product and the solid phase to bring about its separation from the contaminants in the mixture. Structurally the process is similar to ion exchange but the nature of the ligand in the matrix establishes the fundamental difference between them. The ligands differ in specificity: some will bind only one type of protein while others will bind a whole class of proteins. The importance of this process is that if well designed it can be highly specific. The main problems to be solved relates to the matrix itself. It is necessary that the ligand remains active after it is coupled to the matrix and a poor construction will cause the loss of the ligand which will contaminate the product. This is a strong hindrance if the product is a biopharmaceutical substance and the ligand is a biologically derived substance, as, for example, antibody. Affinity matrices are usually much more expensive than those used in ion exchange. Therefore, the design of matrices plays an important role in the enhancement of the process. At present, the most promising affinity methods, with large-scale potential, are those based on commercial textile reactive dyes, immobilized on supports with hydroxyl groups, and those based on metal chelates, a technique known as IMAC (immobilized metal affinity chromatography). A further development is the attainment of new ligands obtained from new insights into protein–synthetic ligand interactions: key binding centres associated with generic groups of proteins should ideally be identified [21] (see also Chapter 13).

This proliferation of options mainly reflects the difficulty of the technological problem that must be solved. None of the alternative chromatographic methods has yet caused a revolution. New technologies that might serve to replace column chromatography have not emerged. A commercially significant solution to the complexity of downstream processing is still lacking [18].

8.5 Process Synthesis and Integration

An important strategy for process integration is process synthesis. The essential principle of engineering design is generating and evaluating methods in order to achieve a specified task. For this goal the normal procedure is to divide the work into three stages: *synthesis, analysis* and *evaluation.* The first stage consists of the translations of design heuristics into real methods and their assembly. For the selection and design of an optimal sequence in a multistep separation and purification process, the designer must choose between alternative operations and seek the optimal sequence with maximum yield and a minimum number of steps. At this point the engineer must keep in mind specifications and constraints. These are not only the common physicochemical parameters but also the strong restrictions imposed by quality control and regulatory legislation. The analysis task evaluates by calculation or comparison similar purification schemes and finds out whether they satisfy other important constraints and conditions. The final stage, evaluation, creates the trial-and-error background that reveals the true nature of design.

The synthesis stage in process design utilizes, as intensively as possible, mathematical models and equations that provide the necessary amount of information to show whether a particular piece of equipment is adequate for the proposed operation. However, the designer is responsible for the final decision beause this kind of judgement is almost exclusively based on the designer's expertise. The knowledge for designing a process must include an extensive definition of the product characteristics, the transport parameters in the process (energy and mass balances) and, for complex mixtures from which the product shall be isolated, properties of contaminants and how they interact with the product when subjected to particular processing conditions.

Downstream processing in protein production contains two classes of actions: protein recovery and purification/isolation. The design of operations belonging to the first group follows, in general lines, the traditional reasoning found in most fermentation systems. Substantial difficulties appear when attempting to solve the problems presented in the second group (purification). There is a lack of design equations and mathematical models as well as thermodynamic databases on the materials to be separated for choosing the adequate optimal sequence with a minimum number of steps, which are, usually, chromatographic. Otherwise other approaches would lead to facing the reality of an almost infinite number of combinations to evaluate, as shown in Figure 8.4.

The use of *expert systems* for dealing with the problem of protein purification is a powerful tool for overcoming the above-mentioned questions [22]. Expert systems are intelligent computer programs that aim to give advice or to solve problems that are difficult enough to require significant human expertise for their solution, by means of logical deductions, in a simulation of a human expert's reasoning. The core of the expert system is a knowledge base,

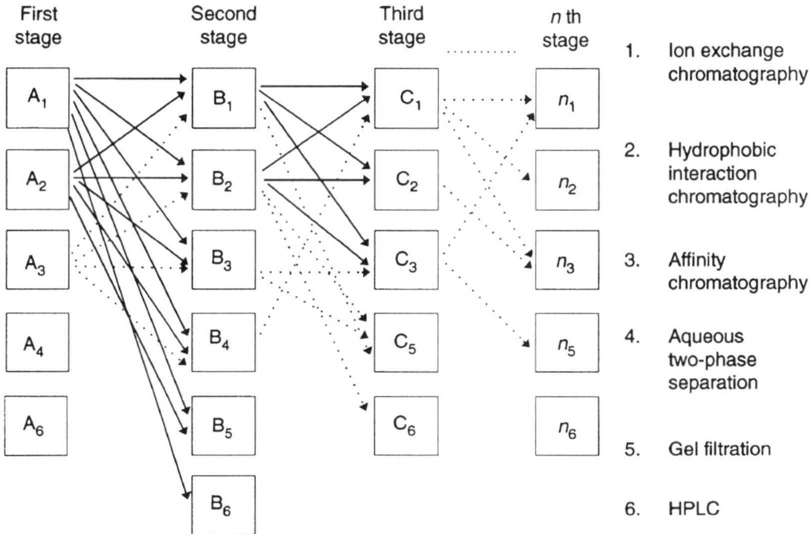

Figure 8.4 The combinatorial characteristics of choosing the sequence of operations for protein purification

which contains not only a structured set of objects but the rules that allow that simulation and lead to a rational solution for the problem. The architecture of an expert system is presented in Figure 8.5 and the functionality and flow of information within the system is shown in Figure 8.6

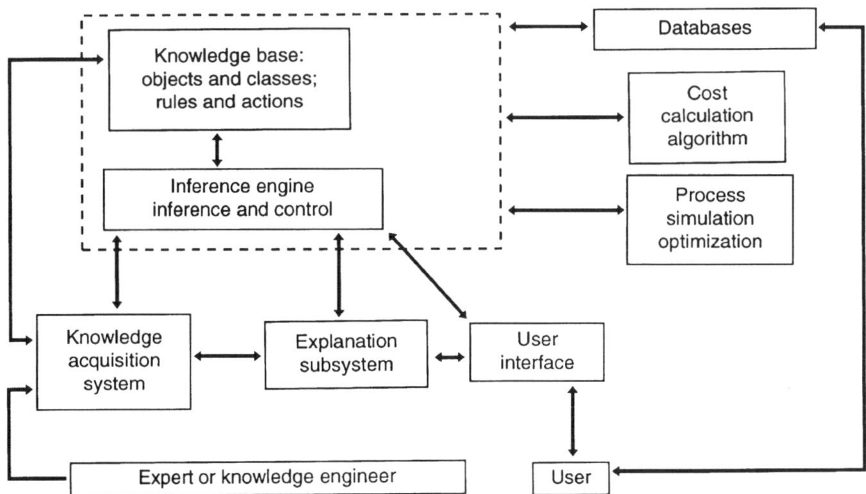

Figure 8.5 The architecture of an expert system

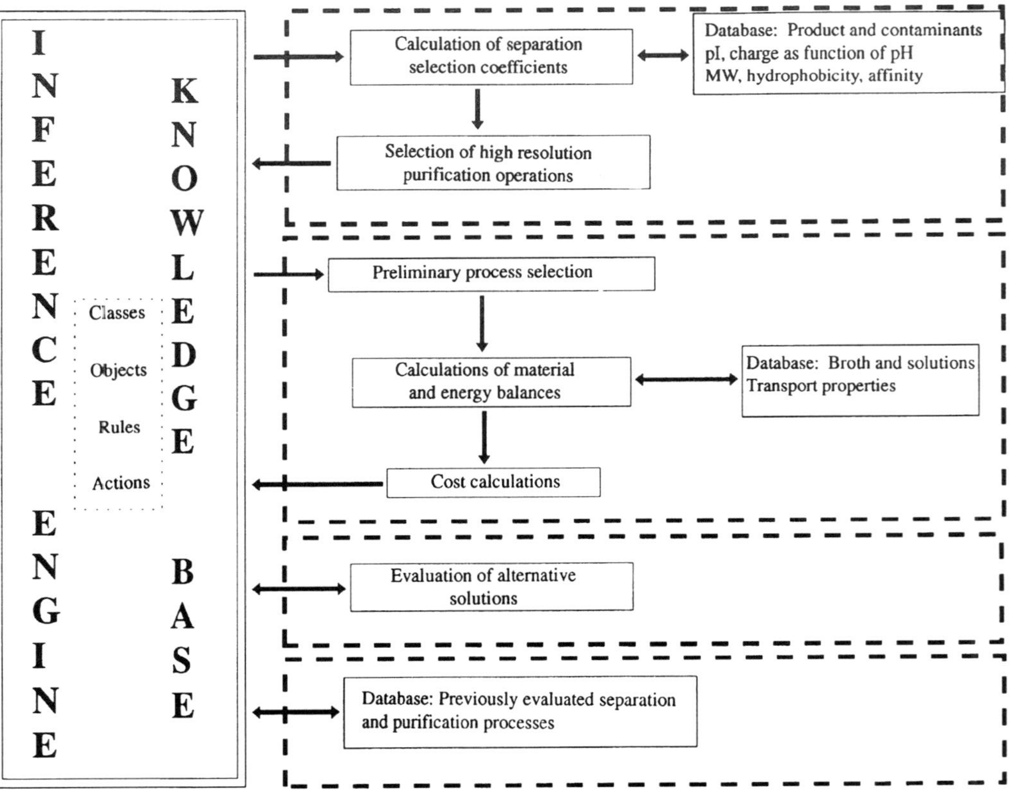

Figure 8.6 The flow of information between the expert system and the external applications: integration for achieving a complete solution for the choice of methods and design of downstream processing of proteins

Using a commercially available shell (Nexpert Object), a knowledge base was built [23]. It contains around six hundred logical rules to help find the best sequence of separation operations for obtaining a defined product on a large scale. The process is divided into two sections: recovery and purification. For recovery the rules are based on relatively well established knowledge of the fermentation industry. The specific know-how related to recovery, for which this knowledge is lacking, was dealt with by consulting industrial experts.

A rationale for the selection of a sequence of processes for purification was developed. It consists of a system that compares extensive data on the product to be purified and data on the main contaminant proteins found in the different expression systems used for biosynthesis. Definitions of parameters that translate the rules-of-thumb employed to guide the selection are presented. The deviation factor is a measure of the difference between individual properties of the product and the corresponding properties of each of the protein 'contaminants'. Efficiency measures how good a process is when exploiting the differences amongst properties. A concentration factor is included in the analysis. These variables define the selection separation coefficient and a comparison of its value determines the best possible method for the separation step. Data on the contaminants are obtained from databases included in the programme. Table 8.2 presents the data from a cell homogenate of *E. coli*.

After each separation step, there is a reduction in the amount of contaminant proteins. The number of steps must be sufficient to achieve a defined level of purity. The determination of the amount of 'contaminants' eliminated is done using an algorithm developed for this purpose, based on simplified interpretations of chromatograms indicating new concentrations after each step. A comparison of the concentration of the product after the separation with the defined level of purity indicates whether an additional step is necessary.

The system can select among three chromatographic processes: size exclusion, hydrophobic interaction and ion exchange. However, if the range of possible operations of ion exchange at each pH is considered, the actual number of alternatives is 22. Affinity chromatography is included as an option given to the user during the selection.

The use of a tool like this will result in a simplification of the whole process, mainly due to the possibility of comparing what would be the consequences of the utilization of different biological expression systems in terms of an easier downstream process.

8.6 Integration and Environment

Besides the fact that process integration is usually regarded as a tool for improving process performance and productivity, it will usually have an important effect towards decreasing the negative impacts on the environment.

Table 8.2 Concentration, molecular weight, hydrophobicity and charge at different pHs for the main proteins in *E. coli* (Data from Woolston [24])

				Charge[e] (coulomb per molecule $\times$ 1E25)										
		g/litre	Da	[c]	pH 4	pH 4.5	pH 5	pH 5.5	ph 6	pH 6.5	pH 7	pH 7.5	pH 8	pH 8.5
Contaminants	pI[a]	Weight	Molecular weight[b]	Hydro-phobi-city[d]	qA	qB	qC	qD	qE	qF	qG	qH	qI	qJ
1	4.67	11.29	18370	0.71	1.94	0.25	-0.80	-1.41	-1.76	-1.97	-2.15	-2.33	-2.45	-2.67
2	4.72	7.06	85570	0.48	2.35	0.29	-1.17	-2.17	-2.83	-3.24	-3.50	-3.63	-3.68	-3.64
3	4.85	4.63	53660	0.76	1.83	0.67	0.04	-0.30	-0.49	-0.65	-0.85	-1.90	-1.34	-1.50
4	4.92	5.58	120000	1.50	3.29	1.38	-0.03	-0.69	-1.07	-1.34	-1.73	-2.30	-2.85	-2.75
5	5.01	4.83	203000	0.36	4.08	1.83	0.04	-1.17	-1.92	-2.46	-3.07	-3.90	-4.98	-5.65
6	5.16	2.48	69380	0.36	5.22	3.17	1.02	-0.72	-1.90	-2.60	-3.05	-3.46	-3.90	-4.24
7	5.29	7.70	48320	0.48	3.96	3.16	1.12	-0.58	-1.36	-1.34	-1.00	-0.95	-1.59	-2.84
8	5.57	6.80	93380	0.93	10.90	5.81	2.78	0.77	-0.81	-2.18	-3.32	-4.12	-4.45	-4.31
9	5.65	7.53	69380		1.09	0.55	0.26	0.10	-0.03	-0.12	-0.21	-0.28	-0.32	-0.32
10	6.02	6.05	114450	0.63	10.40	5.94	3.15	1.51	0.56	-0.05	-0.53	-0.99	-1.43	-1.72
11	7.57	3.89	198000	0.06	0.33	0.03	0.05	0.05	0.05	0.05	0.05	-0.69	-0.97	-1.57
12	8.29	1.48	30400		5.17	4.22	3.20	2.25	1.46	0.87	0.50	0.30	0.20	0.08
13	8.83	0.83	94670		11.70	7.94	5.39	3.73	2.66	1.97	1.50	1.13	0.80	0.51

[a] Measured by isoelectric focusing using homogeneous polyacrylamide gel.
[b] Molecular weight was measured by gel filtration using an LKB Ultropac column.
[c] Hydrophobicity is expressed as the concentration (M) of ammonium sulphate at which the protein eluted (higher values represent lower hydrophobicity).
[d] Hydrophobicity was measured by hydrophobic interaction chromatography using a Poros phenyl PH column in HPLC and a gradient elution from 1.5 to 0.0 M $(NH_4)_2SO_4$ in 20 mM tris buffer.
[e] Charge was measured by electrophoretic titration curve analysis.

In this case, the recycling of by-products will be considered as a case of process integration. Bioprocessing, especially when seen in conjunction with agriculture, offers a wide variety of possible by-product recycling. Recycling processes in this way usually need less technical effort, and hence less area for installations, than in traditional process technologies.

A classic example of integrated ecological bioprocessing is the production of ethanol, used as fuel for vehicles in Brazil. There is a recycling of the stripped bagasse, which is used as fuel for running the plant, carbon dioxide is a secondary product and the waste from the distillation column, a highly pollutant mixture, can be used as a fertilizer and, hence, is pumped directly to the sugar cane plantation. In general, renewed efforts have to be made to optimize bioprocesses to make biotechnology more compatible with the requirements for sustainable development, leading to ecological bioprocessing as a new paradigm for biotechnology.

8.7 References

1. E. W. Leser and J. A. Asenjo (1994) 'Protein recovery, separation and purification. Selection of optimal techniques using an expert system', *Memórias do Instituto Oswaldo Cruz*, **891**, 99–109.
2. R. E. Shamel and M. Keough (1994) 'Trends in biopharmaceutical product development and commercialization', *Genetic Engng News*, **14**(1), 6.
3. D. T. Moir and J.-I. Mao (1990) 'Protein secretion systems in microbial and mammalian cells', in J. A. Asenjo (Ed.), *Separation Processes in Biotechnology*, Marcel Dekker, New York, pp. 67–94.
4. V. Nagarajan, M. Chen and H. Kane (1994) 'Extracellular protein production in *Bacillus subtilis*', presented at the Recombinant DNA Technology III Conference, Deauville, France, 16–21 October 1994.
5. P. Ferrer, L. Hedegaard, L. Halkier, I. Diers, D. Savva and J. A. Asenjo (1994) 'Molecular cloning and high-level expression in *Bacillus subtilis* of a lytic β-1,3-glucanase gene from *Oerskovia xanthineolytica* LLG 109. A β-1,3-glucanase able to selectively permeabilize the yeast cell wall', presented at the Recombinant DNA Technology III Conference, Deauville, France, 16–21 October 1994.
6. J. A. Asenjo, A. M. Ventom, R.-B. Huang and B. A. Andrews (1993) 'Selective release of recombinant protein particles (VLPs) from yeast using a pure lytic glucanase enzyme', *Bio/Technol.*, **11** (February) 214–17.
7. A. W. R. H. Teunissen, J. A. van der Berg and H. Y. Steensma (1993) *Yeast*, **9**, 1–10.
8. A. Skerra, L.-O. Essen, T. Schmidt and C. Wardenberg (1994) 'Engineering of antibody fragments produced in bacteria: solutions to the purification problem', in L. Alberghina, I. Frontali and P. Sensi (Eds.), *ECB6: Proceedings of the 6th European Congress on Biotechnology*, Elsevier Science B. V., Amsterdam, pp. 167–70.
9. M. Moo-Young and Y. Chisti (1988) 'Considerations for designing bioreactors for shear-sensitive culture', *Biotechnol.*, **6**(11), 1291–6.
10. J. Merchuk and J. A. Asenjo (1994) 'Design of a bioreactor system: overview', in J. A. Asenjo and J. C. Merchuk (Eds.), *Bioreactor System Design*, Marcel Dekker, New York, pp. 1–12.

11. L. A. M. van der Wielen, A. J. J. Straathof and K. Ch. A. M. Luyben (1993) 'Adsorptive and chromatographic reactors', in M. P. C. Weijnen and A. A. H. Drinkenburg (Eds.), *Proceedings of the Conference on Precision Process Technology, 482nd meeting of EFChE, Delft, The Netherlands*, Kluwer Publishers, Dordrecht, The Netherlands.
12. J. E. Bailey and D. F. Ollis (1986) *Biochemical Engineering Fundamentals*, 2nd edn, McGraw-Hill Book Company, New York.
13. F. J. Castillo (1994) 'Monoclonal antibody production in extended perfusion culture', presented at the Recombinant DNA Technology III Conference. Deauville, France, 16–21 October 1994.
14. B. Griffiths (1990) 'Perfusion system for cell cultivation', in A. S. Lubiniecki (Ed.), *Large-Scale Mammalian Cell Culture Technology*, Marcel Dekker, New York.
15. S. L. Mistry, J. C. Merchuk and J. A. Asenjo (1994) 'Mathematical modelling and computer simulation of continuous aqueous two-phase protein extraction', in D. L. Pyle (Ed.), *Separations for Biotechnology*, Vol. III, Royal Society of Chemistry, London, pp. 321–8.
16. A. Kaul, R. Pereira, J. A. Asenjo and J. C. Merchuk (1995) 'Phase separation studies for PEG/phosphate aqueous two-phase systems', *Biotechnology and Bioengineering*, **48**, 246–256.
17. A. Schmidt and J. A. Asenjo (1994) 'Modelling the partition behaviour of proteins in aqueous two-phase systems', in D. L. Pyle (Ed.), *Separations for Biotechnology*, Vol. III, Royal Society of Chemistry, London, pp. 462–8.
18. C. A. Bisbee (1994) 'Opportunities in and alternative strategies for biomanufacturing success', *Genetic Eng News*, **14**(14), 8.
19. S. Kampe, R. Hjorth and L.-E. Nystrom (1993) 'Characterization of a novel adsorbent for recovery of proteins in expanded beds', presented at the VIth European Congress on *Biotechnology*, Firenze, 13–17 June 1993.
20. H. A. Chase (1994) 'Purification of proteins by adsorption chromatography in expanded beds', *Trends in Biotechnol.*, **12**, 296–303.
21. K. Jones (1991) 'Protein purification: a new approach to affinity chromatography', in G. Subramanian (Ed.), *Preparative and Process-Scale Liquid Chromatography*, Ellis Howard, New York, pp. 236–49.
22. E. W. Leser and J. A. Asenjo (1994) 'The rational selection of purification processes for proteins: an expert system for the downstream processing design', *Ann. New York Acad. Sci.*, **721**, 337–47.
23. E. W. Leser (1995) 'Prot ex: an expert system for selecting the sequence of processes for the downstream purification of proteins', PhD Thesis, The University of Reading.
24. P. Woolston (1995) 'Database for an expert system for the selection of recombinant protein purification processes', PhD Thesis, The University of Reading, 365pp.

9 MEMBRANE PROCESSING IN BIOTECHNOLOGY: A CASE HISTORY

Michael A. Cook

9.1 Introduction

No lesson seems to be so deeply inculcated by the experience of life as that you should never trust experts. If you believe doctors, nothing is wholesome; if you believe theologians, nothing is innocent; if you believe soldiers, nothing is safe. They all require their strong wine diluted by a very large mixture of common sense.

This article attempts to encapsulate some of the factors considered and evaluated on a crossflow membrane based process as part of activities in a process development department. It is thus written on the basis of experiences, of laboratory, pilot plant and production equipment accumulated over a number of years, and not by someone claiming to be an expert!

Crossflow membrane filtration as a practical, reliable industrial process is already more than twenty years old. The separation and isolation of chemical entities present in a complex mixture is a significant problem in many chemical processes, and none more so than in the biotechnology industry. The development of applications has been rather slow but now, as industry demands more selective separations, more control over effluent and waste disposal, better containment and importantly even more careful control over process costs, so crossflow membrane filtration will continue to gain recognition and acceptance.

9.2 The Nature of the Problem

Whilst many topics on crossflow membrane filtration frequently appear in the literature there seems to be little information on the evaluation of processes through scaling up from laboratory based equipment through to full-scale

Downstream Processing of Natural Products. Edited by Michael S. Verrall
©1996 John Wiley & Sons Ltd

commercial operation. The bulk of this discussion relates to a particular problem which was the subject of substantial evaluation through to successful full-scale operation.

The problem was to extract and purify a low molecular weight solute present in a fermentation broth. The product was of high value and had to be recovered in high yield. The initial separation process had to integrate with subsequent downstream processing stages which involved a number of further purification steps.

Typically fermentation broths contain many different classes of materials, including suspended solids, colloidal solids, micro-organisms, cell debris, macromolecular solutes and microsolutes, including both primary and secondary metabolites. The particulate size of this material may range from 10 angstroms up to several mm in diameter.

It is often the case that commercial microbial fermentations have up to 50 grams/litre of dissolved solids and as much as 50 grams/litre of suspended solids present. Just for good measure filamentous organisms, for example, often exhibit high viscosity and non-Newtonian rheology. Additional challenges may result from the fact that the product may be shear sensitive and labile to small changes in pH, ionic strength or temperature.

Let us consider the problem in hand. The first question to ask is why consider membrane filtration in the first place? The established process was rotary drum precoat filtration. Whilst excellent yields can be achieved by careful optimization of operation it was considered that improved yields and particularly reduced operation costs could perhaps be obtained. The cost of precoat material and the perceived increasing difficulties associated with disposal of substantial quantities of waste precoat from rotary vacuum filtration were further incentives to seek improved processing methods. An alternative process stage evaluated was a semi-continuous disc stack centrifuge to remove microbial debris from the fermenter broth. Centrifugation proved to be inefficient because of significant product yield losses under simple operation or the need to repeat centrifugation steps to increase yield and improve process stream clarity suitable for subsequent downstream processing steps.

Whole broth extraction was also considered but the partition coefficient of the compound in question was poor and, at such low concentrations and with large volumes to treat, this process was not considered practical at the time.

9.3 Initial Experiments

Initial experimentation on the fermentation broth with a proprietary flat sheet crossflow ultrafiltration system was an ignominious failure. After a few tantalizing drips of permeate it became obvious that all the flow channels had been effectively blocked. Attempts at increasing crossflow velocities by removing 75% of the membrane plates had little effect. On dismantling the rig it was evident that the flow channel thickness was too narrow and thus rapid physical blockage, followed by dead-end filtration, had ensured the early demise of the system. Pretreatment of the broth to remove the offending solid was not deemed commercially viable. However, further evaluation of ultrafiltration permeate obtained from rotary vacuum filtrate showed some significant advantages to subsequent downstream processing stages.

Some time later a $1.4\,m^2$ hollow fibre unit with $1.1\,mm$ i.d. fibres was acquired to continue tests. These cartridges were available in a small range of different membrane polymer materials. A unique advantage of the hollow fibre membrane system is the ability to reverse the flow through the membrane lumen. This can be a significant help towards keeping the membrane clean and can even remove physical blockage. The self-supporting nature of the membrane also allows cleaning with no permeate flow and this particularly aids the restoration of good flux during cleaning. Not only were reasonable flux rates achieved and product transmission was 100% but importantly the permeate appeared to be well suited to further downstream processing in an integrated sequence of stages. This integrated process became an attractive option worthy of further investigation.

Considerable attention then focused on evaluation of the hollow fibre system with a view to commercial operation. Factors such as membrane material, lumen i.d., molecular weight cut-off, membrane cartridge length, operational pressure, runtime and membrane area to process volume ratio, pH, temperature and cleaning were all thoroughly investigated and evaluated. Some factors were easy to resolve; some required compromise.

Initially just cellulose and polysulfone membrane materials were available and these were evaluated for flux rate and cleaning efficacy. Subsequently a wider range of membranes, many with surface modification to give a range of hydrophilic–hydrophobic properties, has become available.

The lumen i.d. was also a variable. At the high crossflow velocities required to achieve good fluxes and reasonable concentration levels energy consumptions became an important factor. Too narrow a membrane caused high pressure drops and a risk of fouling. Wide diameter lumens had a restriction of maximum operating pressure yet allowed a slightly higher level of concentration.

A range of molecular weight cut-off membranes were evaluated. Generally the higher the molecular weight cut-off the higher the flux. On the other hand, it was identified early on that a tight specification had to be set on molecular weight cut-off because of effects of downstream processing yields. Transmission of product, even with 'tight' membranes, did not seem a problem at this stage, but it certainly featured heavily in later investigations.

Membrane cartridges were supplied in a variety of lengths. A really short cartridge reduced pressure drop across the cartridge and maximized flux per unit area of cartridge. However, a compromise must be sought otherwise a system with a very large number of short cartridges would result in substantial equipment costs and energy costs would become disadvantageous.

For our product it was noted that the higher the pressure the better the flux, with no decline in product transmission. However, the membranes were, at that time, limited to 2.8 bar. At 3.5 bar substantially increased fluxes were achieved but membrane life could not be guaranteed.

Many membrane systems operating on electrophoretic paint, water purification, milk processing and fruit juice seem to be able to run for days, weeks or even months without significant reductions in flux. We observed a rapid and continuing decline in flux, even on recycle, after only a few hours. It was clearly evident that the longer the membranes were on broth processing, not only did the flux decline but also, on prolonged operation cleaning became progressively more difficult. Too short a runtime and the downtime for cleaning became a significant overhead and cost. Too long a processing time and average fluxes were low and cleaning times extended.

An important factor also related to runtime in a batchwise operation is the membrane area to process volume ratio. To maximize the yield of product in the permeate the concentration factor and diafiltration level need to be established by trial and error. With hollow fibre systems only a small amount of concentration was achievable. Thus to obtain the required yield significant levels of diafiltration were required. The operation of concentration and diafiltration needs to be completed in the designated runtime. The volume of broth therefore needs to be optimized for a set membrane area.

Even the pH of operation was a compromise since flux rates are improved at a pH where the product was more unstable.

Optimization of processing temperature was also a difficult compromise. Flux rates can be significantly improved by increasing operating temperature resulting from a decrease in fluid viscosity affecting the boundary layer thickness. On the other hand, product stability can be significantly decreased by increasing the temperature. Additionally at very high crossflow velocities on commercial systems pumping hundreds of cubic metres of broth per hour significant energy is expended in providing the pressure and crossflow velocity. Significant costs can then be incurred in providing heat exchange capacity to continue processing at, say 10 or 15 °C.

9.4 Process Design

Having identified and optimized a number of key factors for the membrane processing and knowing the proposed demand of the product we were in a position to make reasonably reliable estimates at membrane area and rig numbers. To maximize flux and concentration factors we needed high crossflow velocity and this meant short (300 mm) cartridges run at the maximum design inlet pressure of 2.8 bar. Membrane cartridge costs, energy costs and heat exchange facilities were substantial. It was thus time to reappraise goals and objectives to ensure they were in line with needs.

A 'final' review of a range of available commercial equipment, encompassing flat sheet, hollow fibre and tubular systems had identified a number of new potential systems that had become available. Indeed, during our evaluation of the hollow fibre system we had experimented with a wide range of systems for a variety of commercial uses within our industry.

After a careful survey and application of a risk analysis we selected a new wide channel flat sheet system. This was capable of taking any flat sheet membrane from a range of manufacturers. Membrane systems can be operated in a variety of modes such as batch, semi-batch or continuous. Each system has both advantages and disadvantages which need careful consideration.

Our process under development commenced with a batchwise supply of fermentation broth. A truly continuous supply of broth from the fermentation facilities could not be expected.

Mathematical models are available to calculate the relationship between concentration factor and diafiltration level for a predefined yield at a known level of transmission. After careful consideration it was concluded that a batchwise system was preferred.

For effective and efficient operation the commercial system had to satisfy a number of parameters:

(a) Excessive lysis and contamination by extraneous organisms must be avoided as this can increase membrane fouling and decrease the flux.
(b) Yields were to be 95% +.
(c) The process stage must integrate with further downsteam processing steps.
(d) The process must be cost effective.
(e) It must be reliable.
(f) It must be capable of automation and validation.

Considerable experimentation was then carried out to try and evaluate factors such as the number of aliquots, the size of each membrane rig, plant layout and CIP (clean-in-place) facilities.

One aspect of scale-up that was not observed on the small rigs and therefore not evaluated was the effect of high transmembrane pressure on product transmission. High transmembrane pressures result from the arrangement of commercial flat sheet systems into a combination of membranes in series and in parallel in a rig.

Although it might well be considered ideal to run all membranes in parallel this proves to be totally impracticable on a large scale due to excessive pumping requirements and energy consumption. A number of membranes must be arranged in a section and a number of sections placed in series.

In a pilot plant 1.1 m^2 system with three sections the maximum inlet pressure was 4–6 bar. In the commercial system with nine sections in series and thirty membranes in parallel up to 20 bar inlet pressures were observed. Such high inlet pressures resulted in high transmembrane pressures but did not affect flux rates adversely. Indeed, it is usually advantageous, in terms of flux rate and process economics, to operate at the highest pressure—and crossflow velocity commensurate with optimal flux and the ease of cleaning.

On a pilot plant system is was relatively easy to select a membrane with the correct rejection characteristics which allows 100% transmission of the product. In practice at high operating pressures, although the flux rate at the high-pressure end of the rig was most acceptable, it was disturbing to note that the transmission was significantly reduced compared to that at the low-pressure end. This obviously reduces yield or requires an increased level of diafiltration necessary to maintain the yield.

If sections are removed to reduce the number of membranes in series the inlet pressure and thus transmembrane pressure are reduced, but then the membrane area is reduced. If the membrane area is restored by placing more membranes in parallel in the reduced number of sections then the crossflow velocity is reduced with the likelihood of reduced flux, increased fouling and potential physical blockage.

Even from early experimentation it became obvious that highly variable fluxes were obtained during apparently standard reproduction experiments. Often these average fluxes would vary by a factor of two or even more. This effect was finally and confidently attributed to variation in broth characteristics and must also be taken into account on scale-up, particularly when large-scale 'continuous' sequencing of fermentation batches is required. Additionally it is often observed that when a 'bad broth' gives a low flux rate it is often followed by successive 'bad broths', perhaps due to some unforeseen perturbation during fermentation. Furthermore, a bad flux usually results in a reduced efficiency during the cleaning cycle such that successive operations continue to give poor flux values but with marginal improvement noted after each subsequent clean, even when 'good broths' are produced.

It follows therefore that it may well be necessary to evaluate carefully whether it is advisable or even necessary to plan scale-up on the basis of average flux data or worst-case flux data, plus a contingency, especially in the allowed cleaning time. On a commercial plant where broth supplies are all but continuous from carefully sequenced fermentations it is vital to scale-up on a worse-case flux and then build in a safety contingency. A backlog of valuable fermentations is unacceptable, particularly where the product is labile or prolonged storage has a deleterious effect on downstream processing performance. However, it does clearly result in a system whereby, in times of normal processing, a substantial excess of membrane capacity will be present. In these days of optimal design and careful cost control it can be difficult to obtain acceptance for this facet of process design. The consequences of a bottleneck in a membrane plant can, however, be extremely costly, with a failure to achieve yields by cutting short diafiltration or, at worse, draining aliquots of fermentation broth in order to catch up with sequencing.

It must also be remembered that development of the fermentation itself, aimed at improving the product titre or reducing the actual fermentation costs, may adversely impact on ultrafiltration (UF) flux rate performance. On occasions it may even be necessary to seek alternative cleaning procedures or even alternative membranes. Ideally fermentation development is carried out in parallel with UF evaluation to ensure such processing problems do not arise.

9.5 Membrane Fouling

Membrane fouling is often very unpredictable and appears to vary markedly in severity depending on factors such as membrane composition, operating times and the nature of the broth components such as retained solutes, antifoam type, pH, operating temperature and pressure. Under the worst circumstances fouling can cause up to 90% irreversible flux loss within hours.

In most cases the foulants can be removed from the membrane by appropriate cleaning. In all events membrane cleaning reduces productive operating time, consumes energy and uses costly reagents. Undoubtedly harsh cleaning regimes with frequent temperature cycling between the operating temperature at 10° C and the cleaning temperature at 65° C can also lead to a reduction in membrane, seal and support plate lifetime.

Thus, having carried out the concentration and diafiltration operations, the membranes become fouled and fluxes decline to an unacceptable level. It is often reported that there are membrane systems that can run for days, weeks or even months between simple cleaning operations on the membrane. Certainly many systems manage adequately with process cleaning every 24 hours.

In many biotechnology applications the results of membrane fouling are somewhat more daunting. Even under ideal operating conditions a significant flux rate decline is observed in the first half hour. Thereafter the flux continues to fall during concentration. During the diafiltration step the flux rate also continues to fall. Our experimentation demonstrated that the longer an ultrafiltration step was carried out, using larger aliquots, so the more heavily fouled the membrane became and the more difficult the membrane cleaning step, to restore the water flux, became. Hence the need to carefully evaluate the membrane area to broth volume ratio and thus the operation time for the rig. It must also be remembered that for commercial systems operating at higher transmembrane pressures than test systems it is often more difficult to clean membranes.

When cleaning hollow fibre systems it is usually found that flux rates are more easily restored with permeate flows shut off. With flat sheet systems such options are not applicable. However, it is often best to evaluate cleaning operations either just using the feed pump with slow flow and low transmembrane pressure or alternatively it may be advantageous to also use the recirculation pump as used during normal processing with high flow but with the disadvantage of high transmembrane pressure.

For many biological processes the cleaning time can be a significant proportion of the operating time, even to the extent that more time can be spent cleaning than operating on process fluid. Yet such a system can still be commercially viable and practical.

9.6 Equipment Configuration

In a production plant great care must be taken to ensure the efficient displacement of spent process fluids without the excessive use of water. The water usage can be significant and indeed can easily outpace water treatment plant capacity. Hold-up volumes in large rigs can be surprisingly large, yet the efficient displacement flushing is vital for maximum effectiveness of caustic or hypochlorite cleaners.

The full-scale plant needs to be designed for minimum pipe runs to minimize hold-up volumes. This applied not only with regard to the product for maximum recovery but also waste process streams for efficient displacement prior to cleaning. The importance of this factor must be stressed.

For flexibility and also perhaps peace of mind, if such luxuries are allowed, the centrifugal pump should be sized for a moderate over-capacity. Initial operation should be conduced with an orifice plate in place. However, it should be borne in mind that such devices can act as high shear homogenizers.

After a period of operation and evaluation it should be possible to machine down the centrifugal pump impeller diameter and ultimately remove the orifice plate. This is to aim for minimal energy consumption. However, this irrevocable step can only be carried out if there is complete confidence that broth hydrodynamics will not change adversely.

9.7 Summary

It must be said that membrane systems still contain a significant element of 'art' rather than 'science'. For biological systems there is as yet no ideal, universal optimum system. The ideal way to achieve a satisfactory commercial membrane process is by careful planning and evaluation of the many factors involved in determining optimum product flux. Whilst there is no substitute for detailed laboratory investigations it is vital to ensure an integrated approach upstream and downstream right through from laboratory to pilot plant and finally to full-scale production.

10 DISPLACEMENT CHROMATOGRAPHY

Denise M. Wallworth

10.1 Introduction

For the purification of biomolecules, liquid chromatography (LC) in its various forms is well established as the most efficient method. The demands of high purity in downstream purification in a world of major advances in biotechnology has led to a wealth of stationary phases based both on silica and a wide range of polymeric resins. However, in practice, the scale-up of such processes regularly causes problems and limits the acceptance of LC at the process level. Displacement chromatography lends itself to solving the difficulties experienced when products requiring purification have low chromatographic separation factors which, if combined with a stationary phase of low capacity, leads to highly challenging preparative chromatography. Although method development for displacement can be somewhat slower than for elution chromatography, it lends itself to automation well and produces a final product in a far more concentrated form. Since the technique offers much higher loading levels than for elution chromatography, it could also be far more economical for the purification of components that have a high selectivity.

At low alpha values, critical resolution will be rapidly lost as the loading of the sample increases, when 'overload' conditions apply at quite low sample concentrations, resulting in the necessity for peak slicing or recycling techniques. Conventional, elution mode chromatography can only circumvent this problem by increasing selectivity, α, either by changing the solvent conditions (often normal phase conditions offer much improved separation factors), increasing the column dimensions, reducing particle size to increase efficiency or changing to a different type of stationary phase. In many circumstances, none of these changes are possible for natural products, either due to solubility limits or to the cost of small particle size silicas. Alternatively,

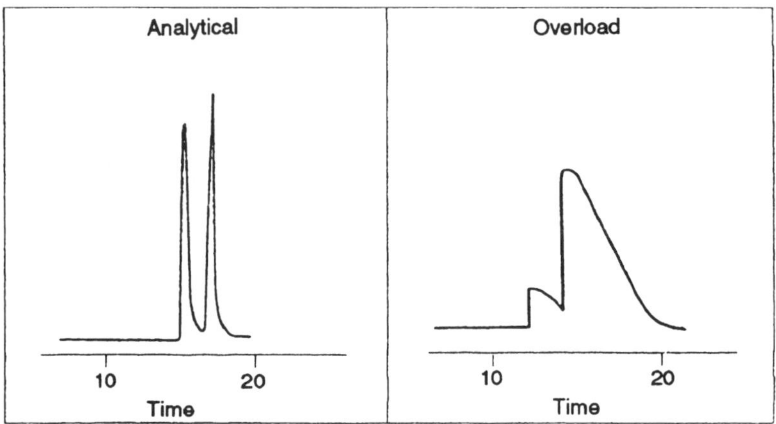

Figure 10.1   The effect of overload conditions in elution mode

heart cutting and extensive recycling can be used, but these can become both laborious and low in yield. In contrast, solute concentrations and column loading orders of magnitude higher than in elution chromatography have been realized in the displacement mode [1–4], even at very low alpha values. When compared to the classical 'touching band' technique [5], the gain is about one order of magnitude [6,7].

10.2 Overload Elution Chromatography

The effect of increasing loading to overload conditions in elution mode chromatography is that peaks become broad and asymmetrical, retention times decrease and loss of resolution is observed (Figure 10.1). This is due to competition for adsorption sites at higher concentrations, resulting in the surface concentration of solute ceasing to be proportional to the solute concentration in the liquid phase. An adsorption (Langmuir) isotherm,

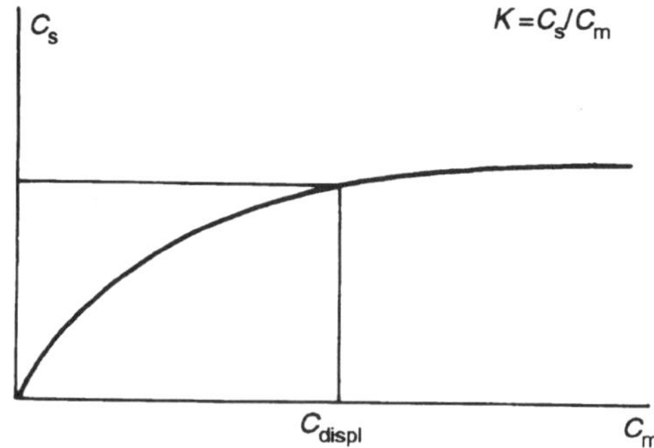

Figure 10.2   Non-linear adsorption isotherm for a typical solute

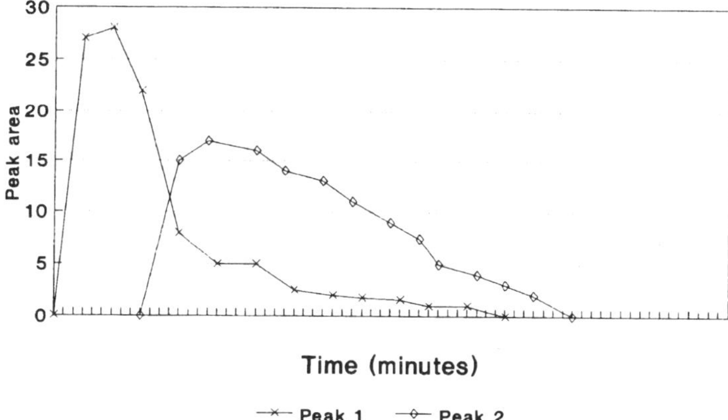

Figure 10.3 Reconstructed chromatogram of an overload elution mode preparative separation [11]

plotting the concentration of adsorbed solute against the concentration of solute in the mobile phase (Figure 10.2), therefore exhibits a distinctive non-linear region at high solute concentrations where competition for the stationary phase surface begins. If the column loadings in the elution mode reach this non-linear region of the isotherm, the peaks begin to develop an approximate triangular shape with the leading edge eluting before the characteristic retention time of the solute, resulting in contamination of both peaks. The contamination can be seen even more clearly if fractions are collected and analysed, and their peak areas plotted against time (Figure 10.3).

The competitive effect is particularly enhanced in a one-to-one binary mixture such as a racemic mixture of an optically active pharmaceutical. Sample molecules in the tailing region of the less retained component fall into the region occupied by the more strongly retained component. Competition forces the tailing molecules to catch up. In this non-linear region of the adsorption isotherm, retention is being strongly affected by local concentrations of one component in the other. In effect, the second component is displacing the first and sharpening its profile. The result is the production of a well-defined band with a roughly rectangular profile (focusing effect).

10.3 Displacement Chromatography

If the concentration of the second component is sufficiently high (equal or greater than that of the first component), sample self-displacement will dominate. For lower second-component concentrations than this, the effect can be the opposite, with a 'tag along' effect dominating [8]. It is simply the ratio of concentrations of the two components that determines whether displacement or 'tag along' operates. However, if a *third* component is

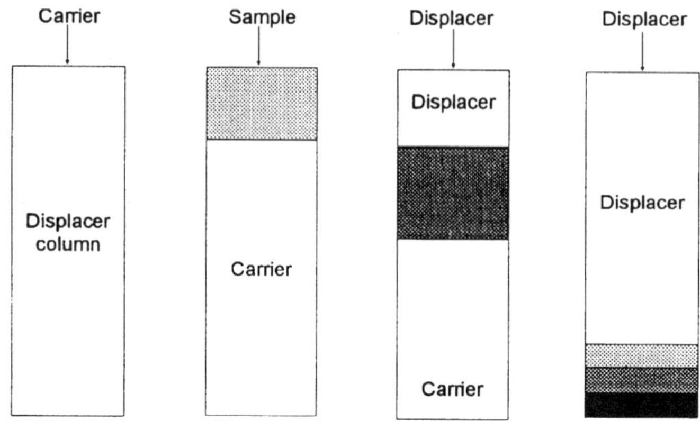

Figure 10.4   Schematic of displacement chromatography method

introduced in high concentration, displacement of each of the first two solutes
can be achieved, and under the right conditions (in particular, a weak mobile
phase) an isotactic train of bands is formed, each band being displaced by the
next eluting component, and each in turn focusing the band ahead of itself.
This third component in displacement chromatography is known as the
'displacer'.

In practice, the column is first equilibrated with the mobile phase (or
'carrier'), a weak solvent that allows the solute of interest to be highly retained.
The sample is then introduced at high concentration, followed by the displacer,
which has the highest affinity for the stationary phase. As the front of the
displacer moves down the column, it displaces the sample components which,
in turn, displace each other according to their adsorption strength. If the
adsorption strengths of the components are sufficiently different and the
column has the necessary efficiency, the components occupy adjacent zones

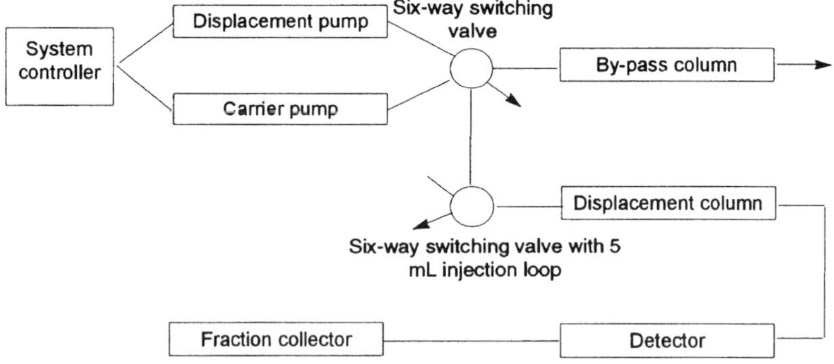

Figure 10.5   Typical equipment requirements for displacement chromatography

and move at the same velocity in the fully developed displacement train (Figure 10.4). Component concentrations in the isotactic train depend only on the respective adsorption isotherms and the concentration of the displacer [9].

The bands produced by displacement chromatography become narrow zones of high concentration, giving an apparent increase in column efficiency and reversing the trend observed in elution mode chromatography of a decrease in theoretical plates with increasing load. Displacement chromatography such as this has been well known for quite some time [10], but was only recently revived by Horvath *et al.* [2]. Interest in this area is now growing steadily, with an increasing interest in chiral applications [11–14], as well as for other isomeric solutes [9,15–17].

10.4 Equipment Required for Displacement Chromatography

A suitably adapted analytical or semi-preparative HPLC system is generally used. Figure 10.5 shows a typical system comprising a computer-controlled LC system incorporating two LC pumps (one for carrier solvent and one for displacer solution), two pneumatically actuated six-way switching valves, a detector and a fraction collector. A column oven may be added for improved column reproducibility and for faster removal of displacer by temperature ramping.

Conditions for successful displacement include the following:

(a) Chromatographic conditions must be available where the solute of interest is well retained.
(b) A displacer must be available which, under the conditions of the separation, is more highly retained than the solute.
(c) The adsorption isotherms of all components must be concave.
(d) Both the displacer and the solute must be sufficiently soluble in the carrier to enable high concentrations to be used, without risk of precipitation.

10.5 Method Development

In order to bring into effect the displacement mode, samples must have a non-linear adsorption isotherm and must be capable of eluting in the non-linear region. In practice, this means optimizing selectivity at high retention values, with k' typically between 3 and 13, using a weak carrier (although it has been reported [17] that for large-scale purifications a k' of 3–7 provides better production rates). The limiting stage of the method development process has always been the choice of displacer since it must by necessity have a higher adsorption on the stationary phase under the chosen conditions. However, a knowledge of the solute and its possible interactions with the stationary phase can in many cases lead to a suitable choice more quickly by selecting a closely related compound that could provide the higher affinity required. Indeed, a similar structure and retention mechanism can insure that effective competition occurs at the stationary phase surface.

The stages of method development are, in summary:

(a) retention and selectivity optimization (Section 10.6),
(b) adsorption isotherm measurements for solute and potential displacer (Section 10.7).

10.6 Retention and Selectivity Optimization

Selectivity, α, is defined in the equation

$$\alpha = k'_1/k'_2$$

where k'_1 and k'_2 are the capacity factors for the two solutes being separated (product and contaminant), given by

$$k' = (t_r - t_0)/t_0$$

and t_r is the retention time of the solute, t_0 the elution time for an unretained solute (equivalent to the void volume of the column). Values less than 1.2 are typically considered low enough to cause problems on scaling-up in elution mode chromatography.

Analytical method development for displacement chromatography needs to be revised to achieve high k' values *and* optimized α values: the mobile phase that achieves this is chosen as the 'carrier' solvent. This may involve an investigation of alternative stationary phases normally rejected because of long retention times. For reversed-phase chromatography of peptides and proteins, the effect of 300 versus 100 Å pore sizes might be studied, or perhaps a move to hydrophobic interaction LC might be explored, as typically carried out on methacrylate phases (see Chapters 11 and 15). Important experiments to be carried out are the influence of buffer concentration, pH and the choice of organic modifier and buffer types, as often subtle changes in these could make major differences to both retention and selectivity. In the development of a displacement purification of the chiral product, 5,10-dideazatetrahydrofolic acid (DDATHF), a potential antineoplastic agent, the influence of pH, buffer concentration and polar organic modifier was investigated, since the compound can carry both a net positive charge and a net negative charge as pH is varied between 2 and 10 [13]. The k' values for the more strongly retained isomer are about half an order of magnitude lower at low pH (Figure 10.6), indicating that pH is a very powerful retention controlling parameter, while selectivity is largely unaffected. In this case, however, a compromise is reached as solubility is much higher in the higher pH eluents, and pH 6 is chosen as the preferred eluent pH. As DDATHF is ionic at pH 6, it is expected that retention and selectivity will also be influenced by buffer concentration (in this case,

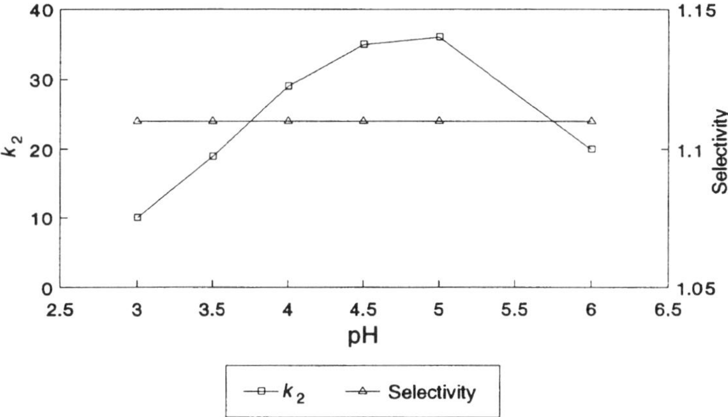

Figure 10.6 The effect of pH on retention and selectivity for DDATHF [13]

citrate buffer). Figure 10.7 shows that citrate concentration is another strong influence on the control of DDATHF retention, but again does not compromise selectivity. It was therefore chosen as the carrier solvent.

Once the mobile phase, or carrier, is selected, its suitability should also be checked on a 250 mm length, 21 mm i.d. preparative column. Using a flowrate of one column volume every two minutes, sample loadings of 10–40 mg (larger if the column diameter is above 20 mm) are evaluated, adjusting the mobile phase further if necessary to achieve the optimum relative retention [17]. The final maximum load is likely to be of the order of several grams on a 21 mm internal diameter column when α is greater than 1.1. The effect of increasing the flowrate at this stage is often an enhancement of production rate, and does not usually affect the yield [17].

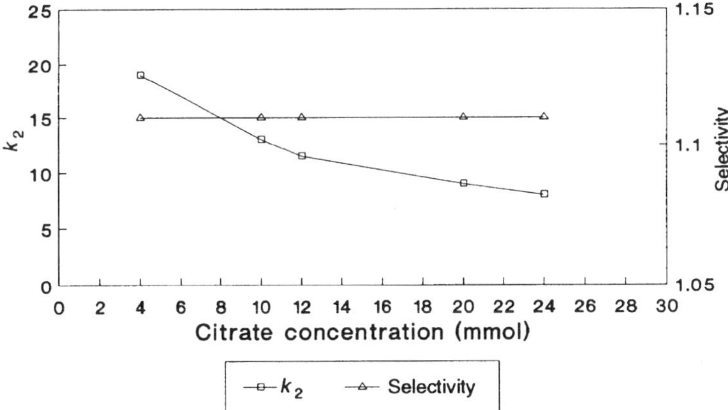

Figure 10.7 The effect of citrate concentration on retention and selectivity for DDATHF [13]

10.7 Adsorption Isotherm Measurement and Choice of Displacer

The success of any displacement chromatographic separation system critically depends on the finding of an appropriate displacer. The next stage in development therefore concentrates on studying the excess adsorption isotherms of the solutes and of potential displacers.

An adsorption isotherm is one way of expressing the relationship between surface-phase concentration of the solute and its concentration in the liquid phase. The relationship is dependent on temperature and is linear as concentration increases, except in preparative chromatography, when column loadings are higher.

Individual adsorption isotherms are determined by running a series of frontal chromatograms at increasing concentration in carrier solvent, recording the breakthrough curve. After conditioning the column with carrier solvent, a low-concentration sample solution is pumped through the column until its front reaches the dummy column (see also Figure 10.5). Rotation of the switching valve diverts the flow on to the column and data collection is started, continuing until the solute detection levels off (the breakthrough volume, V_b). The amount of solute adsorbed on to the stationary phase is then calculated using the equation:

$$\text{Amount adsorbed,} \quad q_i = (V_b - V_0)C_i$$

where V_0 is the void volume of the column and C_i is the concentration of species i in the carrier solvent.

The process is repeated for increasing concentrations of the solute until the non-linear region of the adsorption isotherm, or the solubility limit of the solute, is reached. For each run after the first, the calculation adds the previous run's result, to give the total amount adsorbed:

$$q_j = (V_b - V_0)C_j + q_{j-1}$$

where q_{j-1} is the cumulative number of moles adsorbed through the $j-1$ breakthrough steps.

Potential displacers are evaluated in the same way. The one that exhibits an adsorption isotherm which is above that of the solute is chosen. Initial screening of a series of possible displacers can be carried out, however, by comparing initial breakthrough volumes of each with that of the solute or solutes to be displaced, choosing that displacer which exhibits a value higher than that of the solute [9]. If adsorption behaviour varies with displacer concentration, as is often the case, the concentration of the displacer is chosen at which the isotherm lies above that of the solute to be purified.

In the study of potential displacers for the purification of 1- and 2-naphthol [9], a series of long-chain and symmetrical surfactants was investigated. For the LC column chosen, a β-cyclodextrin silica, the n-hexdecyl derivatives

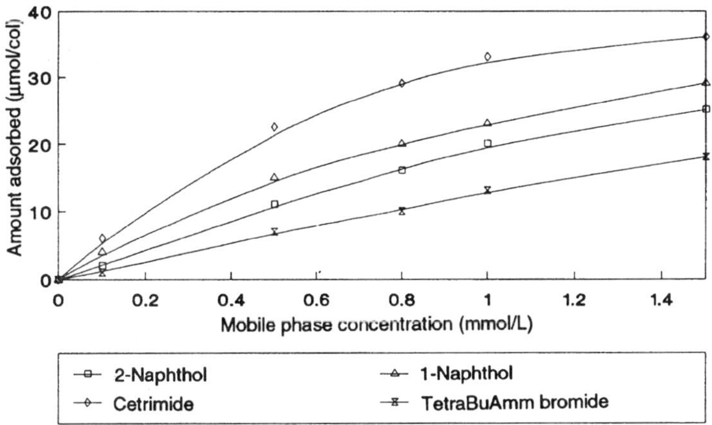

Figure 10.8 Adsorption isotherms for naphthol and its potential displacers [9]

containing an aromatic group showed the highest retention, presumably due to closer interaction with the cavity of the cyclodextrin cavity. It was observed that none of the surfactants exhibited a simple Langmuir isotherm, but tetradecyltrimethylammonium bromide (cetrimide) gave a convex adsorption isotherm that lay above those of the two solutes (Figure 10.8).

10.8 Self-displacement

If the concentration of the second component of a mixture is sufficiently large, and a suitably large sample is injected, it is possible to effect sample self-displacement. Just as is observed for a chosen displacer, the second component causes the concentration and compression of the first band. A mixture of two substituted cyclohexanone epimers, which exhibit an extremely low α of 1.04 on an underivatized silica column, successfully demonstrated that the second eluting epimer could be used to effect preparative displacement of the desired, first eluting epimer [16]. Although the first component tailed into the second, this was found to be at a minimum for a 25:75 ratio of first and second components, especially if a small particle size silica was used. The desired concentration ratio can also be achieved by first harvesting a significant amount of the first eluting compound and returning to the feed. Excellent production rates of the first component were achieved (287 mg from a 1000 mg load on a 250 mm × 21.4 mm i.d. silica column).

A later study [17] utilized a 1:3 mixture of diethyl phthalate and β-tetralone to further demonstrate the self-displacement effect. Although the resolution appeared to be poor, there was in fact only a very narrow mixed band following the first peak. Purity evaluation of the collected fractions of diethyl phthalate revealed that a 98% yield of 95% pure product was obtained: if the purity requirements must exceed 99%, the yield is still acceptable at 67%.

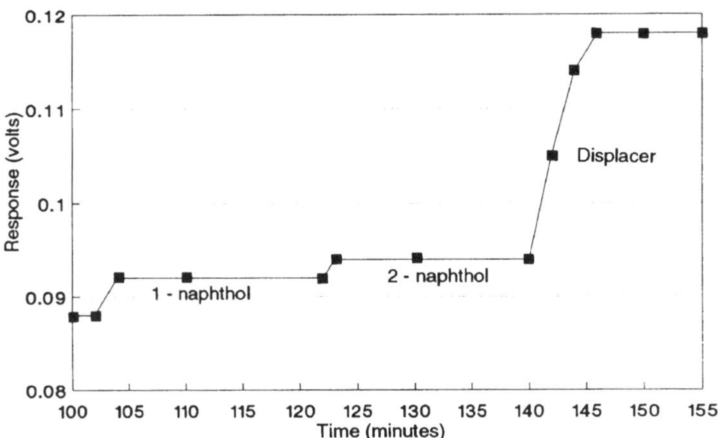

Figure 10.9   Typical displacement [9]

10.9 Displacement Runs

Displacement mode purifications follow equilibration of the column with carrier solvent. Once sample, dissolved in carrier, is loaded into the sample loop, the flow path is switched to the displacer and pumped through the system, linked to the dummy column. When the displacer reaches the dummy column, the sample is injected, displacer flow switched to the displacer column and the detector trace monitored. This method allows for a sharp front for both the sample and displacer [18]. A typical displacement chromatogram is shown in Figure 10.9; each plateau representsd one of the two components of the mixture, followed by the displacer. Fraction collection commences as soon as a slope change is noted. By analysing each of the fractions in turn, a reconstructed chromatogram can be prepared (Figure 10.10). When recovery and production rates are plotted against purity [12,13], it can be seen that displacement chromatography can provide high yields at greater than 95% purity. Since it has been reported that the production rate for a given column is proportional to

$$[(1-\alpha)/\alpha]^2$$

[19], optimizing selectivity can be crucial to success. A small increase in α from 1.21 to 1.24 gave an increase in the production rate of 5% [17].

10.10 Conclusions

Displacement mode preparative chromatography offers several advantages over the traditional elution mode LC. Most importantly, it produces pure fractions at a much higher concentration and in smaller volumes, making recovery by evaporation faster and easier, important for labile products. It provides for higher yields and faster production rates than elution mode

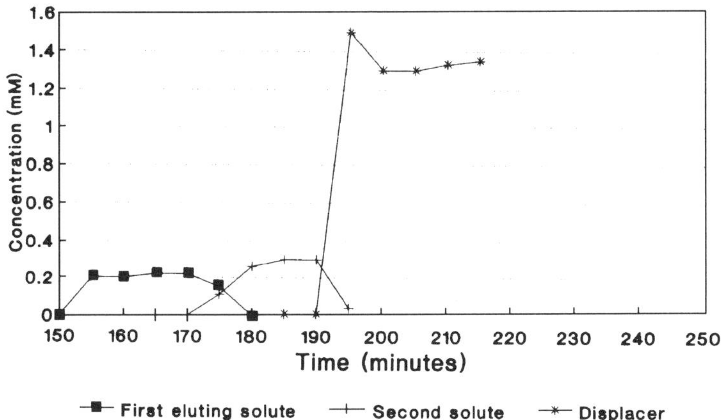

Figure 10.10 Reconstructed chromatogram for a displacement mode purification [13]

chromatography, especially for the product that elutes first. If does, however, have the disadvantage of a more time-consuming and more complex method development stage, but in many cases this is time well spent. The displacement effect is strongest at high concentration [17] and therefore lends itself to the purification of large-volume, high-loading products.

10.11 References

1. Cs. Horvath, A. Nahum and J. H. Frenz (1981) *J. Chromatogr.*, **218**, 365.
2. Cs. Horvath, J. Frenz and Z. El Rassi (1983) *J. Chromatogr.*, **255**, 273.
3. Cs. Horvath (1985) in F. Bruner (Ed.), *The Science of Chromatography*, Journal of Chromatography Library 32, Elsevier, Oxford, p.179.
4. H. Kalasz and Cs. Horvath (1981) *J. Chromatogr.*, **215**, 295.
5. J. H. Knox and H. M. Pyper (1986) *J. Chromatogr.*, **363**, 1.
6. S. Ghodbane and G. Guiochon (1988) *Chromatographia*, **26**, 53.
7. S. Golshan-Shirazi and G. Guiochon (1989) *Analy. Chem.*, **61**, 1368.
8. S. Jacobson, S. Golshan-Shirazi, A. Katti, M. Czok, Z. Ma and G. Guiochon (1989) *J. Chromatogr.*, **484**, 103.
9. G. Vigh, G. Quintero and G. Farkas (1989) *J. Chromatogr.*, **484**, 237.
10. A. Tselius and S. Claesson (1943) *Arkiv Kemi Mineral. Geol.*, **16A**, 18.
11. G. Vigh, G. Quintero and G. Farkas (1990) *J. Chromatogr.*, **506**, 481.
12. G. Farkas, L. H. Irgens, G. Quintero, M. D. Beeson, A. Al-Saeed and G. Vigh (1993) *J. Chromatogr.*, **645**, 67.
13. L. H. Irgens, G. Farkas and G. Vigh (1993) *J. Chromatogr.*, **666**, 603.
14. C. J. Shaw, P. J. Sanflippo, J. J. McNally, S. A. Park and J. B. Press (1993) *J. Chromatogr.*, **631**, 173.
15. G. Vigh, G. Farkas and G. Quintero (1989) *J. Chromatogr.*, **484**, 251.
16. J. Newburger and G. Guiochon (1989) *J. Chromatogr.*, **484**, 153.
17. J. Newburger and G. Guiochon (1990) *J. Chromatogr.*, **523**, 63.
18. L. H. Irgens (1991) PhD Thesis, Texas A&M University.
19. S. Golshan-Shirazi and G. Guiochon (1990) *American Biotechnology Laboratory*, June.

11 NON-IONIC ADSORBENTS IN SEPARATION PROCESSES

Hiroaki Takayanagi, Junji Fukuda and Eiji Miyata

11.1 Introduction

Synthetic adsorbents are organic polymeric materials with hydrophobic adsorption characteristics which can be exploited in downstream processes. They are now widely used in the field of biotechnology and natural products isolation from the laboratory research scale to industrial production. Today a number of natural products are isolated using synthetic adsorbents as purification agents. This chapter describes the fundamental properties of synthetic adsorbents and gives a practical guide for starting research work with these resins.

Synthetic adsorbents are spherical and highly porous materials with a hydrophobic surface. They can adsorb hydrophobic organic compounds on their surface. By this selective function, they can be used for extraction of organic compounds from aqueous solutions such as fermentation broths (the aqueous solution to be treated with a synthetic adsorbent is often referred to as 'liquor' hereafter because the discussion may also be about partially purified solutions). Considering the mass transfer of target compounds, the function of synthetic absorbents is similar to that of solvent extraction. However, the use of synthetic adsorbents has several advantages over conventional solvent extraction processes.

A difference is that adsorbent processes require much smaller amounts of organic solvents. These are often toxic and flammable so their elimination may have practical advantages. Another difference is that adsorbents have a molecular sieving function based on their pore structure so that fractionation by molecular size can take place during the process. In addition, adsorbents can be used either in a contained form (in a column) or in batch mode (in

Downstream Processing of Natural Products. Edited by Michael S. Verrall
©1996 John Wiley & Sons Ltd

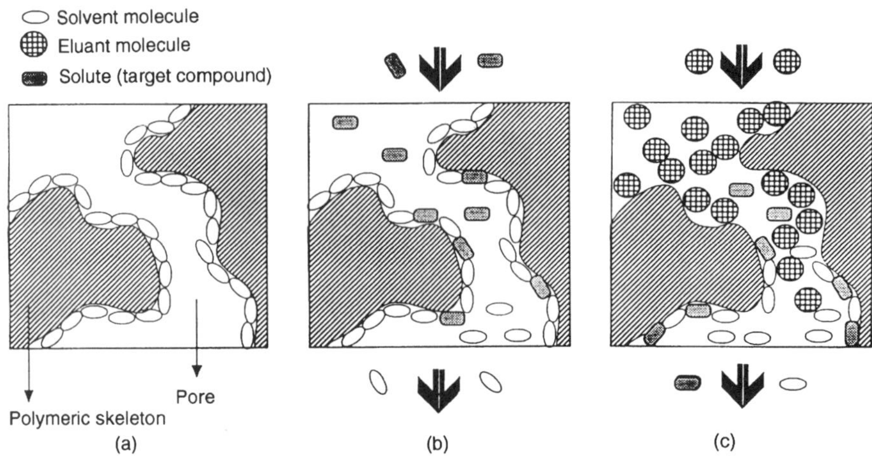

Figure 11.1 Adsorption/desorption on adsorbent surface

suspension). When a large amount of liquor has to be treated and the concentration of a target compound is very low, column processes with an adsorbent may be quite effective.

Synthetic adsorbents can be applied to a variety of compounds so long as they are stable and soluble in the solvents used. A simple explanation of the adsorbent process is shown schematically in Figure 11.1. Fermentation broth usually contains a number of contaminants such as inorganics, proteins and peptides, saccharides and other unwanted organic compounds. When a liquor is fed to a column of adsorbent, hydrophilic (or highly polar) materials such as inorganics and small saccharides do not interact with the adsorbent and pass through the column. High molecular weight compounds (proteins or polysaccharides) cannot diffuse into the adsorbent particles so are not retained. Consequently, only the target compound is adsorbed and it can be recovered using a suitable eluant (desorbing solvent). However, it should be noted that some types of protein can be adsorbed on the resin and may contaminate it (see Section 11.3.5). In many applications the liquor also contains isomers of the target compound or other closely similar compounds and additional purification steps will be required.

Unlike other types of adsorbent such as activated carbon and zeolite (i.e. inorganic solids), synthetic adsorbents show relatively mild adsorptivity and thus adsorbed compounds can be eluted under mild conditions. This also means that the adsorbents can be used repeatedly. In practical processes, these characteristics can give rise to considerable economical advantages.

There have been many theories of the sorption mechanism of synthetic adsorbents; a simple explanation is given here. Figure 11.1 shows a schematic of an adsorption/desorption process on the surface of a synthetic adsorbent.

The surface is covered with the solvent molecules after swelling with the solvent (Figure 11.1a). When an organic compound (solute) is added to the system, some of the solute molecules are adsorbed on the adsorbent surface and release the solvent molecules (Figure 11.1b). The ratio of the solvent concentration on the adsorbent surface to that in the liquid phase is constant (distribution constant) at low concentrations. When a much larger amount of solute is added, the adsorbed amount reaches a saturation value which is called the adsorption capacity. In practical terms the relationship between the adsorbed amount and the solute concentration can be treated as a Langmuir sorption system. In the elution step, another solvent (the eluant) is used to desorb the solute from the adsorbent (Figure 11.1c). For this purpose, a solvent (or mixture) of reduced polarity is used as eluant.

The following sections describe the basic properties and some experimental procedures for synthetic adsorbents, which would be helpful for practical applications. In Section 11.4, some literature applications are briefly reviewed. Additional information can be obtained from the useful handbooks available from the resin manufacturers [1].

11.2 Basic Properties of Synthetic Adsorbents

11.2.1 GENERAL ASPECTS AND PORE CHARACTERISTICS

There are many types of synthetic adsorbent and they are classified according to their chemical and physical nature. Since each type of adsorbent has different adsorption characteristics, it is very important to know their general properties. Basically they are classified in terms of their chemical structure and details of each type of adsorbent are described in the following sections.

The pore character of an adsorbent is an important factor to be considered. As shown in Figure 11.2, synthetic adsorbents have a totally porous structure. The solid part gives an adsorptive surface and the vacant part (pore) functions as a passage for molecular diffusion. In evaluating the pore character of adsorbents, specific surface area and pore radius (pore distribution) are used as the measure.

Specific surface area is usually measured by nitrogen adsorption (BET method), and reflects the contact ratio of the sorbent–liquid interface. Pore radius (pore distribution) is measured by the nitrogen adsorption/desorption method (BJH method). Pore radius is concerned with the size of the diffusion space within an adsorbent particle and reflects the molecular sieving effect of an adsorbent.

11.2.2 POLYSTYRENE RESINS

Among the variety of synthetic adsorbents, crosslinked polystyrene resins (styrenic resins) are most typical. The chemical structure of this type of resin is shown in Figure 11.3. Polystyrene resins have moderate adsorptivity and thus

Figure 11.2 Electron microscopic view of synthetic adsorbent

they can be applied to a wide range of organic compounds. In most cases, polystyrene resins are the type first investigated. The pore characteristics of some such resins are shown in Figure 11.4 and their properties are summarized in Table 11.1.

Polystyrene resins can be classified into two subgroups based on their pore characteristics. One group consists of resins that have relatively large pores, i.e. with a pore radius around 10 nm or larger. Because they have such large pores these resins can adsorb various macromolecular compounds such as proteins. Although some proteins (MW *ca.* 10^5 Da) may be adsorbed on these resins they are often denatured, so this procedure is not usually recommended as a protein purification process.

Resins of another group have smaller pores so that large molecules cannot interact with resin particles. These resins are thus more suitable for the separation of small organic compounds (MW $< 10^3$) having a much larger surface area than those above and consequently a large adsorption capacity for small molecules. They are used for the purification of antibiotics such as cephalosporins. As seen from the table, there is a close relation between adsorption capacity for cephalosporin and specific surface area. It should be noted that resins of high adsorption capacity require a more lipophilic eluent or higher elution volume in the desorption stage.

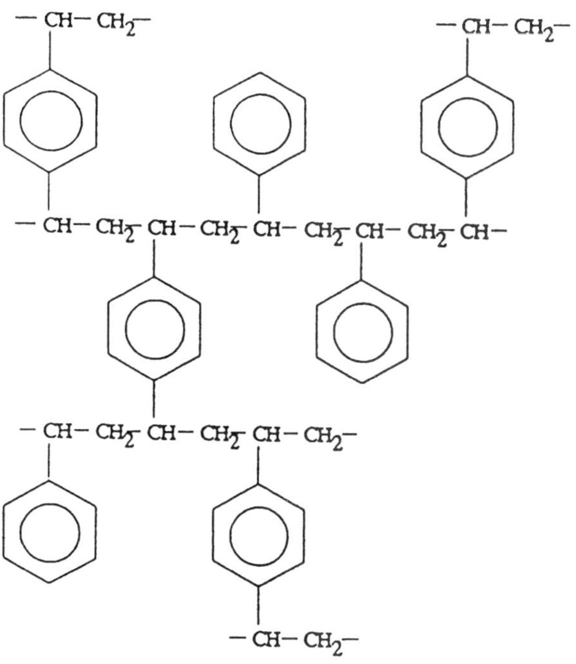

Figure 11.3 Chemical structure of polystyrene resin

11.2.3 MODIFIED POLYSTYRENE RESINS

Polymers with modified aromatic rings (Figure 11.5) provide enhanced adsorptivity. Brominated polystyrene resins offer a larger adsorption capacity than unmodified resins (Table 11.1). Resins of this type are effective for treating a large volume of feed solution or for removing organic impurities from the liquor. The pore characteristics of these resins is shown in Figure 11.6.

Brominated polystyrene resins are unique in having a high density. Because of this property, they can be easily applied to the treatment of dense liquors or to upflow applications.

11.2.4 POLYMETHACRYLATE RESINS

Polymethacrylate resins have a somewhat hydrophilic nature compared with the above polystyrene resins, as one would expect from their chemical structures (Figure 11.7). These resins are suitable for the extraction of relatively hydrophilic organic compounds that are not adsorbed on polystyrene resins. They can also be used for the sorption of highly hydrophobic compounds which are irreversibly adsorbed on polystyrene resins. The properties of methacrylate resins are shown in Table 11.1 and Figure 11.8.

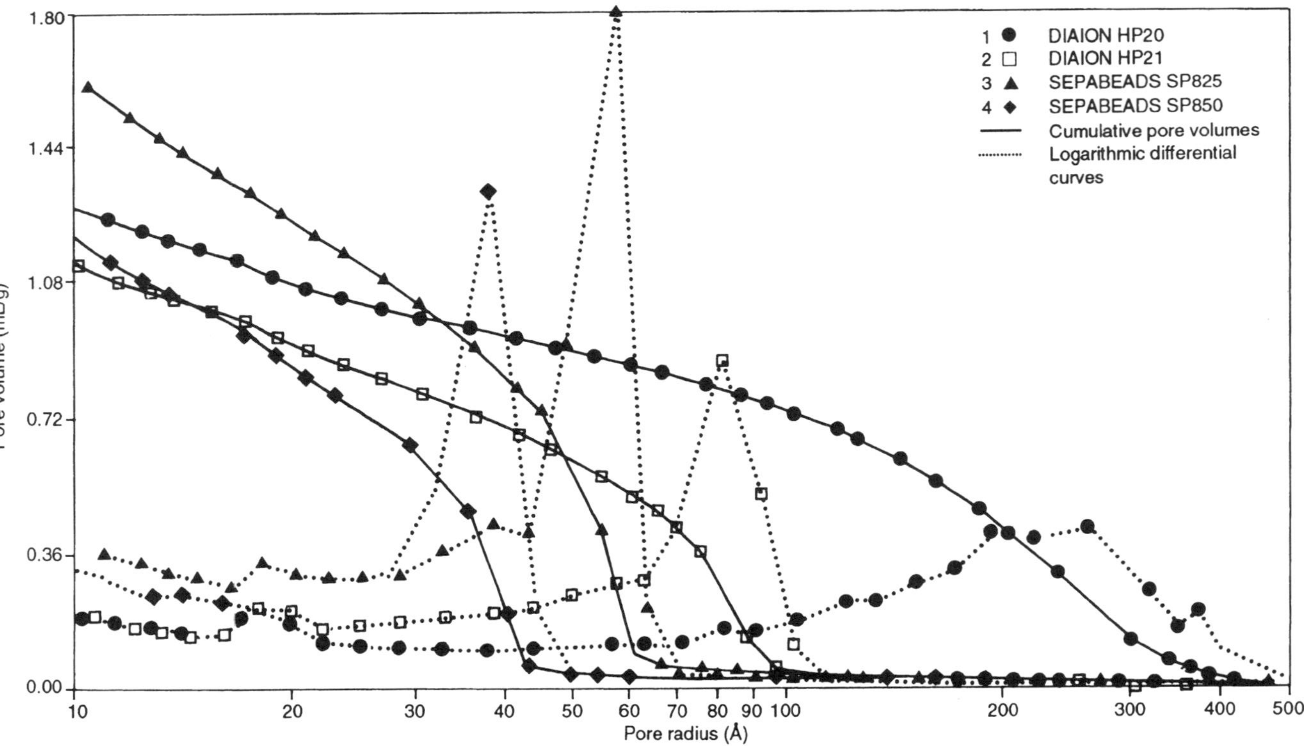

Figure 11.4 Pore characteristics of polystyrene resins

Table 11.1 Basic properties of adsorbent resins

Grade	Type	Moisture content (%)	Specific gravity	Specific surface area (m^2/g)	Specific pore volume (mL/g)	Pore radius (A)	Swelling ratio in solvents				Cephalo-sporin adsorption (g/L-adsorbent)	Application
							H_2O	Methanol	Propanone	Toluene		
DIAION HP20	Crosslinked polystyrene	56	1.01	∼600	∼1.3	200–300	1.00	1.28	1.29	1.32	38	Applicable to a wide range of organic compounds (∼MW 10^4)
DIAION HP21	Crosslinked polystyrene	54	1.02	500–600	∼1.1	80–100	1.00	1.31	1.34	1.35	48	Applicable to a wide range of organic compounds (∼MW 10^4)
SEPABEADS SP825	Crosslinked polystyrene	58	1.01	∼1000	∼1.6	55–60	1.00	1.19	1.18	1.20	76	Specific adsorption for compounds (MW < 10^3)
SEPABEADS SP850	Crosslinked polystyrene	51	1.01	∼1000	∼1.2	35–40	1.00	1.16	1.15	1.19	85	Specific adsorption for compounds (MW < 10^3)
SEPABEADS SP206	Brominated crosslinked polystyrene	50	1.19	500–600	∼1.0	200–300	1.00	1.12	1.15	1.14	101	Applicable to a wide range of organic compounds (∼MW 10^4)
SEPABEADS SP207	Brominated crosslinked polystyrene	50	1.18	∼600	∼1.1	∼100	1.00	1.15	1.15	1.18	119	Applicable to a wide range of organic compounds (∼MW 10^4)
DIAION HP1MG	Crosslinked polymethacry-late	63	1.07	300–400	∼1.2	200–300	1.00	1.05	1.06	1.04		Applicable to a wide range of organic compounds (∼MW 10^4)
DIAION HP2MG	Crosslinked polymethacry-late	61	1.09	400–500	∼1.2	200–300	1.00	1.05	1.06	1.04		Applicable to a wide range of organic compounds (∼MW 10^4)

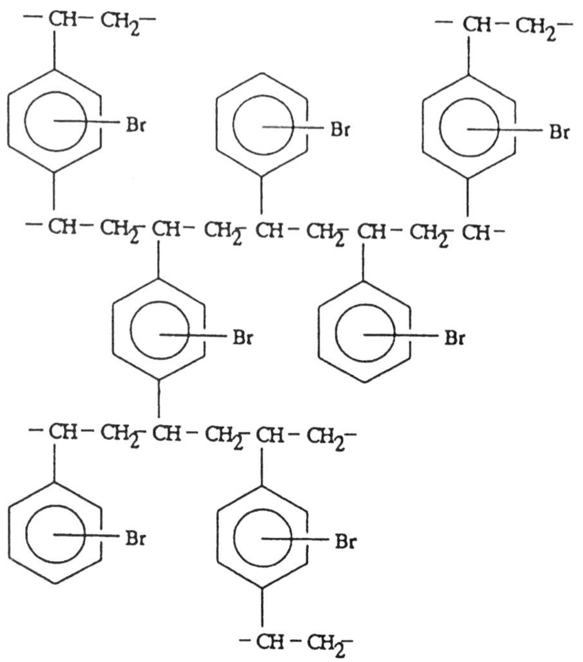

Figure 11.5 Chemical structure of modified polystyrene resin

Table 11.2 Synthetic adsorbent of smaller particle size

Grade	Type	Particle size (μm)	Other properties other than particle size	Application
DIAION HP20SS	Crosslinked polystyrene	ave 100	The same as DIAION HP20 (Table 11.1)	Chromatographic separation of isomers or homologues
SEPABEADS SP20DF	Crosslinked polystyrene	ave 70	The same as DIAION HP20 (Table 11.1)	Chromatographic separation of isomers or homologues

11.2.5 OTHER PROPERTIES

11.2.5.1 Particle Size

Adsorbent particle diameter becomes an important factor when a separation process is to be scaled-up, although it tends to be neglected in laboratory-scale work. It is desirable to use adsorbents of largest particle size in production-scale processes because smaller particles are difficult to handle and often cause trouble in pumping systems. Therefore most synthetic adsorbents are designed

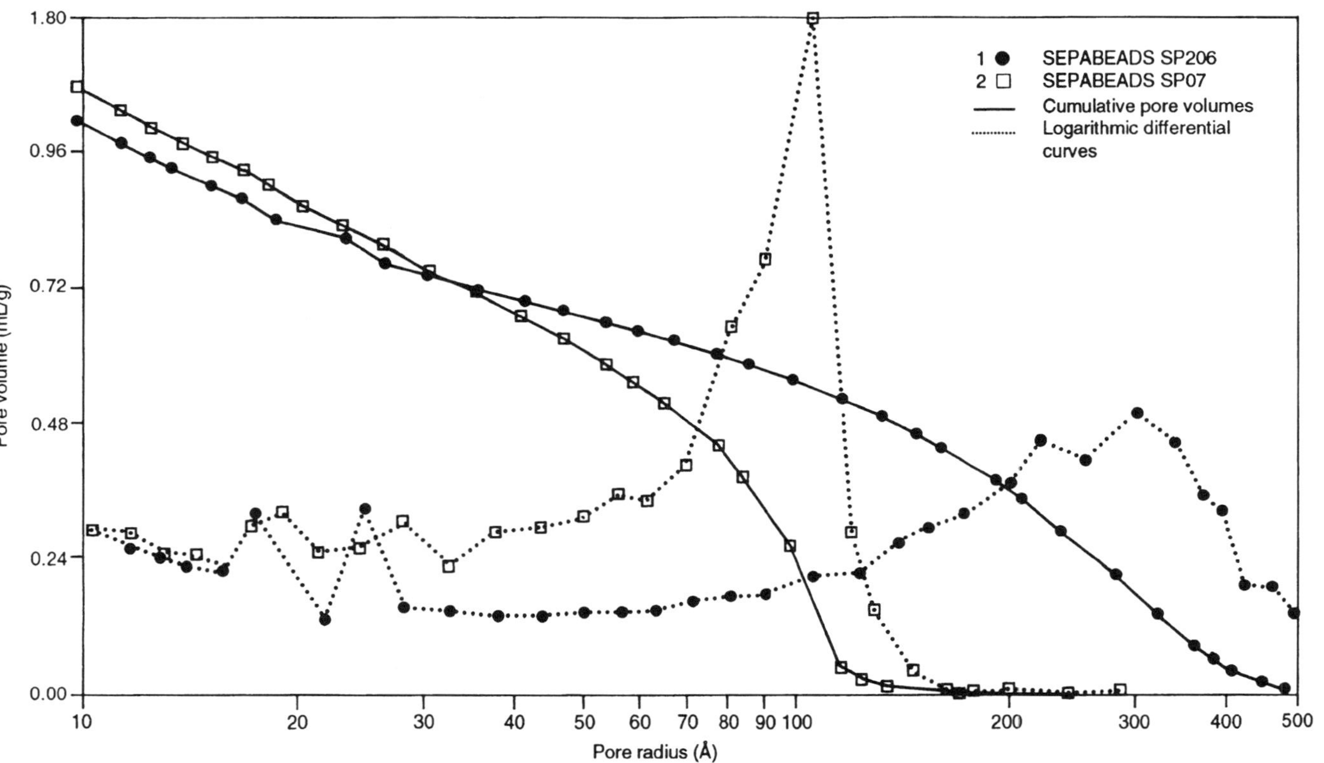

Figure 11.6 Pore charasteristics of modified polystyrene resins

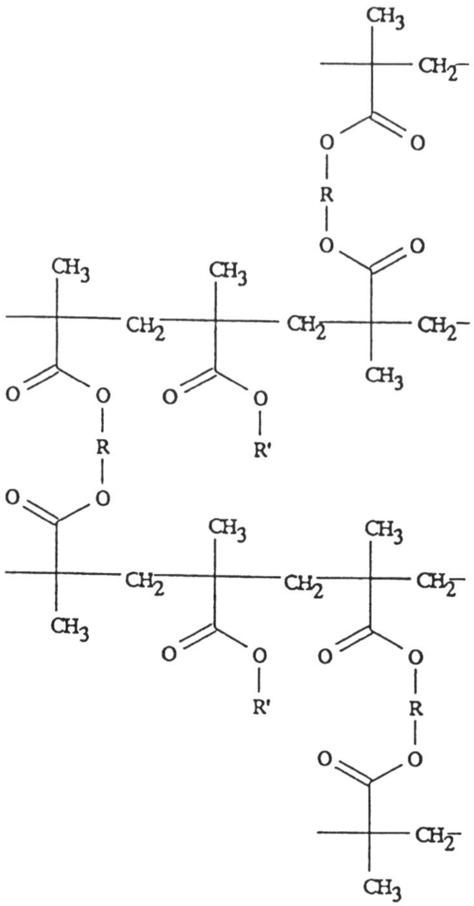

Figure 11.7   Chemical structure of polymethacrylate resin

for production-scale operations and the particle diameter is generally in the range 0.3–1.0 mm.

It sometimes happens that a liquor contains compounds of similar structure to the desired product and it may be difficult to separate them using large particle size resin. In such cases smaller particle sizes are used in chromatographic mode. Table 11.2 shows some examples of such resins. A chromatographic application of this type of adsorbent is described later (Section 11.4.3).

11.2.5.2 Moisture Content

Synthetic adsorbents are generally purchased in their water-swelled state and used without drying. Moisture content is expressed as the weight ratio (%) of

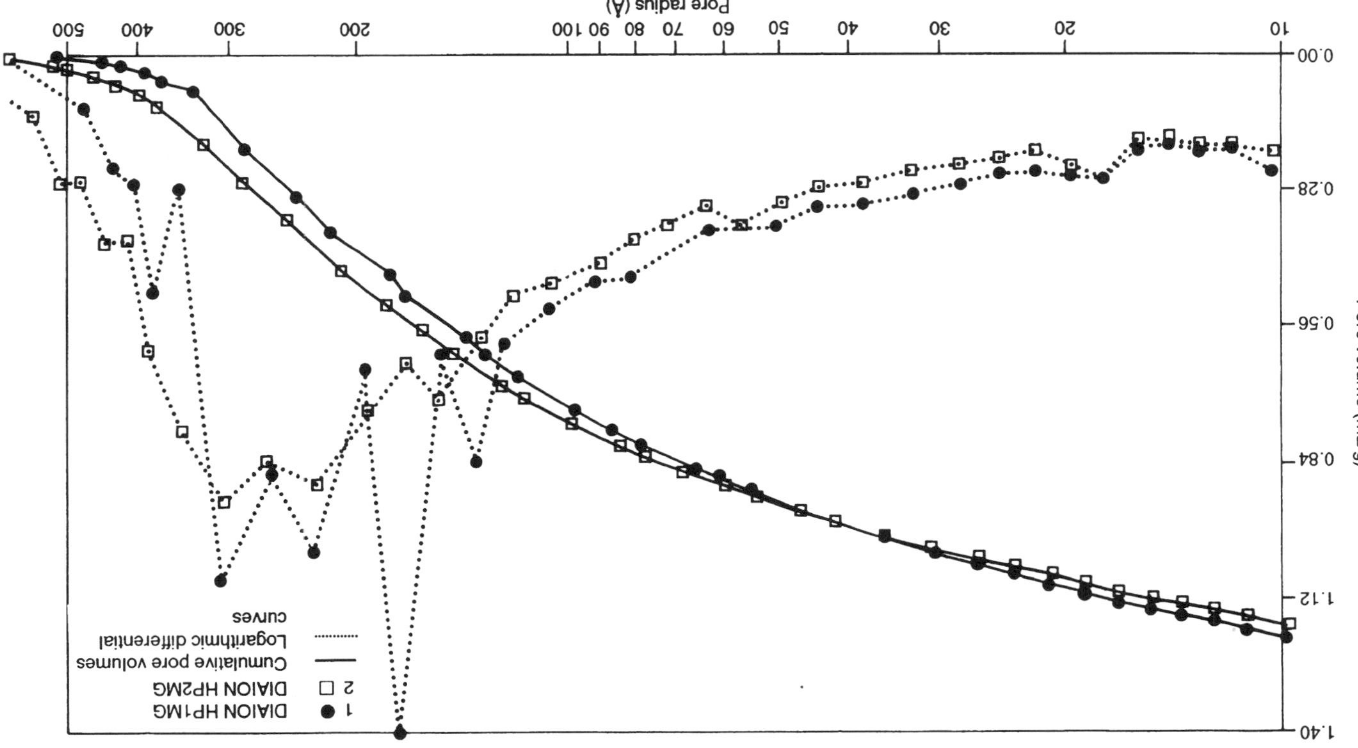

Figure 11.8 Pore characteristics of polymethacrylate resins

water retained in the resin particles. Each adsorbent has its characteristic moisture content and this value has a close correlation to the pore volume. The moisture content changes when the adsorbent is contaminated or has deteriorated and this parameter can be used to assess resin life.

11.2.6 STABILITY OF SYNTHETIC ADSORBENTS

Under the conditions normally met during downstream processing, synthetic adsorbents are very stable materials, but there are several points to note. The following are relevant to both polystyrene and polymethacrylate resins. Generally the resins are chemically stable; they are insoluble in organic solvents and can withstand pH 1 or pH 13 for a certain period of time. However, they are damaged by oxidizing agents (see Section 11.3.5). In downstream processing the temperature is usually $< 100\,°C$; the resins are stable at these temperatures and they can withstand steam sterilization.

11.3 Practical Operations

When synthetic adsorbents are to be applied to a purification process the user has to select an appropriate type of adsorbent and then optimize the conditions before scaling up the process. This section gives a guide for the evaluation of resins on the laboratory scale.

11.3.1 PRETREATMENT

The resins are normally used after being swollen with water. When an adsorbent is partially dried it does not swell to its former value by direct water contact. It is necessary to swell it first with a water-miscible organic solvent and then change to pure water.

It is also important to note that resins sometimes contain small amounts of impurities and degradation products. They should therefore be washed with a suitable solvent such as propanone or methanol and then water. A typical

---

Wash with an organic solvent — propanone
- Suspend a synthetic adsorbent into propanone in a beaker.
- Stir the mixture generously to remove bubbles and adhering solids from adsorbent particles.
  Do not use magnetic stirrer to avoid grinding down the particles.
- Transfer the adsorbent slurry to a column fitted with a glass filter frit and a tap.
- Feed more than 3 BV of propanone on to the column.

Displacement with water
- Wash the column with more than 5 BV of pure water to remove propanone.
- Transfer the adsorbent to a reservoir with water.
  Do not dry the adsorbent before use.

Scheme 11.1 A typical pretreatment procedure for synthetic adsorbent

scheme is shown in Scheme 11.1 When adsorbents of higher specific surface area are tested vacuum degassing is desirable.

11.3.2 TESTING METHODS

11.3.2.1 Batch Tests

Batch tests are convenient and informative for selecting the type of adsorbent or for optimizing certain conditions. In batch experiments, parameters to be examined include adsorbent volume to liquor (phase ratio), pH, temperature and contact time in order to obtain reproducible data. If the target compound has one or more ionizable groups pH often influences adsorption capacity. To maximize adsorption, pH should be adjusted such that the target compound is un-ionized (at the isoelectric point for amphoteric compounds). A typical process for adsorption testing is shown in Scheme 11.2.

- Measure a certain volume of adsorbent in a graduated cylinder with water.
- Filter the adsorbent and remove free water by centrifuge.
- Transfer the adsorbent into an Erlenmeyer flask and add a measured volume of test liquor.
- Seal the flask and shake it gently in an incubator.
- Analyse the concentration of the target compound in the supernatant after several hours.

Scheme 11.2   Batch test procedure

In this process, the capacity (adsorbed compound per unit volume of adsorbent) is calculated from

$$C = 1000 \; (W_t - W_s)/V$$

where C is the adsorption capacity (g/L adsorbent), W_t is the amount (g) of target compound in the liquor, W_s is the amount of target compound in the supernatant after contact and V is the volume (ml) of the resin tested. It is desirable to collect data on time dependence of the adsorption process in order to evaluate the kinetics of adsorption. For the initial stage of testing a number of adsorbents are evaluated in parallel and then some are selected for further examination. Tests of the desorption stage can be carried out in a similar way. However, it is often convenient to use a column system for this stage.

11.3.2.2 Column Test

Having selected a few adsorbents in batch tests, they should be evaluated in column mode. At this stage the actual tests willl depend on the specific process or compound but the following points should be considered:

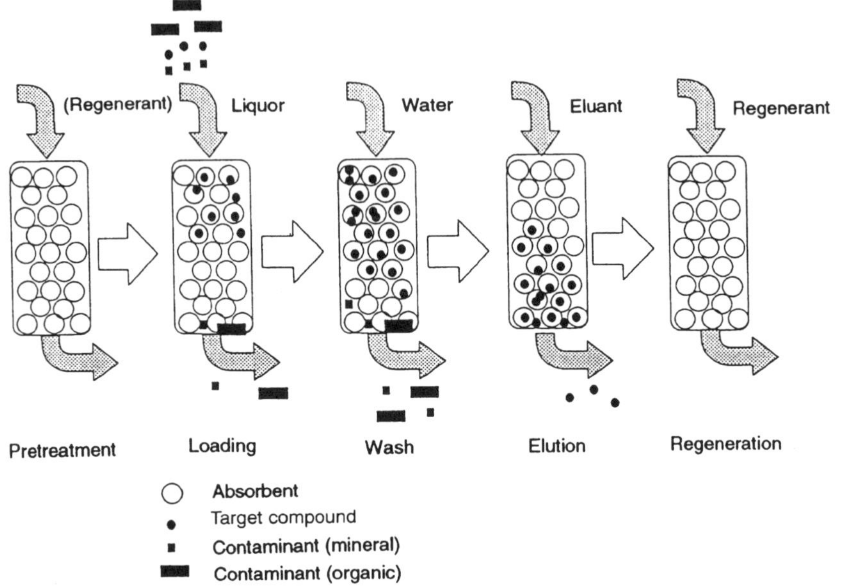

Figure 11.9 Typical operation cycle

(a) effect of the type of eluant on recovery,
(b) effect of the type of eluant on the elution profile (chromatogram),
(c) effect of loading on recovery,
(d) effect of flowrate (if adsorption or desorption rate is low).

When the target compound is neutral a non-ionizable, a water-miscible organic solvent such as lower alcohols or propanone is used as eluant. However, an aqueous buffer or buffer/solvent mixture may be preferable for ionizable (i.e. more water-soluble) compounds. Having selected an adsorbent, operating conditions are evaluated.

11.3.3 OPERATING CONDITIONS

A typical column purification process consists of several steps, as shown in Figure 11.9. The apparatus and the solutions should be thermostatted. In the column conditioning step the resin is swelled with water and the pH is adjusted to that of the feed liquor. Column loading consists of feeding the liquor at a constant flowrate. The volume of feed solution is set so that the amount of target compound is less than the adsorbent capacity of the column. The effluent (percolate) should be monitored to ensure that the target compound does not leak from the column.

After loading, the column is still filled with the feed liquor and contains contaminants in the intraparticle voids. Therefore the column should be washed with water (or pH-adjusted water) before elution. Only sufficient washwater to displace the residual feed liquor is required, e.g. 0.5–1.0 bed volume.

During elution a suitable eluant is passed through the column to recover the target compound. To accelerate this process, gradient elution may be beneficial. It may be possible by careful selection of elution conditions to separate impurities quite similar to the target compound.

To re-use the column it is washed with water and re-equilibrated at the correct pH. After several cycles the resin will probably require rejuvenating to recover the original adsorptivity (see Section 11.3.5).

11.3.4 CYCLE TESTS

When an operating condition has been established, the next stage is to estimate the average throughput of the purification process and the life of the adsorbent. Cycle tests are done by repeating the operations described in Section 11.3.3. In most downstream processing applications, the liquor contains a number of contaminant compounds which accumulate on the adsorbent surface. Therefore the purification capacity of the process will decrease over a period of time. The desired average purification capacity of the process determines the number of cycles between rejuvenation processes.

11.3.5 REGENERATION AND REJUVENATION

11.3.5.1 Regeneration Methods

The adsorbent is regenerated to be reloadable in each cycle as described in Section 11.3.3, the regeneration condition depending on the eluant used. Should the target compound be eluted chromatographically and be followed by slower running components, eluant should be supplied until all components have eluted. A multistep regeneration process may be appropriate in such cases. After elution of the target, a less polar solvent is used to complete the elution and then aqueous regenerant is used.

11.3.5.2 Rejuvenation Methods

When a column is used repeatedly the adsorbent is gradually contaminated by irreversible adsorption of contaminants such as proteins. These contaminants cannot be removed by normal elution conditions so resin rejuvenation is eventually necessary. Rejuvenation consists of treating the resin with much stronger conditions such as sodium hydroxide/propan-2-ol. The effect is dependent on temperature, concentration of agents, contact time, etc., and must be judged from the degree of contamination. In cases of severe fouling an oxidizing agent such as sodium hypochlorite can be used, but with great caution, as it may degrade the resin.

11.3.6 PRESERVATION OF ADSORBENTS

It is sometimes necessary to store used resins for some months. Although the resins are stable during long-term storage there are several precautions required to keep them in a usable condition. Generally they should be kept in the swollen state in a cool place out of direct sunlight so as to avoid heating or ultraviolet irradiation. They should not be frozen as the formation of ice crystals may break the pore structure of the resin. The resins should be in closed containers to retain moisture and minimize microbial contamination. Although the resins are not biodegradable, microbial growth will compromise performance. One way to minimize microbial contamination is to store in concentrated sodium chloride solution. Preservatives such as sodium azide are effective but toxicity and explosion hazards should be considered.

11.4 Examples of Applications

There is much literature on the application of synthetic adsorbents in downstream processing. This section mentions some laboratory-scale data and examples from recent papers and patents.

11.4.1 SEPARATION OF CEPHALOSPORIN DERIVATIVES

A fermentation broth was applied to Sepabeads SP825 or SP850 and cephalosporin or its derivatives were separated as shown in Figure 11.10.

11.4.2 REJUVENATION OF A SYNTHETIC ADSORBENT (HP20)

As noted previously, rejuvenation of adsorbents is usually necessary in practical applications. An example of rejuvenation is given in Table 11.3. Since some agents used in these processes often damage the resin itself, the conditions should be carefully set according to the degree of contamination.

Table 11.3 Effect of rejuvenation condition on surface area of adsorbent (DIAION HP20)

Sample resin	Condition	Specific surface area (m^2/g)
New HP20		~ 700
Used resin		96
After treatment	99% methanol (3 bed volume)	246
	75% methanol (3 bed volume)	141
	95% 2-propanol (3 bed volume)	406
	95% propanone (2 bed volume)	563
	75% 2-propanol and 4% NaOH (4 bed volume)	562
	4% sodium hypochlorite (5 bed volume)	246
	4% sodium hypochlorite with air bubbling (4 bed volume)	527

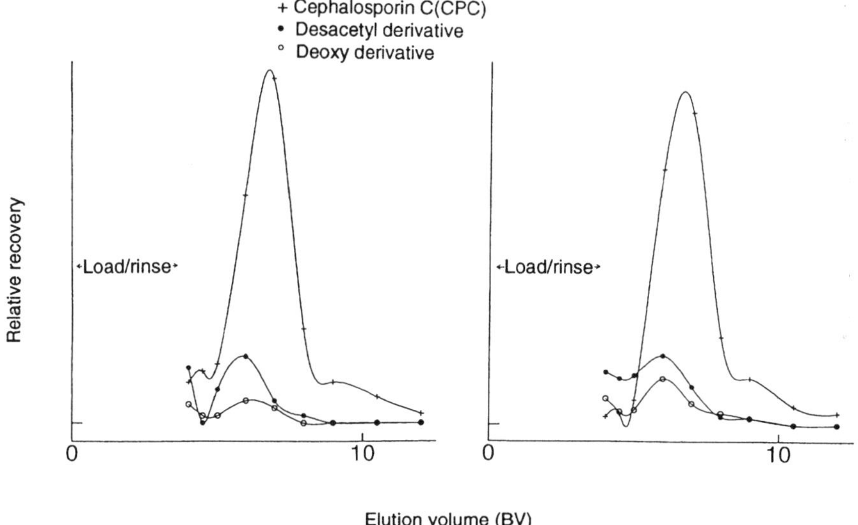

Figure 11.10 Separation of cephalosporin derivatives on SEPABEADS SP825 (left) and SP850 (right) columns. Filtered broth was loaded onto the column (*ca.* 40 g CPC/L–adsorbent). The column was rinsed with H_2O and then eluted with 50 mM sodium acetate

11.4.3 OTHER APPLICATIONS IN THE LITERATURE

11.4.3.1 Antibiotics

Thienamycin was recovered from fermentation broths by the use of a polystyrene resin [2]. Glycopeptide antibiotics (actaplanin and vancomycin) were isolated on a polystyrene resin [3]. Separation of cephalosporin derivatives was achieved using polystyrene resin [4]. A method of recovery of mitomycin C from fermentation broths was disclosed using a brominated polystyrene resin in combination with reversed-phase chromatography [5].

11.4.3.2 Vitamins

Processes for the purification of vitamin B_{12} or related compounds were disclosed, using crosslinked polystyrene resin as adsorbent [6]. Separation of D-biotin was carried out on a polystyrene resin [7].

11.4.3.3 Peptides and Proteins

Polystyrene resin was used for the isolation of an angiotensin-converting enzyme inhibiting peptide from hydrolysed milk casein [8]. Serrapeptase was

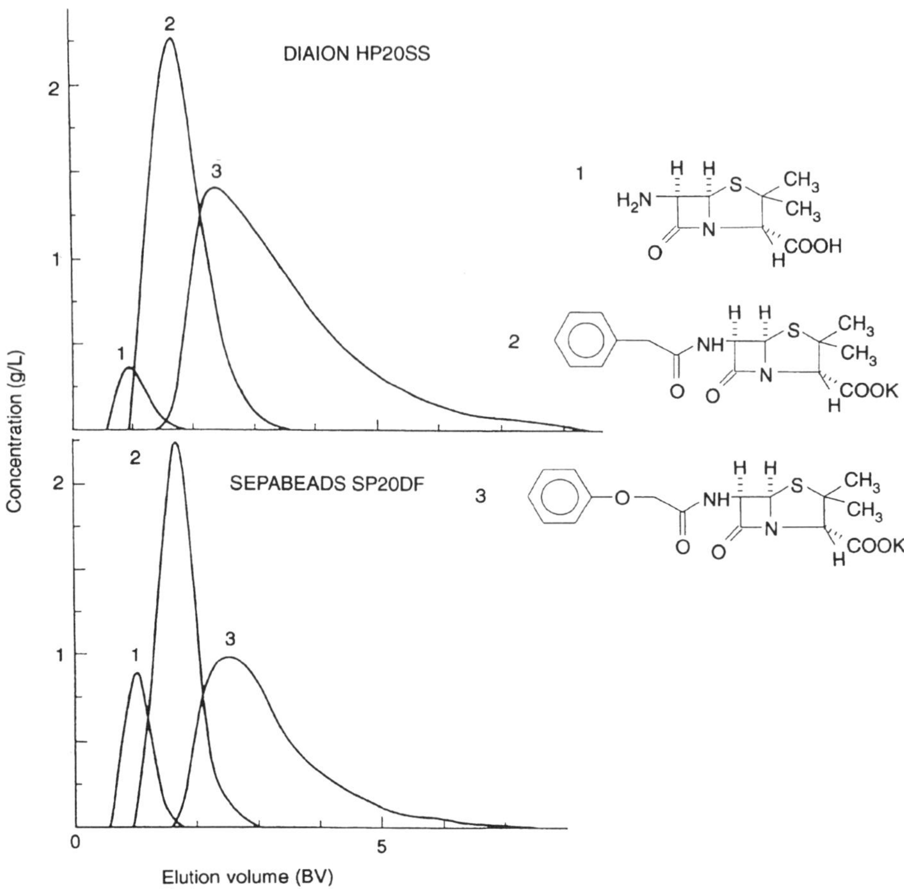

Figure 11.11 Chromatographic separation of penicillin derivatives on columns of synthetic adsorbent (small particles). Column: 500×20 mm i.d.; mobile phase, 50% methanol in 50 mL phosphate buffer (pH 8); flowrate, 30 mL/min (SV = 11.5); temperature, 10°C. Solute: (1) 6-aminopenicillanic acid 0.025 g, (2) penicillin G 0.25 g and (3) penicillin V 0.25 g; sample size, 50 mL (0.32 BV)

purified on polymethacrylate resin [9]. Polymethacrylate resin was used for the concentration of tissue plasminogen activator produced in a bioreactor [10].

11.4.3.4 Saccharides

A plant derived sweetener (from *Stevia*) was purified by polystyrene resin [11]. A method has been described for the production and purification of cyclodextrin and its derivatives using polystyrene resins [12].

11.4.3.5 Amino Acids

Brominated polystyrene resin or polymethacrylate resin was used for the separation of phenylalanine [13,14].

11.4.3.6 Nucleosides

Brominated polystyrene resin was used for the purification of 2′,3′-dideoxyinosine or 2′,3′-dideoxyadenosine from a fermentation [15,16].

11.4.3.7 Other Organic Compounds

The recovery of nicotine alkaloids from plant culture was reported using polymethacrylate resin [17]. Ergot alkaloids from culture filtrate were separated on polystyrene resin [18]. The recovery of prostaglandins from a culture has been patented [19]. Polystyrene resin was used for the purification of *cis,cis*-muconic acid [20]. Arachidonic acid and isomyristic acid were separated on polystyrene resin as their methyl esters [21]. Gibberellic acid was isolated using polystyrene resin [22].

11.4.4 APPLICATION OF SMALLER PARTICLE ADSORBENTS

A chromatographic separation of penicillin derivatives has been examined in the author's laboratory with the following results. A large amount of three penicillanic acids (a synthetic mixture) was chromatographed using a preparative-scale column of Diaion HP20 or Sepabeads SP20DF. The compounds were successively separated under a simple isocratic elution condition (Figure 11.11).

11.5 References

Many of the references cited below are to Japanese patents. To assist the reader in accessing this literature the corresponding *Chemical Abstracts* (*CA*) reference is provided.

1. Mitsubishi Kasei Corporation (1992) *Diaion: Manual of Ion Exchange and Synthetic Adsorbent, Parts I and II.*
2. V. P. Gullc and L. R. Treiber, US Patent 4 198 338.
3. M. H. McCormick and G. M. Wild, US Patent 4 440 753.
4. W. Voser and K. Weiss (1980) *J. Chromatogr.*, **201**, 287.
5. S. Iwata and M. Shiomi, DE Patent 4 042 156 [*CA* 117:190268y].
6. K. Yamanaka and S. Hikami, Japan Patent 02286093 [*CA* 114:227506r].
7. M. Terasawa, S. Nara and H. Yugawa, Japan Patent 05078369 [*CA* 119:7341w].
8. R. Sugai, T. Shinai, H. Uchiwa and U. Murakami, Japan Patent 04341193 [*CA* 118:146252k].
9. K. Katamoto, T. Kuramoto and Y. Matsuda, Japan Patent 61215330 [*CA* 106:83035].

10. S. Mitsuda, Y. Matsuda, N. Kobayashi, A. Suzuki, Y. Itagaki, E. Kumazawa, K. Higashio and G. Kawanishi (1991) *Bioprocess Engng.*, **7**, 137.
11. R. Ise and I. Harada, Japan Patent 54132599 [*CA* 92:109391d].
12. Y. Sawaguchi, M. Aikawa, K. Kikuchi and K. Sasaki, Japan Patent 02255095 [*CA* 114:141647a].
13. M. Otani, C. Sano and I. Kusumoto, European Patent 147 974.
14. B. McCulloch and W. Goodman, US Patent 5 071 560.
15. T. Tanabe, M. Ootani and T. Yugawa, Japan Patent 01165390 [*CA* 111:230677r].
16. M. Ootani, T. Tanabe and T. Yugawa, Japan Patent 01098496 [*CA* 111:132624b].
17. K. D. Green, N. H. Thomas and J. A. Callow (1992) *Biotechnol. Bioengng.*, **39**, 195.
18. S. Takano and T. Sasaki, Japan Patent 02167091 [*CA* 113:170483].
19. I. Nakajima, K. Suzaki and M. Tsuji, European Patent 424 915.
20. N. Yoshikawa and K. Ota, Japan Patent 02145537 [*CA* 113:130760k].
21. Y. Noguchi, M. Kaji and T. Funada, Japan Patent 63112694 [*CA* 111:113739b].
22. R. Heropolitanski (1978) *Przem. Ferment. Owocowo-Warzywny*, **22**, 7 [*CA* 90:49625].

12 ION EXCHANGE CHROMATOGRAPHY AND SECONDARY ADSORPTION EFFECTS

Peter R. Levison

12.1 Introduction

Ion exchange chromatography is routinely carried out in the industrial bioprocessing market and for low-pressure chromatography is typically based on polysaccharide matrices derivatized with the appropriate functional group [1,2]. The theory underlying the principles and practice of ion exchange chromatography is outside the scope of this article and has been described elsewhere [3].

In an ion exchange separation there are several key stages which influence the effectiveness of the procedure and ultimately its economic success (Table 12.1). Of the stages summarized in Table 12.1, attention is generally paid to the efficiency of the adsorption/desorption process and product purity, etc. The aim of this chapter is, however, to examine practical aspects of ion exchange chromatography and, to this end, the areas to be covered include feedstock preparation, media preparation, media fouling and media regeneration. The scope of the article presupposes that media selection and mobile phases have been identified and that the process will take place in a column. Many of the requirements for batch chromatography are similar to those for a column and a comparison of both techniques has been made elsewhere [4,5].

12.2 Feedstock Preparation

Since the ion exchange chromatography step is typically part of an overall downstream process, the feedstock is generally the product of the previous

Table 12.1 Stages of chromatographic separation

1. Feedstock preparation — clarification, dilution/concentration, buffer adjustment, etc.
2. Media preparation — equilibration, column packing, etc.
3. Adsorption
4. Washing
5. Desorption
6. Regeneration — clean-in-place, sanitization, re-equilibration

upstream/downstream process. Consequently, it is easy to overlook the preparation of the feedstock and develop a process based entirely on the product as recovered upstream. However, since ion exchange chromatography relies on the ionic properties of the adsorbate and the adsorbent [3], it is necessary to ensure that the pH of the feedstock is sufficiently far from the pI of the adsorbate to facilitate adsorption. This would typically require a pH of at least $pI + 1$ for anion exchange and at least $pI - 1$ for cation exchange. Furthermore, in order for the ion exchanger to interact effectively with the adsorbate the pH should be at least $pK_a - 1$ for anion exchange and $pK_a + 1$ for cation exchange. The pK_a for the ionizable groups of anion and cation exchangers can be obtained by titration, or from the media manufacturer. If the pH of the product from the upstream process is outside these limits then it may need adjustment to render it suitable as a feedstock for the ion exchange process.

While ion exchange is pH dependent, it is, as the name implies, an exchange of ions and thus is dependent on the ionic strength of the mobile phases. Ionic strength is routinely measured as conductivity and can be calibrated as such. Protein is typically desorbed from ion exchangers by addition of, for example, 0.5 M NaCl. This clearly increases conductivity and proteins desorb. However, the feedstock during adsorption also has a conductivity and it is important that this be reduced as much as possible, in order to facilitate adsorption and minimize spontaneous desorption. We have investigated the influence of mobile-phase composition on adsorption efficiency using hen egg-white proteins on Whatman QA52 [6]. In this work we used 0.025 M tris/HCl buffer, pH 7.5, which has a conductivity of ~ 1.8 mS. A 10 mg/mL solution of hen egg-white proteins in this buffer has a conductivity of ~ 2.7 mS and this gave good protein-binding capacity. If, however, protein concentration is increased to 30 mg/mL, the conductivity increased to ~ 5.6 mS, which significantly reduced the effectiveness of the adsorptive step. It is therefore recommended that the chromatographer assess the effect of feedstock concentration on the ion exchange step and where necessary dilute the product from the upstream process in order to improve the capture efficiency of the adsorbate in a single contacting operation. While such a measure may increase the overall process time, the associated increase in productivity should outweigh this inconvenience.

12.3 Media Preparation

Bulk ion exchange media are supplied as either dry or wet powders, or slurried in a preservative, typically dilute ethanol or bacteriostatic solution. In each case the media must be equilibrated such that all preservatives etc. are removed and it is in a similar mobile phase to that of the feedstock. The chemical manipulations required to fully equilibrate the media will generally be provided by the manufacturer and a medium can be regarded as equilibrated when the pH and conductivity of the mobile phase within the medium are the same as that used for equilibration. It is, however, important that the bulk media is handled in a manner such that attrition does not occur, which can affect the performance of the media. In our experience in handling non-regenerated cellulose adsorbents, a typical slurry consistency suitable for stirring and pumping would be 30% (w/v) for microgranular products and 15% (w/v) for fibrous products, prepared by adding media to buffer and not buffer to media. We would typically use a tank with a height to diameter aspect ratio of 2:1 equipped with a variable-speed agitator fitted with turbine impellers of either open type or centre disc design fitted with back slope blades [7]. These are marketed as retreat curve impellers by, for example, Corning [8]. When stirring slurries of ion exchange media, it is important to operate the agitator at a speed sufficiently low as to prevent mechanical shear of the particles. A useful measure of stirring is to calculate a tip speed for the stirrer. This is calculated as follows:

Tip speed $= 2\pi r v$

where $r =$ radius of impeller
        $v =$ rotation speed (rev/min)

We have found that for ion exchange celluloses a tip speed of 3.5 metres per second should not be exceeded.

Slurries of ion exchange media are often transferred from vessel to vessel and again high-shear systems should be avoided. Pumps of either peristaltic or rotary lobe design are preferred for the handling of cellulosics and centrifugal pumps, for example, should not be used since they can result in disruption of the matrix.

12.4 Column Packing

There are several column designs available for large-scale chromatography from a range of manufacturers that are constructed from various materials including glass, acrylic and stainless steel. The two major concepts underlying column design relate to the use of either axial or radial flowpaths for the mobile phases. The features and benefits of these column configurations have been reviewed previously [9,10]. A widely used column design involves axial flow using an adjustable bed support to vary bed volume. Generally, the column packing protocol would reflect the column manufacturer's instructions but a general protocol may be as follows:

(a) Transfer the slurry of ion exchanger into the empty column barrel, fitted with an extension tube (if necessary).
(b) Fit the upper flow adaptor.
(c) Consolidate the bed at constant pressure (typically 0.13–0.34 bar above the operating pressure).
(d) Depressurize the column.
(e) Remove the extension tube and residual packing buffer.
(f) Bring the flow adaptor into contact with the relaxed bed.
(g) Compress the bed using the flow adaptor to the height observed during the initial bed consolidation step.

We have investigated the sensitivity of this protocol with particular emphasis on step (g) described above. An Amicon G450 × 500 column (45 cm i.d.) was

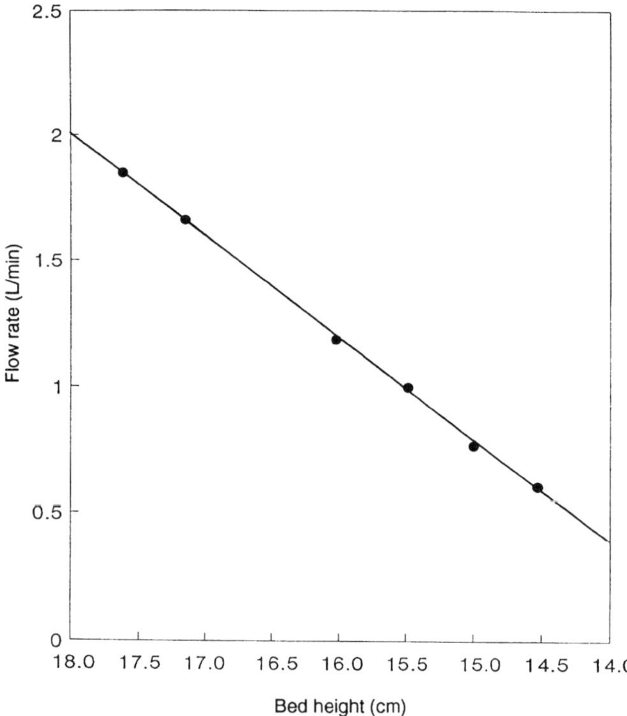

Figure 12.1 The influence of overcompression of a bed of Whatman DE52 using the flow adaptor of an Amicon 45 cm i.d. column on the flowrate at a constant inlet pressure of 0.68 bar using 0.025 M tris/HCl buffer, pH 7.5

packed with a 30% (w/v) slurry of the anion exchange cellulose DE52 in 0.025 M tris/HCl buffer, pH 7.5, using a packing pressure of 0.68 bar to give a consolidated bed height of 17.6 cm. The procedure in step (g) was carried out and the flowrate at 0.68 bar measured. The bed was then progressively overcompressed by up to 3 cm using the flow adaptor, thereby increasing the packing density of the column. During this study flowrate was measured at 0.68 bar. The results of this study are represented in Figure 12.1. These data demonstrate that overcompression of the column bed using the flow adaptor has a significant effect on column pressure/flow performance and a seemingly small difference in bed height, i.e. 1 cm will give rise to reductions in flowrate of 25–35% at constant pressure. The results of this study demonstrate the sensitivity of what might, at the outset, be considered a robust operation, namely column packing. It is therefore essential that a reliable column packing protocol be devised and implemented, especially where this becomes a routine operation, in order that the chromatographic step becomes a reproducible procedure.

12.5 Media Fouling

Having prepared the feedstock and packed the column of pre-equilibrated ion exchange medium, the chromatographer has very little control over the adsorption process, especially when one is applying a large mass of protein. The basis of the interaction between an ion exchange medium and the protein adsorbate molecule is ionic and it follows that the adsorbed protein molecule could act as a stationary phase to facilitate secondary adsorption. Such secondary adsorption could arise from ion exchange interactions, chiral interactions or protein–protein interactions arising from the biological specificity of the protein, its hydrophobicity, local changes in tertiary structure, etc. [11]. Furthermore, the matrix may interact directly with proteins by, for example, hydrogen bonding or via an affinity interaction as exhibited by lectins [11]. Lipids in the feedstock may interact with an ion exchanger through the polar head group, leaving the non-polar tail as a site for hydrophobic interactions to take place. In summary, the scope for secondary adsorption to occur by various intra-feedstock interactions is high and generally unpredictable. It may be possible to control such interactions by addition of surfactants or detergents such as Tween or Triton, or the addition of denaturants such as urea or guanidine hydrochloride. These measures may reduce, but are unlikely to eliminate, these secondary adsorptive effects.

If the ion exchange group itself is considered then it could give rise to secondary adsorptive effects under certain conditions. For example, the tertiary amine diethylaminoethyl (DEAE) group, a widely used anion exchange functionality, has the structure:

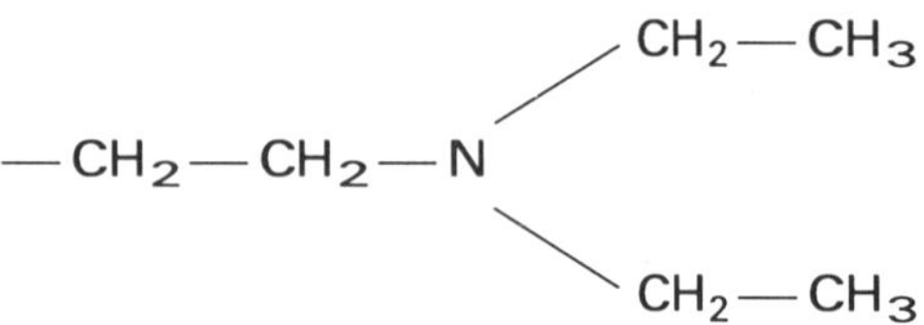

The tertiary amine has a pK_a of ~ 9.5 and if, for example, desorption is effected by increasing the pH to > 9.5, i.e. deionizing the amine, then the functional group more closely resembles a hydrophobic interaction group rather than an ion exchange group. This could result in a secondary adsorption occurring subsequent to ion exchange desorption via a hydrophobic interaction.

Furthermore, if desorption is effected by using a high concentration of chloride as counter-ion, e.g. 0.5 M NaCl, then assuming 75% (v/v) voidage for a DEAE anion exchanger with a small-ion capacity of 1 milliequivalent/dry g, the chloride ions would be in excess on an electrical charge basis. Under these conditions the positively charged nitrogen atom would be unlikely to interact electrostatically with proteins, and consequently secondary adsorptive effects could occur, for example, by hydrophobic interactions as discussed above.

Given that media fouling could occur for any of the reasons given above, i.e. feedstock composition or ion exchanger effects, then the nature of the fouling is extremely difficult to characterize and quantify. In general there are three effects that may become apparent with a fouled column, namely discoloration of the medium, binding capacity reduction and chromatographic performance reduction, each of which may be manifest by an increase in the pressure drop required to maintain flow. These three effects will be discussed in more detail.

12.5.1 MEDIA DISCOLOURATION

Media fouling is often observed qualitatively by an adsorption of coloured material from the feedstock near the top of the bed which concentrates during use giving a noticeable coloured band. In many instances this material is difficult to elute and may necessitate replacement of the top of the bed. In order to avoid this fouling, it may be desirable to undertake further feedstock preparation using, for example, a small guard column of the ion exchanger to remove the colourants. Although media discoloration is aesthetically unattractive, whether such discoloration affects the ion exchange separation in this or subsequent runs should be established, and the implications of running a 'coloured' column several times on process validation should be considered in the overall context of process design and economics.

12.5.2 CAPACITY REDUCTION

Media fouling may be accompanied by a concomitant reduction in protein-binding capacity. If this is the case then fouling becomes quantifiable and the

Table 12.2 Recovery of IgG following chromatography of goat serum on Whatman QA52. For full details see Levison *et al.* [12]

Separation number	IgG recovery (% w/w)
1	49
2	42
3	36
4	27

effectiveness of the clean-in-place (CIP) protocol for column regeneration is easy to demonstrate. We have demonstrated that under certain conditions in the purification of immunoglobulin G (IgG) from goat serum using Whatman QA52, a serious media fouling occurred. By either a modification of the loading conditions and/or a periodic CIP with 0.5 M NaOH, column performance was restored and maintained [12]. In this study, we observed a significant reduction in IgG recovery which we could attribute to fouling of the QA52. These data are summarized in Table 12.2.

In a similar experiment we investigated the re-use of Whatman DE52 for a process-scale egg-white separation [13]. Following five consecutive loadings of hen egg-white (~ 1.8 kg) on to a 25 L column of DE52, protein capacity remained constant at ~ 1.38 kg. Under these conditions protein capacity is not a useful indicator of column fouling and a chromatographic performance test is a useful means of demonstrating fouling.

12.5.3 CHROMATOGRAPHIC PERFORMANCE TESTS

When process-scale ion exchange chromatography is performed and the column effluent is monitored by absorption at, for example, 280 nm, the resulting chromatogram is not as 'elegant' or detailed as perhaps the inexperienced person may expect. For example, the chromatography of 200 L of 9.24 mg/mL hen egg-white proteins on a 25 L column of Whatman Express-Ion™ Exchanger Q (45 cm i.d. $\times$ 16 cm) is represented in Figure 12.2(a). The protocols used in this type of study have been reported previously [14].

Under identical conditions we applied 10 L of the same feedstock to the column of Express-Ion Q. These results are represented in Figure 12.2(b) and the resulting chromatogram gives good separation of the components present in egg-white. In the former case we have overloaded the column, whereas in the latter case we are using $\sim 5\%$ of the total protein-binding capacity of the bed. Under conditions of high protein loading Beer's law is not obeyed and small differences in absorbance can give rise to large differences in protein capacity — hence the broad peak shape seen in Figure 12.2(a). We have described this effect previously [15] and if chromatographic performance is to be used as an indication of media fouling then it is suggested that process-scale chromatograms be interpreted with caution.

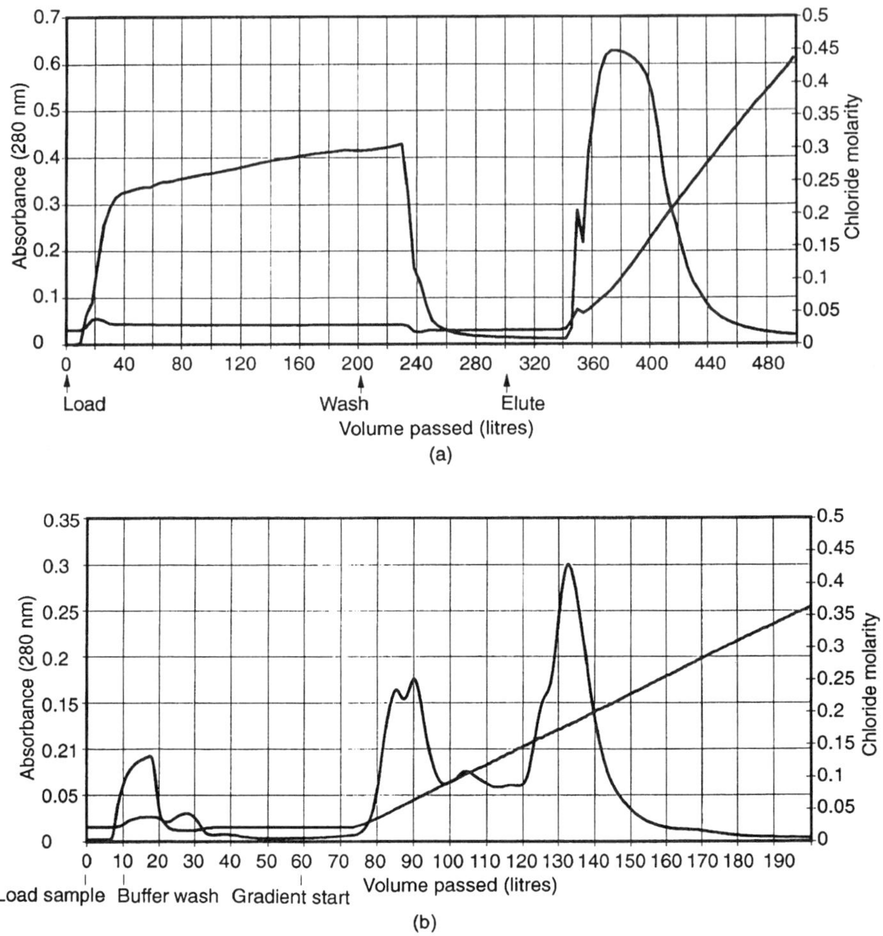

Figure 12.2   The chromatography of 9.24 mg/mL hen egg-white proteins on Whatman Express-Ion Q in a 25 L column (45 cm i.d. × 16 cm), with (a) process loading and (b) analytical loading using 0.025 M tris/HCl buffer, pH 7.5 at a flowrate of 150 cm/h

It is clear that an analytical loading of the same feedstock used for process chromatography gives a 'typical' chromatographic profile (Figure 12.2b). By performing an analytical chromatogram before and after process loadings and/ or a CIP regime, it is possible to assess the effect, if any, that the process loading has on the chromatographic performance of the media and the effectiveness of the CIP on media regeneration. We have used this approach in validating a CIP regime after the process-scale chromatography of hen egg-white proteins on Whatman DE52 [15] and Express-Ion D [14].

In order to demonstrate the applicability of such an approach to monitor media fouling, we investigated the chromatography of hen egg-white proteins

Table 12.3   Process-scale chromatography of egg-white proteins on Express-Ion C

Stage of study	Chromatographic step
1	100 g analytical load — 150 cm/h
2	9 kg preparative load — 112 cm/h
3	100 g analytical load — 150 cm/h
4	CIP — 150 cm/h
5	100 g analytical load — 150 cm/h

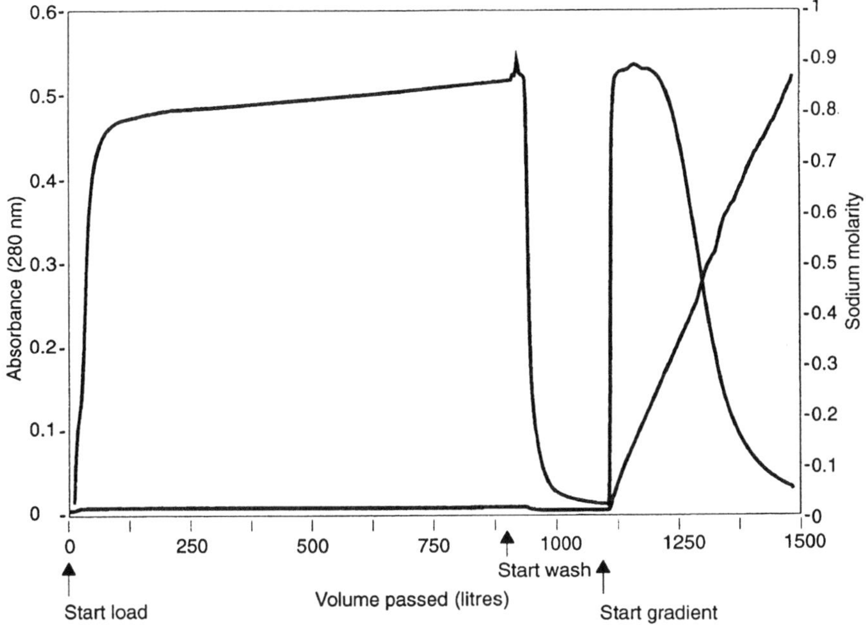

Figure 12.3   The process-scale chromatography of 10.25 mg/mL lysozyme-depleted hen egg-white proteins on Whatman Express-Ion C in a 25 L column (45 cm i.d. × 16 cm) using 0.01 M sodium citrate buffer, pH 5.0 at a flowrate of 112 cm/h

Table 12.4   Protein mass balance data for the process-scale chromatography of hen egg-white proteins using Express-Ion C

Stage of chromatography	Total protein (g)		
	Feedstock	Mobile phase	Adsorbed
Loading	9229	7486	1743
Wash	—	153	1590
Elution	—	1540	50

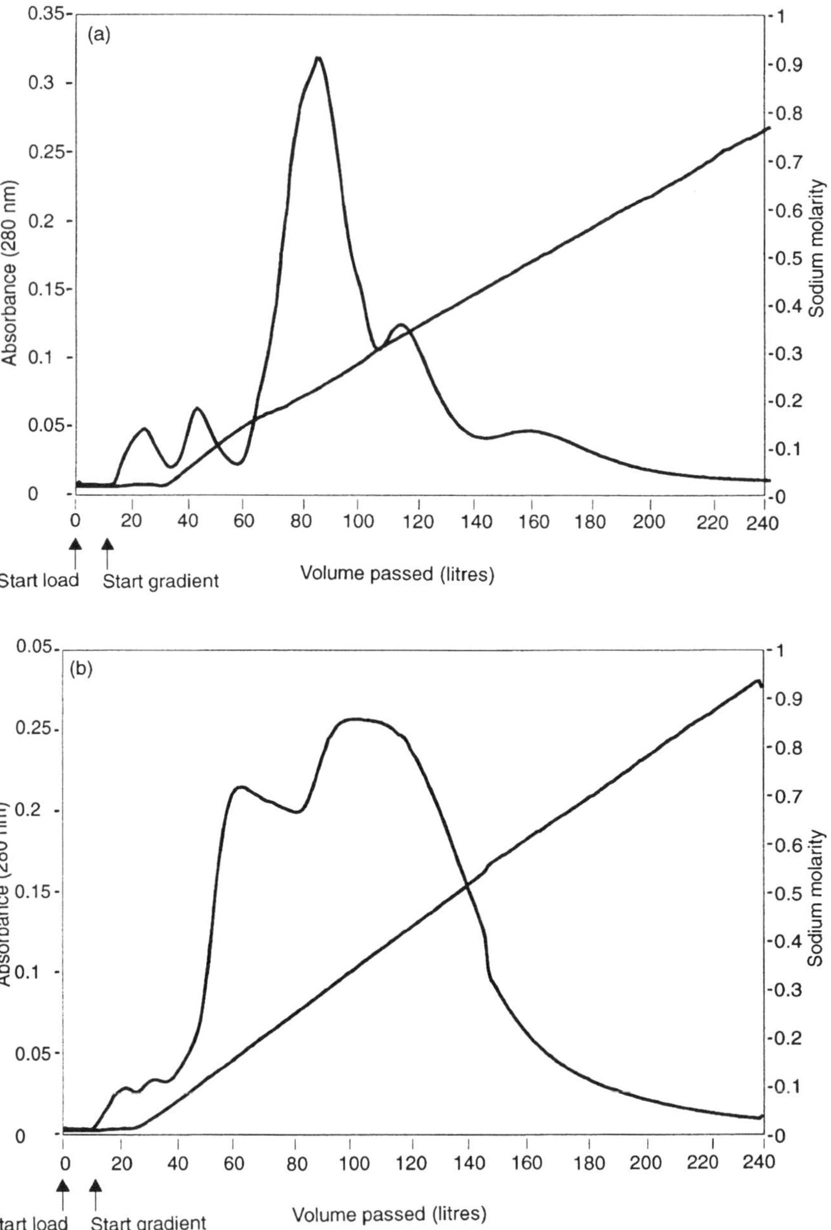

Figure 12.4. The analytical-scale chromatography of 10.25 mg/mL lysozyme-depleted hen egg-white proteins on Whatman Express-Ion C in a 25 L column (45 cm i.d. × 16 cm) using 0.01 M sodium citrate buffer, pH 5.0 at a flowrate of 150 cm/h. (a) Run 1 — before process-scale chromatography, (b) run 3 — after process-scale chromatography, pre-CIP and (c) run 5 — after CIP

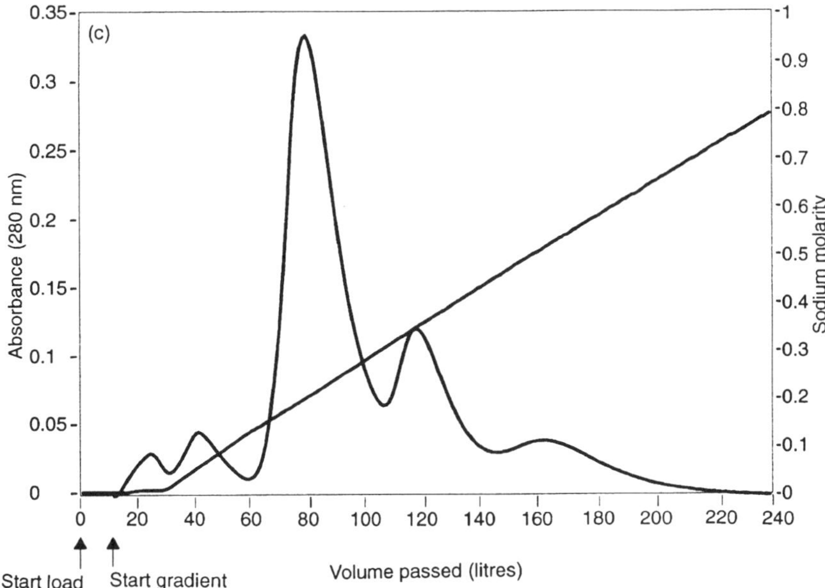

Figure 12.4 (*continued*)

using the cation exchange cellulose Whatman Express-Ion C. In this study the feedstock contained 10.25 mg/mL total protein and the lysozyme component was reduced by pretreatment with Whatman SE-cellulose/D529. The strategy for this process-scale chromatography is summarized in Table 12.3. In this study, the feedstock 10 L or 900 L, for analytical or preparative loadings respectively, was adjusted to pH 5.0 with citric acid and loaded on to the 25 L column (45 cm i.d. × 16 cm) of Express-Ion C previously equilibrated with 0.01 M sodium citrate buffer, pH 5.0. After washing with equilibration buffer, proteins were eluted using a gradient of 0–1 M NaCl in 0.01 M sodium citrate buffer, pH 5.0. The CIP protocol involved storage overnight in 0.5 M NaOH. The process-scale chromatogram is represented in Figure 12.3 and the protein mass balance data is summarized in Table 12.4. The data in Table 12.4 reveal that of the 9.23 kg protein applied to the column only 1.57 kg is bound and, furthermore, desorption gave a recovery of 96.8%. Using such a high protein loading, one might anticipate media fouling to have occurred.

The analytical chromatogram (run 1) before the process-scale chromatogram is represented in Figure 12.4(a). Following the process loading the analytical separation (run 3) of egg-white proteins is significantly impaired (Figure 12.4b). This is most likely due to fouling of the Express-Ion C following a severe protein overload. Following a CIP regime the analytical separation (run 5) of egg-white protein is improved and the chromatogram (Figure 12.4c) is similar to that obtained prior to the process loading (Figure 12.4a).

12.6 Column Regeneration

On the basis of the data reported above media fouling is not always apparent in terms of a drop in protein capacity but may be manifest by altered resolving power of the medium which could affect purity of the eluted product. In order to restore column performance it is necessary to implement a column regeneration (CIP) procedure. Since the effectiveness of a CIP regime is dependent on the nature of the fouling, and this is feedstock and thus application dependent, there is no universal CIP procedure. On the basis of our investigations and experiences it is felt that conditions that are effective for media regeneration may be as follows:

(a) ionic strength of up to 2 M,
(b) elevated pH up to 2 M NaOH,
(c) elevated temperatures of up to 50 °C,
(d) reduced pH down to 1 M HCl,
(e) use of alcohols, including ethanol and propan-2-ol,
(f) use of detergents, including Triton and Tween,
(g) use of denaturants, including urea and guanidine hydrochloride.

These conditions often used in combination may be useful starting points for a CIP procedure.

12.7 Summary

In this short overview of process-scale ion exchange chromatography, we have highlighted several areas of the chromatographic process that can have significant effects on the effectiveness and efficiency of the process. These areas include aspects of feedstock preparation, media handling, column packing, and identifying and rectifying media fouling. It is suggested that these aspects be considered during the early stages of process development such that scale-up becomes a relatively straightforward operation.

12.8 References

1. D. Freifelder (1982) *Physical Biochemistry*, W. H. Freeman, San Francisco.
2. E. F. Rossomando (1990) 'Ion-exchange chromatography', *Methods Enzymol.*, **182**, 309–17.
3. P. R. Levison (1993) 'Cellulosics as ion-exchange materials', in J. F. Kennedy, G. O. Phillips and P. A. Williams (Eds.), *Cellulosics: Materials for Selective Separations and Other Technologies*, Ellis Horwood, Chichester, pp. 77–84.
4. P. R. Levison, S. E. Badger, D. W. Toome, M. L. Koscielny, L. Lane and E. T. Butts (1992) 'Economic considerations important in the scale-up of an ovalbumin separation from hen egg-white on the anion-exchange cellulose DE92', *J. Chromatogr.*, **590**, 49–58.
5. P. R. Levison (1993) 'Techniques in process-scale ion-exchange chromatography', in G. Ganetsos and P. E. Barker (Eds.), *Preparative and Production Scale Chromatography*, Marcel Dekker, New York, pp. 617–25.

6. P. R. Levison, D. W. Toome, S. E. Badger, B. N. Brook and D. Carcary, (1989) 'Influence of mobile phase composition on the adsorption of hen egg-white proteins to anion-exchange cellulose', *Chromatographia*, **28**, 170–8.

7. N. H. Parker (1964) 'Modern theory and practice on the universal operation . . . mixing', *Chem. Engng*, June, 166–220.

8. Corning (1991) *Pilot Plant Reactors — Technical Data*, Corning Ltd, Stone, UK.

9. L. Lane, M. L. Koscielny, P. R. Levison, D. W. Toome and E. T. Butts (1990) 'Chromatography of hen egg-white proteins on anion-exchange cellulose at high flow rates using a radial flow column', *Bioseparation*, **1**, 141–7.

10. P. R. Levison, S. E. Badger, D. W. Toome, E. T. Butts, M. L. Koscielny and L. Lane (1991) 'Influence of column design on chromatographic performance of ion-exchange media', in A. Huyghebaert and E. Vandamme (Eds.), *Upstream and Downstream Processing in Biotechnology*, Vol. III, Royal Flemish Society of Engineers, Antwerp, pp. 3.21–3.28.

11. L. Stryer (1981) *Biochemistry*, W. H. Freeman, San Francisco.

12. P. R. Levison, M. L. Koscielny and E. T. Butts (1990) 'A simplified process for large-scale isolation of IgG from goat serum', *Bioseparation*, **1**, 59–67.

13. P. R. Levison, S. E. Badger, D. W. Toome, D. Carcary and E. T. Butts (1990) 'Performance of anion-exchange cellulose at process-scale', in D. L. Pyle (Ed.), *Separations for Biotechnology*, Vol. 2, Elsevier Applied Science, London, pp. 381–9.

14. P. R. Levison, S. E. Badger, D. W. Toome, M. Streater and J. A. Cox (1994) 'Process-scale evaluation of a fast-flowing anion-exchange cellulose', *J. Chromatogr.*, **658**, 419–28.

15. P. R. Levison, S. E. Badger, D. W. Toome, D. Carcary and E. T. Butts (1989) 'Studies on the reuse of anion-exchange cellulose at process scale', in R. de Bruyne and A. Huyghebaert (Eds.), *Downstream Processing in Biotechnology*, Vol. II, Royal Flemish Society of Engineers, Antwerp, pp. 2.11–2.16.

13 AFFINITY CHROMATOGRAPHY

Steven J. Burton

13.1 Introduction

Protein extracts from living organisms are invariably very complex, whether obtained from whole cell lysates or cell culture supernatant, and virtually all require a significant degree of purification if a homogeneous protein is to be obtained. Chromatography has played a central role in protein purification where techniques such as ion exchange and gel permeation chromatography have traditionally formed the mainstay of preparative protein fractionation methods. However, owing to its high selectivity, affinity chromatography is rapidly superseding traditional protein purification methods, particularly in instances where very high levels of purification are required.

In its simplest form, affinity chromatography relies on the ability of proteins and other biological macromolecules to interact with certain compounds (ligands) in a selective and reversible manner. If a ligand is bonded on to a solid support matrix, the resulting affinity adsorbent may be used to selectively bind the complementary protein to the exclusion of contaminating proteins and other molecules (Figure 13.1). Once protein binding has taken place, the adsorbent is washed to ensure removal of residual contaminants and the purified protein subsequently eluted by adjusting the column conditions to diminish the protein's affinity for the immobilized ligand. The end result is normally a concentrated solution of highly purified protein.

The inherent simplicity and flexibility of this approach has led to a diverse array of separation technologies which utilize affinity ligands, though a packed column of adsorbent particles remains the most favoured format for production purposes. The starting point for affinity purification is normally a clarified solution containing soluble protein. Affinity chromatography (also known as bioselective adsorption chromatography) offers several advantages when used for preparative applications. In addition to the very high levels of

Downstream Processing of Natural Products. Edited by Michael S. Verrall
©1996 John Wiley & Sons Ltd

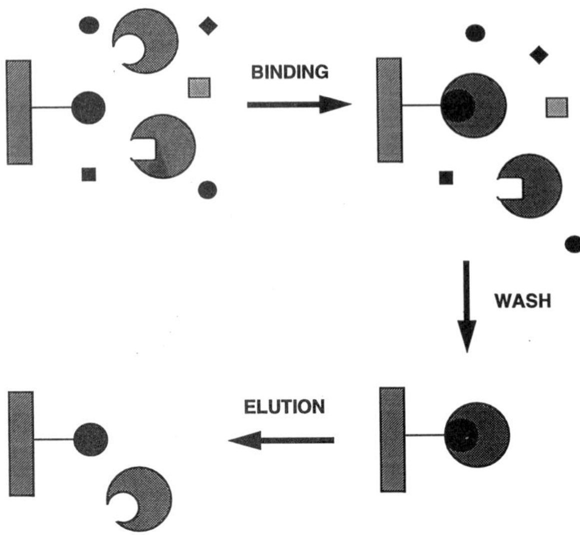

Figure 13.1   Schematic diagram of affinity chromatography

purification that are possible ($> 10^6$-fold in some instances [1]), the purified
protein is frequently concentrated as well. Furthermore, recoveries of purified
protein and protein activity are often greater for affinity separations owing to
the relatively mild conditions that are commonly (but not always) required for
binding and elution, the stabilizing effect of the protein-ligand complex and the
overall reduction in the number of process steps and concomitant reduction in
process losses. This latter point is particularly relevant since reduction of the
overall number of processing steps can have a dramatic effect on process
economics.

In some instances affinity chromatography can provide one-step purification
methods, though more usually it is used in combination with other techniques.
In a survey of 100 published papers on protein separations, affinity
chromatography was used in 60% of the examples studied [2]. Of the processes
that used affinity chromatography, 45% resulted in maximal purification.
Rather ironically, the literature abounds with examples where affinity
chromatography has been used as a final polishing step to remove residual
contaminants, as opposed to a major purification step at the front end of the
process where maximum benefit is gained from the technique's resolving
power. This situation has often been enforced owing to the relatively high cost
and low stability of affinity adsorbents, especially those based on natural
biological ligands, which are frequently very expensive and susceptible to both
chemical and biological (enzymic) degradation. However, in instances where
stable affinity adsorbents with suitable binding properties are available,
relatively large columns have been used to perform the principal purification
step [3, 4]. Judicial choice of affinity adsorbent is therefore of vital importance.

13.2 Preparation of Affinity Adsorbents

Affinity adsorbents may be considered to consist of three basic elements: the affinity ligand, the support matrix and the bonding chemistry used for ligand immobilization. All three components have a major influence on the properties of the composite affinity adsorbent. Some of the more important parameters include selectivity/product purity, capacity, recovery, ease of binding and elution, stability and flowrate.

13.2.1 AFFINITY LIGANDS

Almost any compound may be used as an affinity ligand provided that it binds selectively and reversibly to the protein in question and can be immobilized to a solid support without impairing its protein binding properties. Historically a wide variety of compounds have been used (Table 13.1). The specificity of the protein–ligand interaction ultimately determines the degree of purification that can be achieved. Thus highly specific antibody–antigen interactions with dissociation constants (K_d's) in the range 10^{-8}–10^{-10}M have the potential to deliver higher degrees of purification as compared to immobilized co-enzymes, which typically have K_d's in the region of 10^{-6}M. However, very tight binding can be disadvantageous in that the bound protein may be difficult to elute without causing permanent damage. Similarly, very weak binding is disadvantageous as proteins are merely retarded and soon appear in the column eluate. Consequently there is an optimal K_d range of 10^{-4}–10^{-8}M.

There is also an optimal concentration of immobilized ligand which, in practice, can vary enormously depending upon the nature of the ligand and its

Table 13.1 Examples of compounds used as affinity ligands

Ligand	Target protein
Metal ions	Proteins with an abundance of His, Trp and Cys residues
Reactive dyes	Most proteins with ligand binding pockets
m-Aminophenylboronic acid	Glycoproteins
Lectins	Glycoproteins
Carbohydrates	Lectins
Nucleic acids	Nuceleases, polymerases and other DNA and RNA binding proteins
Nucleotides	Nucleotide dependent enzymes
Amino acids	Proteases
Glutathione	Glutathione-s-transferase/GST fusion proteins
Heparin	Clotting factors, antithrombin III
Protein A/protein G	Immunoglobulins
Antibodies	Antigens
Antigens	Antibodies
Hormones/drugs	Receptor proteins

affinity for the target protein. Too little ligand results in low capacities whilst too much ligand can also serve to reduce capacity owing to steric hindrance and increased non-specific binding (particularly for highly charged or very hydrophobic ligands). In the case of dye–ligand adsorbents, immobilized ligand concentrations of 2–5 µmol/ml are often optimal, whilst aminophenyl-boronic acid, structurally similar to the sulfonated aromatic rings of reactive dyes, must be present at concentrations an order of magnitude higher for efficient binding of glycoproteins [5]. The optimal ligand loading, normally ascertained by experimentation, is achieved either by controlling the extent of matrix activation or limiting the amount of ligand in the coupling mixture.

For small ligands bound in deep clefts, as in the case of *p*-aminobenzamidine binding to trypsin, the ligand must be immobilized to the support matrix by a spacer arm to allow interaction with the binding site to take place [6]. Spacer arms are normally 6–12 units (CH_2 or O) long, the longer length chains often incorporating hydroxyl groups to increase hydrophilicty. For optimally immobilized affinity ligands, capacities approaching 50 mg protein per ml adsorbent are possible [7].

Ligand stability is particularly important if the adsorbent is to be used for industrial applications. Ideally the ligand should be sufficiently robust to withstand chemical immobilization, subsequent treatment with buffers across a wide pH range and reagents commonly used for column cleaning and sanitization (e.g. ethanol, urea, ethanoic acid and sodium hydroxide). In general natural compounds (such as antibodies and co-factors) are not as stable as synthetic organic compounds which are now commonly used in industrial-scale separations. Synthetic dye ligands are a good example since these molecules have excellent stability, reasonably good selectivity and have been applied to a very wide range of protein separations, often on an appreciable scale [3, 8].

13.2.2 SUPPORT MATRICES

Unlike conventional high-performance liquid chromatography, plate theory does not really apply to affinity chromatography and selection of support matrices for use in affinity separations is based on rather different criteria. Of particular importance are ligand accessibility, the absence of non-selective binding interactions and an abundance of functional groups which can be utilized for ligand immobilization. A support matrix with large open pores is advantageous since this increases the probability of the target protein encountering the immobilized ligand and adopting an appropriate orientation for ligand binding. A neutral hydrophilic surface is also important since any hydrophobic or ion exchange interactions associated with the matrix surface can promote the non-selective binding of contaminants (such as unwanted proteins, pigments and nucleic acids) and so reduce the extent of purification and column capacity for the target protein. Background hydrophobic or ionic interactions can also result in adsorption of the immobilized affinity ligand on to the matrix surface, making the ligand unavailable for protein binding. This situation can be encountered when relatively hydrophobic ligands are attached to coated polymer beads possessing a rigid hydrophobic core.

A wide variety of support matrices for use in column chromatography are now available commercially, including materials based on polysaccharides (e.g. dextran, agarose and cellulose), synthetic polymers (e.g. acrylamide, hydro-xyethylmethacrylate, styrenedivinylbenzene) and various inorganic compounds (e.g. silica and glass). Crosslinked beaded agarose is one of the most widely used support materials for the construction of affinity adsorbents since it has an open porous network, very low non-specific binding properties and is readily derivatized using a variety of chemical procedures. In direct comparative trials, affinity adsorbents based on agarose often exhibit substantially higher protein-binding capacities compared to those based on synthetic polymer beads [7]. Agarose is also relatively inexpensive and resistant to concentrated sodium hydroxide, enabling the use of 1M NaOH as a column cleaning, sanitization and depyrogenation agent [9].

In a move towards faster protein separations, columns containing small particles (5–20 µm) operated at high linear flowrates (> 10 cm/min) have received much attention in recent years. However, whilst significant improvements in productivity are possible for high-performance ion exchange chromatography using small ridged particles and very high flowrates, the same is not always true of affinity separations owing to the slower binding kinetics which are often encountered for affinity systems (particularly when the protein is only weakly bound to the immobilized ligand) and the lower adsorbent capacities that are often observed when 'hard' synthetic polymer beads are compared to 'soft' agarose beads. Consequently, affinity columns in down-stream processes of any significant scale currently tend to use larger-particle-sized (100 µm) soft gels in combination with short-column bed lengths to allow higher flowrates. Radial flow columns are ideally suited for this purpose [10] (see also Chapter 14).

13.2.3 COUPLING CHEMISTRIES

Selection of the optimal bonding chemistry for a particular combination of affinity ligand and support matrix is of vital importance if stable, high-capacity adsorbents are to be produced. A wide variety of bonding chemistries have been developed, each having its strengths and weaknesses, which must be considered in combination with the nature of the ligand to be immobilized and the support matrix.

With the exception of reactive dye ligands, most ligands cannot be reacted directly with the support matrix and some form of matrix activation chemistry is required to either facilitate reaction with the ligand or to enable the introduction of particular functional groups which can subsequently react with

Table 13.2   Activation techniques for hydroxylic support matrices

Activating agent	Reactive group on ligand	Reference
Cyanogen bromide	Primary amines	15, 17
Tresyl chloride	Primary amines, thiols	57
Tosyl chloride	Primary amines, thiols	58
Epichlorohydrin	Primary amines, thiols, hydroxyls	59
1,4-Butanedioldiglycidyl ether	Primary amines, thiols, hydroxyls	21
1.1'-Carbonyldiimidazole	Primary amines, hydroxyls,	60
Cyanuric chloride	Primary amines, hydroxyls, thiols	61
Divinylsulfone	Primary amines, hydroxyls	62
2-Fluoro-1-methylpyridinium toluene-4-sulfonate	Primary amines, thiols	63

the ligand. In the case of hydrophilic 'soft gels' such as agarose and cellulose, almost all matrix activation techniques involve reaction with primary or secondary hydroxyl groups on the surface of the matrix (Table 13.2). Such procedures cannot be adopted for materials that lack hydroxyl groups such as silica, PTFE (polytetrafluoroethylene), nylon or polyacrylamide, and alternative approaches to matrix activation must be used [11–13]. In general, the procedures used to activate support matrices result in the introduction of functional groups which are reactive towards nucleophiles such as primary amines, primary hydroxyls and thiols. Virtually all proteins and the majority of compounds commonly used as affinity ligands contain these functionalities and so a wide variety of coupling chemistries can be considered. In cases where the ligand does not possess amine, hydroxyl or thiol groups, alternative functional groups can be used. For example, carboxyl groups can be coupled to immobilized primary amines by carbodiimide-mediated condensation [14].

One of the first activation chemistries to be developed involves the use of cyanogen bromide to introduce reactive cyanate ester groups on to the surface of a hydroxylic support matrix [15, 16]. Cyanogen bromide activated supports have been widely used for the immobilization of compounds containing primary amino groups, particularly proteins. One reason for the popularity of cyanogen bromide coupling chemistry is its good reactivity, allowing ligand immobilization under relatively mild conditions (pH 8.0–10.0; 4–37 °C). However, the resulting imidocarbonate bond is notoriously unstable and prone to hydrolytic cleavage at both acid and alkaline pH [17]. In view of the requirement for low leakage adsorbents, particularly in the production of health care products, the use of cyanogen bromide coupling is declining.

A variety of active ester coupling techniques have been developed in which active esters (such as *N*-hydroxysuccinimide esters) are reacted with amine-containing compounds to form amide bonds [18]. Reactions with active esters can be performed under relatively mild conditions and the resulting bonds are usually more stable than for cyanogen bromide immobilization, though bond breakage and ligand leakage is still encountered at extremes of pH [19].

Some of the most stable coupling methods are those that yield secondary amine, ether or thioether bonds (e.g. epoxide, tosyl chloride, divinylsulfone). Whilst such chemistries often have excellent bond strengths and are often resistant to 1M NaOH and 1M HCl, more aggressive conditions are sometimes required for efficient immobilization (particularly in the case of epoxide bonding). Consequently, the lability of the affinity ligand often dictates which bonding chemistry can be used; however, where a choice is available it is always preferable to choose the method that yields the most stable bond.

Larger ligands may have many exposed groups capable of reacting with activated supports and, as a consequence, are oriented randomly following immobilization. If a particular ligand orientation is required, the bonding chemistry must be selected to favour attachment at a single point. In the case of antibodies and other glycoproteins, immobilization can be effected through carbohydrate moieties rather than lysine amino groups by use of sodium periodate oxidation followed by coupling to either hydrazide agarose or amino agarose, the latter in the presence of sodium cyanoborohydride [20]. Small primary amino- or thiol-containing ligands requiring immobilization via a spacer arm may be conveniently coupled to supports activated with 1,4-butanedioldiglycidyl ether which provides a flexible, stable and hydrophilic spacer chain [21].

To achieve the desired immobilized ligand concentration, either the extent of matrix activation or the concentration of ligand in the coupling reaction mixture should be limited. If the ligand is relatively inexpensive and freely available then it is probably better to control the extent of matrix activation and derivatize all active groups using an excess of ligand. Conversely, if the ligand is expensive or scarce, it is preferable to use a higher level of matrix activation and limit the amount of ligand in the coupling mixture. However, in the latter case, the surplus active groups that remain after ligand immobilization should be inactivated or blocked by reacting with a small, hydrophilic compound such as ethanolamine or 2-mercaptoethanol. The nature of the blocking group will vary depending of the activation chemistry used and the stability of the immobilized ligand.

13.3 Use of Affinity Chromatography in Downstream Processing

Affinity chromatography may be used at any point in a purification process, either as an initial capture/concentration step, an intermediate high resolution step or as a final polishing step. Furthermore, affinity adsorbents may be used to either bind the target protein to the exclusion of contaminants (positive binding) or to selectively bind certain contaminants and allow the target protein to pass through unretarded (negative binding). In most cases affinity adsorbents are packed into columns and used in conventional chromatographic mode. This approach enables residual contaminants to be quickly flushed from the column and the bound protein eluted in a minimal volume. However, when the crude protein mixture is heavily contaminated with particulates (e.g. in the case of direct binding from unclarified cell culture media or fermentation broth), it is advantageous to perform the binding and washing steps with the adsorbent in the form of a stirred slurry (batch mode) or liquid fluidized bed [22]. Affinity separations may also be performed with ligands attached to hydrophilic polymers and used in aqueous two-phase affinity partitioning [23] or attached to water immiscible polymers and used to form affinity emulsions [24].

13.3.1 ADSORBENT SELECTION

Selection of an affinity adsorbent for a particular separation is largely dictated by the availability of immobilized affinity ligands with appropriate protein-binding properties. In many instances ligand selection will be obvious since the protein to be purified will have documented binding sites with affinities for particular compounds. Thus immobilized substrates, cofactors, activators, inhibitors and reaction products may be used to purify enzymes; immobilized hormones and agonists may be used to purify receptor proteins etc. Various structural features may also be used as the basis of an affinity separation. For example, glycoproteins can be purified via their carbohydrate groups using immobilized lectins [25] or aminophenylboronic acid agarose [5]. Proteins rich in surface histidine, tryptophan or cysteine residues can be purified by immobilized metal ion affinity chromatography [26]. Immobilized antibodies are commonly used for protein purification by 'immunoaffinity chromatography' [27]. Since antibodies (both monoclonal and polyclonal) can be raised against most proteins, this technique can be considered to be a universal affinity technique. However, raising suitable antibodies can be a very expensive, difficult and time consuming task, and, once immobilized, antibodies are generally unstable, difficult to clean and have short operational lifetimes. Furthermore, since antibodies normally bind antigens relatively tightly (typical K_d's of 10^{-8}–10^{-10}), it can be difficult to elute proteins without inflicting permanent damage on the bound protein and/or immobilized antibody.

As an alternative to naturally occurring biochemical ligands, stable synthetic ligands such as immobilized dyes are becoming increasingly popular [28–30]. These compounds fulfil the role of general affinity ligands, one ligand structure having the ability to bind several different and often unrelated proteins. A very wide range of proteins can be purified with immobilized dye affinity adsorbents, the size and shape of the dye coupled with the nature and distribution of charge, hydrophobic and hydrogen bonding groups determining which proteins are bound [29, 30]. In the absence of prior knowledge, the best immobilized dye ligand for a particular protein separation may be determined by screening a suitably diverse array of adsorbents for appropriate protein binding properties [28]. Screening kits containing prepacked columns of adsorbent are now available commercially (e.g. PIKSI kits from Affinity Chromatography Ltd). The screening procedure is relatively quick and has the advantage that little if any structural information on the target protein is required for development of an optimized purification protocol. This makes dye ligand adsorbents ideal for the purification of previously unknown proteins or for the purification of proteins with no obvious ligand-binding sites. Furthermore, in view of their low cost and excellent stabilities, they are ideally suited to use in large-scale separation processes [3, 4].

13.3.2 BINDING AND ELUTION

Protein binding is normally performed with a clarified (0.45 µm filtered or centrifuged) extract adjusted to the pH, ionic strength and polarity required to encourage interaction of the target protein with the adsorbent but exclude the binding of contaminants. The packed column of adsorbent is generally pre-equilibrated in an appropriate buffer, the composition of which can vary considerably depending on the nature of the protein–ligand interaction, the composition of the applied mixture and the stability of the target protein. In many instances, slow flowrates in the region of 10–50 cm/h have to be used for binding owing to diffusional limitations and kinetic limitations. Temperature can have a variable effect on binding and a protein's affinity for the immobilized ligand can often be modulated by temperature changes, lower temperatures tending to promote binding in the case of predominantly ionic interactions and lessen binding in the case of predominantly hydrophobic interactions. In most cases adequate binding can be achieved at ambient temperature, though lower temperatures (4 °C) may have to be used for unstable proteins.

Elution of bound proteins may be performed by selective or non-selective methods. With non-selective elution, the column conditions are adjusted to neutralize predominant binding interactions. Consequently, proteins bound predominantly by ionic interactions are eluted by changes of pH and/or the addition of salts to raise the ionic strength, whereas proteins bound largely by hydrophobic interactions are eluted by the addition of polarity-reducing agents such as ethylene glycol. Other elution techniques include changes of temperature [31], pressure [32] and electric fields [33], though such techniques are not commonly employed.

Selective elution with a competing ligand normally provides higher levels of purification, particularly if more than one protein has bound to the adsorbent. For maximal effect, the competing ligand should interact with the same site on the protein as the immobilized ligand and should have a higher affinity. In some instances it is possible to fractionate a mixture of bound proteins by use of competing ligands with differing selectivities to the immobilized ligand [34]. The cost of suitable competing ligands can be prohibitively expensive and further processing steps may be required to remove the competing ligand from the purified protein. For these reasons, non-selective elution techniques are often chosen for process scale applications.

It is good practice to clean affinity adsorbents prior to re-use, particularly if very crude mixtures are passed through the column. Failure to clean a column adequately can render it useless. Various cleaning regimes may be employed, strongly alkaline solutions such as 1M NaOH being amongst the most effective, though not all adsorbents are sufficiently robust to withstand this treatment [9, 35]. Alternative methods such as the use of high and low pH buffers, high salt concentrations, organic solvents or chaotrope solutions (such as 8M urea) may also be effective in regenerating affinity adsorbent columns. Many of these treatments are also effective sanitization and depyrogenation agents, particularly sodium hydroxide. Where column hygiene is of paramount importance (in the purification of therapeutic products for example) a column sanitization/depyrogenation step must be performed prior to use. To prevent microbial growth during long-term storage a bacteriostatic solution such as 20% ethanol should be employed. Storage at 4 °C is also beneficial in this respect.

13.3.3 EXAMPLES

Affinity chromatography has been used extensively for the purification of proteins and other biological molecules and a thorough review of these applications is beyond the scope of this chapter. For further information on the applications of affinity chromatography the reader should consult one of the many comprehensive review articles on this subject [36–38].

One of the main applications of affinity separations has been the isolation of proteins and enzymes which are present in very low concentrations and require extensive purification to reach homogeneity. For this reason intracellular enzymes are commonly purified by affinity chromatography. A variety of affinity ligands have been used, including immobilized substrates, inhibitors, cofactors activators, transition state complexes and synthetic dye ligands which mimic the binding of natural compounds. In one example, morphine dehydrogenase from *Pseudomonas putida* was purified 1216-fold in two steps by affinity chromatography on Mimetic™ Orange 3 A6XL and Mimetic™ Red 2 A6XL [39]. This application is notable in that it represents the first recorded purification of this particular morphine dehydrogenase, which constitutes less than 0.1% of the total cell protein, and was purified to homogeneity with an 84% recovery of activity. A related enzyme, morphinone reductase, was isolated from the same source in a single step using Mimetic™ Yellow 2 A6XL, which provided a 90-fold purification factor and a homogeneous product [40] (Figure 13.2). Other examples of enzyme purification using synthetic alternatives to natural products include trypsin purification from bovine pancreas using immobilized *p*-aminobenzamidine, an

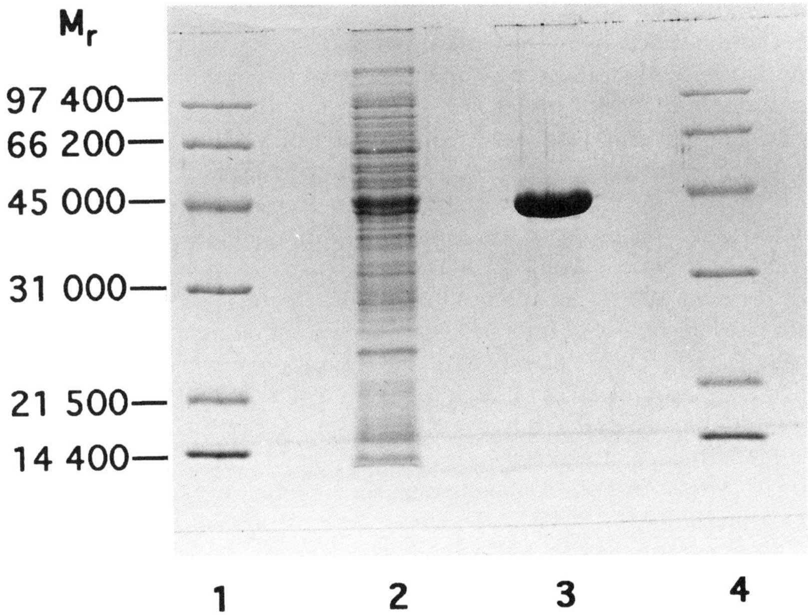

Figure 13.2 Purification of *P. putida* Morphinone reductase by affinity chromatography on Mimetic Yellow 2 A6XL. SDS polyacrylamide gel electrophoresis showing (1) molecular weight markers, (2) crude cell extract, 16 µg protein, (3) purified morphinone reductase, 1.5 µg, (4) molecular weight markers. (Reproduced by permission of the Biochemical Society and Portland Press from Ref. 40)

analogue of arginine which is the natural substrate for this enzyme [6]. Alkaline phosphatase from calf mucosa has also been purified to homogeneity in a single step using an immobilized phosphate analogue incorporated into a synthetic dye ligand [41]. This adsorbent is now available commercially.

Many plasma proteins are now of significant medical and commercial importance and the isolation of highly purified plasma proteins is now performed on an appreciable scale, both from mammalian plasma and from genetically engineered organisms. Affinity chromatography is ideally suited to the purification of plasma proteins since it provides the high levels of purification required, as an adsorption technique it often concentrates the target protein and can be operated on a very large scale, and there is an abundance of compounds that may be used as affinity ligands for plasma protein purification. Some examples used in commercial operations include the purification of plasminogen on immobilized lysine (Figure 13.3), the isolation of antithrombin III with immobilized heparin [42] and the purification of human serum albumin with immobilized CI Reactive Blue 2 [43]. Other blood proteins such as clotting factors and haemoglobin have also been purified by affinity chromatography.

Immunoglobulins are an important group of plasma proteins which are commonly purified by affinity chromatography. Early methods relied on the natural affinity of antibodies for their antigens. Although elution of antibodies from immobilized antigens can be difficult due to the tight binding involved, this technique is still widely used for isolation of a particular antibody from a polyclonal antiserum [27, 44]. Where the purification of particular antibody types (IgG, IgM, etc.) from other contaminating proteins is required, certain bacterial cell wall proteins which bind immunoglobulins selectively, such as protein A (isolated from *Staphylococcus*) and protein G (isolated from *Streptococcus*), are frequently used as affinity ligands [45, 46]. Protein A tends to bind a wide variety of immunoglobulins from a number of different sources, though not all IgGs are bound. Protein G, on the other hand, binds most IgGs from most plasma sources, though it has little or no affinity for other classes of immunoglobulins. Both protein A and protein G are relatively expensive and unstable, which has encouraged the development of a variety of synthetic alternatives [47, 48]. It is anticipated that synthetic ligands will eventually replace protein A/G in view of their lower cost and increased chemical and biological stability.

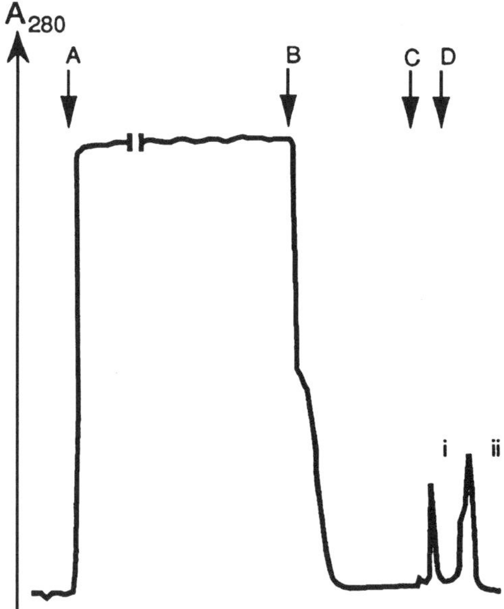

Figure 13.3. Purification of plasminogen from human plasma on Lysine Agarose 6XL. Column: 2.6 cm dia. × 18.6 cm (100 ml bed. vol.); flow rate: 1.0 ml/min; applied solutions: (A) 1 litre human plasma diluted 1:1 with 50 mM potassium phosphate buffer, pH 8.0; (B) 50 mM potassium phosphate buffer, pH 8.0; (C) 0.3 M potassium chloride in 50 mM potassium phosphate buffer, pH 8.0; (D) 0.2 M aminocaproic acid. Eluted peaks: (i) salt eluted contaminants; (ii) plasminogen (Reproduced by permission of Affinity Chromatography Ltd)

Advances in genetic engineering techniques over the last decade allow the introduction of affinity labels or 'tails' into genetically engineered proteins by use of gene fusion. The resulting fusion protein can then be isolated by virtue of the affinity tail, which is subsequently cleaved to yield the native protein. A variety of affinity tails have been used including protein A [49] and glutathione-s-transferase [50], which allow isolation of fusion proteins with IgG agarose and glutathione agarose respectively.

Biotinylation reagents provide another mechanism for introducing affinity tags on to proteins as an aid to selective isolation from complex mixtures. Reagents such as N-hydroxysuccinimidobiotin (NHS-biotin) provide a convenient means of attaching biotin groups to surface amino groups on the protein [51]. Once biotinylated, the protein may be selectively captured using immobilized avidin or streptavidin [52]. In view of the very high affinity between biotin and avidin ($K_d = 10^{-15}$M) binding is almost irreversible. In some instances this can be advantageous as it provides a means of 'immobilizing' sensitive compounds under very mild conditions. However, by use of immobilized monomeric avidin which binds biotin with a reduced affinity [53], it is possible to elute biotinylated proteins and thereby provide a means of purification. The biotin–avidin system has found numerous applications where biotinylated proteins are removed rapidly and with high efficiency and selectivity from biological fluids [54]. For a more comprehensive discussion of affinity tails see Chapter 7.

13.4 Future Prospects

One of the major limitations to increased use of affinity chromatography in downstream processing is the availability of suitable affinity ligands. Rather than using natural compounds and known analogues as affinity ligands, it is now possible to design synthetic ligands that are targeted at particular sites on a protein [55]. Increasingly sophisticated molecular modelling packages and a wider availability of protein structural data enable affinity ligands to be modelled and optimized without the need for extensive chemical synthesis and evaluation. A number of entirely new affinity ligands have been developed using this approach which offers great promise for the development of novel adsorbent materials for the purification of proteins which are not currently amenable to purification by affinity techniques [30, 55, 56].

13.5 References

1. R. H. Allen and P. W. Majerus (1972) *J. Biol. Chem.*, **247**, 7209–17.
2. J. Bonnerjea, S. Oh, M. Hoare and P. Dunnill (1986) *Biotechnology*, **4**, 954–8.
3. Y. D. Clonis (1987) *Biotechnology*, **5**, 1290–3.
4. J.-C. Janson (1984) *Trends in Biotechnol.* **2**, 31–8.
5. M. J. Benes, A. Stambergova and W. H. Scouten (1993) in T. T. Ngo (Ed.), *Molecular Interactions in Bioseparations*, Plenum Press, New York, pp. 313–22.

6. Hixson H. F. and A. H. Nishikawa (1973) *Arch. Biochem. Biophys.*, **154**, 501.
7. D. J. Stewart, D. R. Purvis, J. M. Pitts and C. R. Lowe (1992) *J. Chromatogr.* **623**, 1–14.
8. M. D. Scawen and T. Atkinson (1987) in Y. D. Clonis, A. Atkinson, C. J. Bruton and C. R. Lowe (Eds.), *Reactive Dyes in Protein and Enzyme Technology*, Macmillan Press Ltd, London, pp. 51–85.
9. G. K. Sofer and L. E. Nystrom (1989) *Process Chromatography — A Practical Guide*, Academic Press Ltd, London, pp. 93–105.
10. V. Saxena and M. Dunn (1989) *Biotechnology*, 7, 250–5.
11. J. K. Inman and H. M. Dintzis (1969) *Biochemistry*, **8**, 4074–82.
12. C. R. Lowe, M. Glad, P.-O. Larsson, S. Ohlson, D. A. P. Small, T. Atkinson and K. Mosbach (1981) *J Chromatogr.*, **215**, 303–16.
13. D. J. Stewart, P. Hughes and C. R. Lowe (1989) *J. Biotechnol.*, **11**, 253–66.
14. D. G. Hoare and D. E. Koshland (1967) *J. Biol. Chem.*, **242**, 2447–53.
15. R. Axen, J. Porath and S. Ernbach (1967) *Nature*, **214**, 1302.
16. J. Porath, R. Axen and S. Ernbach (1967) *Nature*, **215**, 1491.
17. C. R. Lowe (1979) in T. S. Work and E. Work (Eds.), *Laboratory Techniques in Biochemistry and Molecular Biology*, Vol. 7, North-Holland, Amsterdam, pp. 267–522.
18. M. Wilcheck, T. Miron and J. Kohn (1984) *Methods in Enzymol.*, **104**, 3.
19. M. Wilcheck and T. Miron (1987) *Biochemistry*, **26**, 2155–61.
20. D. J. O'Shannessy (1990) *J. Chromatogr.*, **510**, 13–21.
21. L. Sundberg and J. Porath (1974) *J. Chromatogr.*, **90**, 87–98.
22. H. A. Chase and N. M. Draeger (1992) *J. Chromatogr.*, **597**, 129–45.
23. P. O. Hedman and J.-G. Gustafsson (1984) *Anal. Biochem.*, **138**, 411–15.
24. G. E. McGreath, H. A. Chase, D. R. Purvis and C. R. Lowe (1992) *J. Chromatogr.*, **597**, 189.
25. I. E. Liener, N. Sharon and I. J. Goldstein (1986) *The Lectins: Properties, Functions and Applications in Biology and Medicine*, Academic Press, San Diego.
26. E. Sulkowski (1985) *Trends in Biotechnol.*, **3**, 1–7.
27. G. W. Jack and H. E. Wade (1987) *Trends in Biotechnol.*, **5**, 91–5.
28. S. J. Burton (1992) in A. Kenny and S. Fowell (Eds.), *Methods in Molecular Biology*, Vol. 11, *Practical Protein Chromatography*, The Humana Press Inc., Totowa, New Jersey, pp. 91–103.
29. Y. D. Clonis, A. Atkinson, C. J. Bruton and C. R. Lowe (1987) *Reactive Dyes in Protein and Enzyme Technology*, Macmillan Press Ltd, London.
30. C. R. Lowe, S. J. Burton, N. P. Burton, D. J. Stewart, D. R. Purvis, I. Pitfield and S. Eapen (1990) *J. Molecular Recognition*, **3**, 117–22.
31. M. J. Harvey, C. R. Lowe, D. B. Craven and P. D. G. Dean (1974) *Eur. J. Biochem.*, **41**, 335–40.
32. W. C. Olson, S. K. Leung and M. L. Yarmush (1989) *Biotechnology*, **7**, 30–5.
33. M. R. A. Morgan, N. A. Slater and P. D. G. Dean (1978) *Anal. Biochem.*, **92**, 144–6.
34. R. L. Easterday and I. M. Easterday (1974) *Adv. Exp. Med. Biol.*, **42**, 123–33.
35. E. Boschetti, X. P. Duteil, C. Nguyen and Y. Moroux (1993) *Chimicaoggi*, March/April, 29–35.
36. P. G. D. Dean, W. S. Johnson and F. A. Middle (1985) *Affinity Chromatography: A Practical Approach*, IRL Press Ltd, Oxford.
37. G. T. Hermanson, A. K. Mallia and P. K. Smith (1992) *Immobilised Affinity Ligand Techniques*, Academic Press, London.
38. C. R. Lowe and P. D. G. Dean (1974) *Affinity Chromatography*, John Wiley, London.

39. N. C. Bruce, C. J. Wilmot, K. N. Jordan, L. D. Gray-Stephens and C. R. Lowe (1991) *Biochem. J.*, **274**, 875–80.
40. C. E. French and N. C. Bruce (1994) *Biochem. J.*, **301**, 97–103.
41. N. M. Lindner, R. Jeffcoat and C. R. Lowe (1989) *J. Chromatogr.*, **472**, 227–40.
42. R. Ektorp (1980) in J. M. Curling (Ed.), *Methods of Plasma Protein Fractionation*, Academic Press, London, pp. 175–88.
43. R. Hanford, W. Maycock and L. Vallet (1978) in *Chromatography of Synthetic and Biological Polymers*, Vol. 2, Ellis Horwood, Chichester, pp. 111–19.
44. H. A. Chase (1984) *J. Biotechnol.*, **1**, 67–80.
45. B. Akerstrom, T. Brodin, K. Reis and L. Bjorck (1985) *J. Immunol.*, **135**, 2589–92.
46. H. Hjelm, K. Hjelm and J. Sjoquist (1972) *FEBS Lett.*, **28**, 73–6.
47. K. L. Knudsen, M. B. Hansen, L. R. Henriksen, B. K. Andersen and A. Lihme (1992) *Anal. Biochem.*, **201**, 170.
48. T. T. Ngo and N. Khatter (1992) *J. Chromatogr.*, **597**, 101–9.
49. T. Moks, L. Abrahmsen, B. Osterlof, S. Josephson, M. Ostling, S.-O. Enfors, I. Persson, B. Nilsson and M. Uhlen (1987) *Biotechnolgoy*, **5**, 379–82.
50. D. B. Smith (1993), *Methods in Molecular and Cellular Biology*, **4**, 220–9.
51. J. M. Becker, M. Wilcheck and E. Katchalski (1971) *Proc. Natl. Acad. Sci. USA*, **68**, 2604–7.
52. E. A. Bayer and M. Wilcheck (1980) *Meth. Biochem. Anal.*, **26**, 1–45.
53. N. M. Green and E. J. Toms (1973) *Biochem. J.*, **133**, 687–98.
54. M. D. Savage, G. Mattson, S. Desai, G. W. Nielander, S. Morgensen and E. J. Conklin (1992) *Avidin-Biotin Chemistry: A Handbook*, Pierce Chemical Co., Rockford.
55. C. R. Lowe, S. J. Burton, N. P. Burton, W. R. Alderton, J. M. Pitts and J. A. Thomas (1992) *Trends in Biotechnology*, **10**, 442–8.
56. C. R. Lowe, S. J. Burton, J. C. Pearson and Y. D. Clonis (1986) *J. Chromatogr.*, **376**, 121–30.
57. K. Nilsson and K. Mosbach (1981) *Biochem. Biophys. Res. Commun.*, **102**, 449–57.
58. K. Nilsson and K. Mosbach (1980) *Eur. J. Biochem.*. **112**, 397–402.
59. J. Porath and N. Fornstedt (1970) *J. Chromatogr.*, **51**, 479–89.
60. G. S. Bethel, J. S. Ayers, W. S. Hancock and M. T. W. Hearn (1979) *J. Biol. Chem.*, **254**, 2572–4.
61. T. Lang, C. J. Suckling and H. C. S. Wood (1977) *J. Chem. Soc. Perkin I*, 2189–94.
62. J. Porath and L. Sundberg (1972) *Nature New Biol.*, **238**, 261–2.
63. T. T. Ngo (1986) *Biotechnology*, **4**, 134–7.

14 THE USE OF RADIAL FLOW CHROMATOGRAPHY FOR THE PURIFICATION OF BIOMOLECULES

Denise M. Wallworth

14.1 Introduction

Although well accepted as the technique of choice, the use of low-pressure chromatography in the final stages of purification is by no means perfected. Resins based on cellulose or agarose matrices shrink and swell with changing buffer strength, causing a reduction in flow through the column. Current resins that have been crosslinked for additional strength, albeit at the expense of ion exchange capacity, go some way to solving the problem, but flowrates can still remain quite low when these resins are used in conventional, axial columns, making a process discouragingly slow. In addition, when a purification developed at the bench scale is scaled up several hundredfold, unavoidable changes in column dimensions for axial columns at the process level frequently give results that differ from that predicted [1, 2]. Since the cost of the isolation and purification steps of recombinant proteins can represent as much as 80% of the total production cost [3], any improvement in the time these take would represent a distinct advantage.

14.2 Column Requirements

Essentially, a chromatography column is a simple container for the resin that will perform the purification. Nonetheless, it has to fulfil a number of requirements in order to do this successfully and in a manner that takes maximum advantage of the resin properties. It must allow a uniform flow

Downstream Processing of Natural Products. Edited by Michael S. Verrall
©1996 John Wiley & Sons Ltd

pattern through the bed and minimize any band spread of products as they exit
out of the column. It is also essential that it allows the uniform application of
sample to the column bed in order to fully utilize the resin volume. Sanitization
of the column must be easy to perform and validate, and, finally, the column
must be safe to use at the pressures required.

Traditionally, axial flow columns have been utilized for protein purification:
samples are applied and separated down the length of a vertical column. Figure
14.1 shows a typical design which incorporates a piston pressed against the top

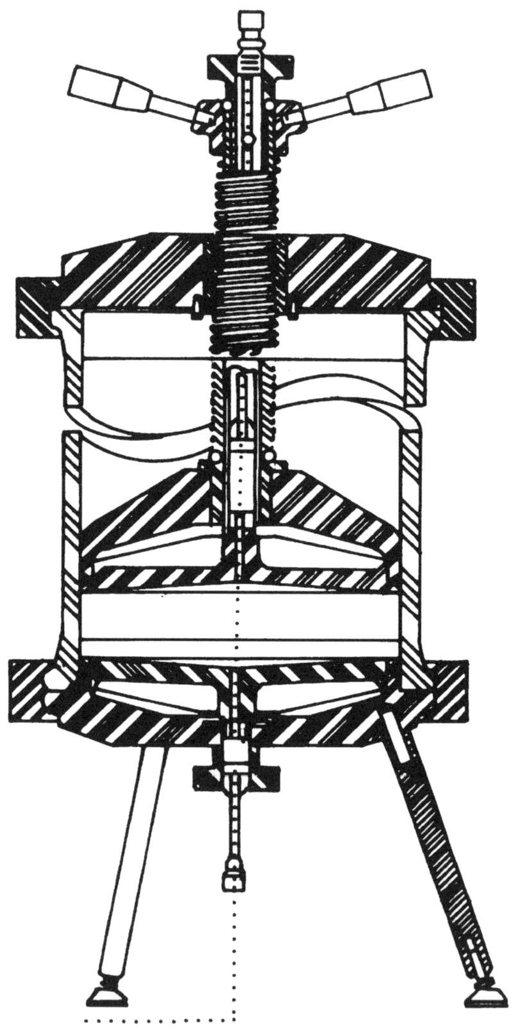

Figure 14.1 Schematic of axial flow column

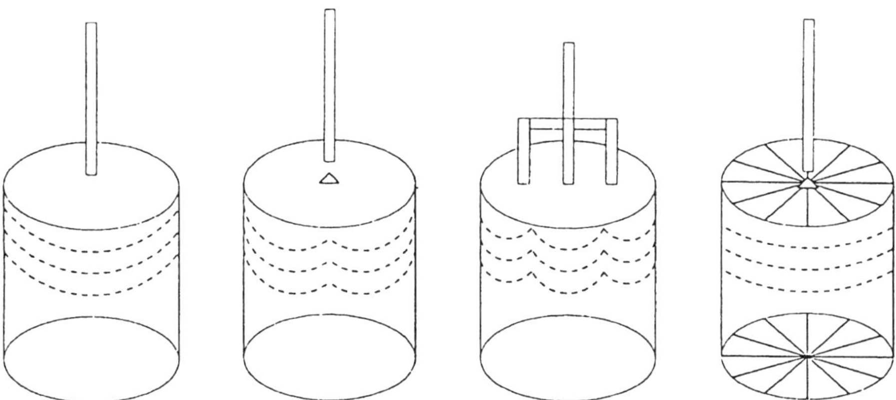

Figure 14.2   Distribution plate designs

of the resin bed to ensure an evenly packed bed and to remove dead spaces which develop during resin compression.

14.2.1 SAMPLE INTRODUCTION

Clearly, a non-uniform application of sample will result in a compromise in separation performance due to inconsistencies and the inefficient use of the resin bed. All axial column designs therefore incorporate a distribution plate at the top of the column, often of the type shown in Figure 14.2, where horizontal flow channels distribute the flow to the outer diameter of the column. This has to be carefully designed, taking into account the maximum and minimum flowrates that the column will be subjected to. At extremely low flowrates, the sample may go through the centre of the column before it has had time to flow outwards, and at extremely high flowrates, the diffuser may not adequately handle outward flow, with the result that the liquid flows preferentially through the centre. The same priorities also apply to the column outlet, in that inefficient fluid collection can result in product mixing or dilution. All of these issues become even more critical in very wide, process axial columns.

14.3 Radial Flow Column Design

Most of these problems can be eliminated to a large extent by the use of radial flow technology. The design utilizes two concentric cylindrical porous frits which hold the resin between them (Figure 14.3). Eluent and sample flow from the outer cylinder to the inner cylinder, across the radius of the column, which is now the effective bed height. The outer cylindrical frit is therefore the inlet of the column, and as such has an extremely large surface area in contact with the resin, making radial flow chromatography ideally suited to adsorptive-type separations.

The flow path of a radial flow column begins at the column inlet at the top of the column (Figure 14.3). Capillary channels take the flow rapidly to the perimeter of the column where it fills the capillary channel that lies just outside the outer frit. When sample is applied to a radial flow column, the entire surface area of the outer frit is utilized, giving an even distribution. Flow then proceeds through the frit and across the resin bed. After passing through the inner frit (the equivalent of the bottom frit in an axial column), the flow path proceeds down the capillary channel to the exit port of the column. The design makes radial flow columns suitable for ion exchange, affinity, hydrophobic interaction, reversed phase and any other adsorption–desorption type of separation. However, they are generally unsuitable for bed-depth-dependent isocratic separations such as size-exclusion chromatography.

Figure 14.3 also shows two additional ports at the top of the column. These are designed for packing the column: a packing manifold is attached at this point and a slurry of resin in buffer is pumped into the column (see Section 14.4). To empty the column, the reverse procedure is followed, although in pilot and process columns, further ports on the base of the column are available for rapid column emptying into a waste or recycling tank.

In the commercially available product [4], the columns are manufactured from acrylic, polyethylene, polycarbonate or stainless steel, the choice depending on the eluent conditions, the sanitizing procedures to be used and their compatibility with the biological molecules to be purified. Inner and outer frits are normally manufactured from either porous polyethylene or sintered stainless steel.

14.3.1 FLOWRATES

A consequence of the high inlet surface area (outer frit) is that exceptionally high flowrates and low column back pressures are possible, often as high as one or two column volumes per minute. This can be especially important in the treatment of large volumes of dilute products from some bioprocesses, in that the loading of such products on to a process column is often the rate-determining step. In addition, for those products that have a high binding constant, it is possible to desorb by reversing the flow through the column, thereby reducing product dilution and speeding up purification further. For unstable products, this technique would potentially increase the yield by reducing time spent on the column. The initial step in the purification of the enzyme uridine phosphorylase, used to catalyse the synthesis of a number of pyrimidine nucleosides, involves the break-up of the cells and removal of cell debris by diafiltration. This and other enzymes from *Escherichia coli* were purified without this step by utilizing a radial flow column packed with Q Sepharose [5]. Because of the high flowrates possible through the radial

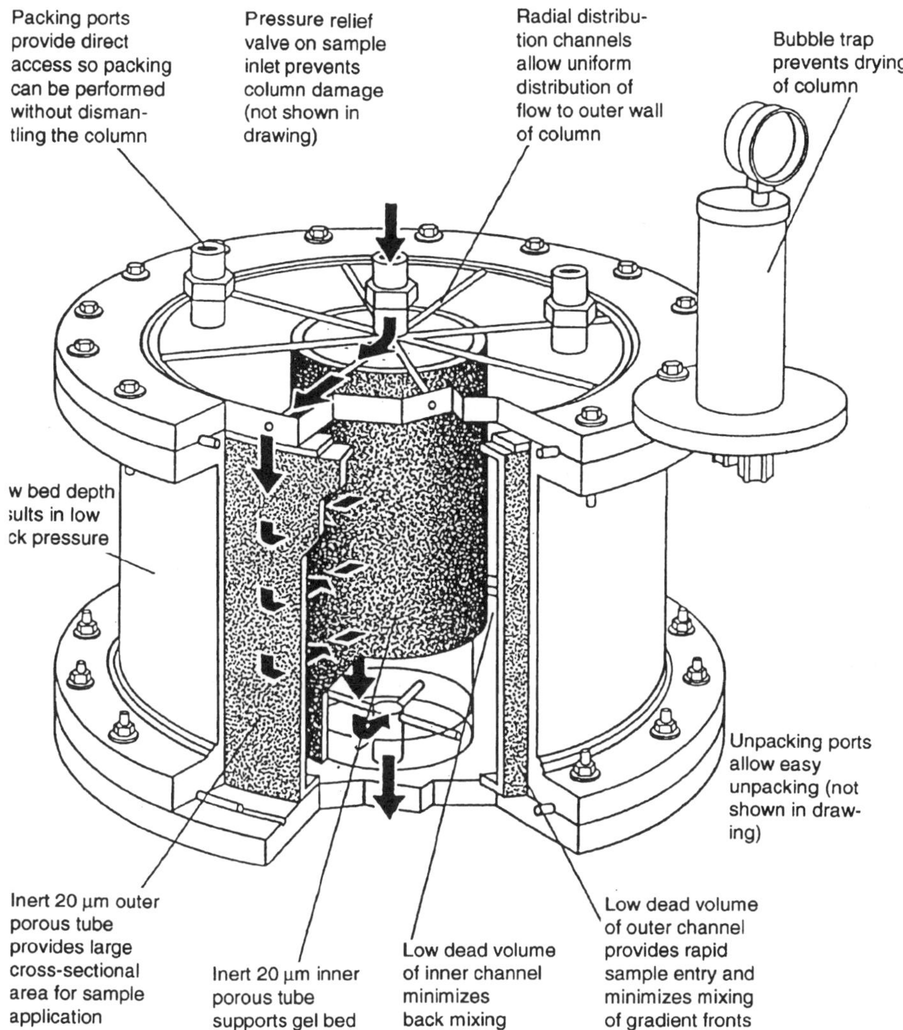

Figure 14.3 Schematic of radial flow column design

column, it was possible to force cell debris through the resin without clogging the column. No chanelling was evident under these conditions and, over some 60 cycles, no diminution in resin-binding capacity or stability was observed.

One of the observations that can be made about this column design is that because the inner frit surface area is much smaller than that of the outer frit, the eluent velocity will increase across the column. In practice, this does not affect the chromatographic process. Since chromatographic performance is dependent on a constant residence time, t_R, defined by

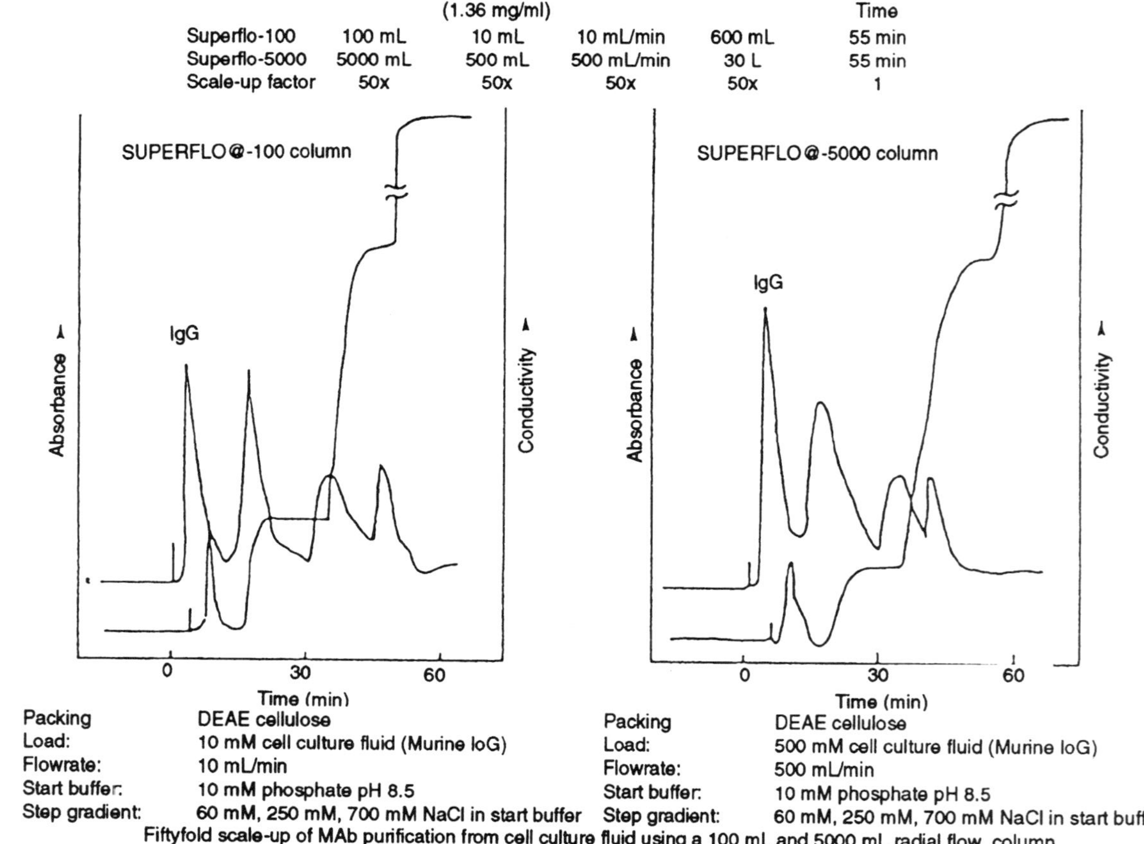

Figure 14.4 MAb scale-up using radial flow columns

$$t_R = V/Q$$

where V represents volume and Q the volumetric flowrate, across the column it can be concluded that since flowrate and volume are constant over a given bed segment, the residence time will also be constant, and so will be the binding efficiency. Hence linear velocity, generally referred to when using axial columns for product purification, does not exist for radial flow columns. Design equations that use simple volumetric flowrate as an alternative to linear velocity have been verified experimentally and shown to be both useful and recommended in the scale-up predictions for radial flow columns [6].

For axial columns, scale-up essentially means increasing the column diameter, with the aim of maintaining the same linear flowrate between laboratory and production columns. More often, a compromise is reached between an increase in column diameter and an increase in column length. For radial flow columns, however, the bed height is kept constant and so larger columns are simply longer, giving exactly linear scale-up. In the fiftyfold scale-up of an antibody purification from cell culture fluid [7], the scale-up was performed by increasing flowrate, sample size and elution volume fifty times. Two columns, 100 mL laboratory scale and 5 litre pilot scale, were packed with DEAE cellulose, and 10 mL or 500 mL samples respectively were applied. Flowrate was also proportionately scaled up from 10 to 5000 mL/min. A three-step gradient using 60, 250 and 700 mM NaCl in 10 mM phosphate buffer (pH 8.5) was employed in both cases and the resulting chromatograms, shown in Figure 14.4, exhibit an almost identical profile.

In a further study [8], a tenfold scale-up of a prorenin purification on QAE and ConA Sepharose packed into a radial flow column was shown to be linear, and activity and recovery levels were maintained. However, an additional result was that the yield increased 75-fold and it was concluded that a reduction in losses caused by on-column denaturation, proteolysis or irreversible binding sometimes seen on axial columns had been reduced due to the short bed depth and shorter residence times.

14.3.2 PRESSURE DROP–FLOWRATE RELATIONSHIP
   IN RADIAL FLOW COLUMNS

Three columns packed with Sepharose CL4B resin, having column volumes of 100 mL, 500 mL and 1.5 L, were evaluated for column pressure drop over a range of flowrates [8]. The pressure flowrate curves were generated and compared with axial columns of corresponding sizes. It was observed that flowrates up to three times faster were possible in the radial flow column when compared to the axial column at the same column back pressure. In a further study using process radial flow columns [9], the pressure drop across a series of columns from 10 to 100 L packed with Sepahrose Fast Flow adsorbent was

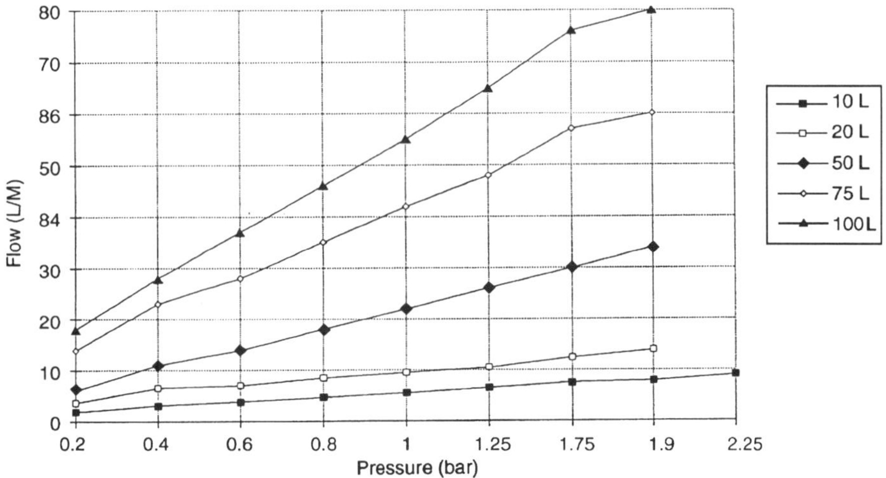

Figure 14.5 Pressure–flowrate curves for radial flow columns

measured (Figure 14.5). In general, these are found to be significantly less than those expected for equivalent axial columns.

14.4 Packing Radial Flow Columns

One of the other consequences of the design of the radial flow column is that it need not be taken apart for packing and unpacking. Whilst not a critical factor for laboratory columns, it becomes very important at the process level that columns be packed faster and under highly sanitary conditions, and then subsequently unpacked as a slurry straight into a disposal or recycling tank. Further, the safety aspects of removing a large and heavy inlet plate and the manual removal of spent resin can be avoided.

To pack a radial flow column, the column is placed in a closed-loop system as shown in Figure 14.6(a). The packing manifold is attached to the top of the column and to the pump, a peristaltic pump which has both forward and reverse flow capabilities. The first stage in the process is that the column is filled with packing buffer (Figure 14.6b), a buffer that has the highest ionic strength in the process to be used. For some resins that have a high shrinkage, this buffer can be increased to 0.5M higher than the highest ionic strength that the column will experience in the process. This stage of the packing process appears to be easiest if the column is operated in reverse, pumping buffer into the column from the outlet port. Once all the air is dispelled from the column, the flowpath is switched to allow the resin slurry to be pumped into the column through the packing ports (Figure 14.6c). Generally, a 25–30% slurry of the resin in packing buffer is used [9], at a flowrate of one half-column volume per

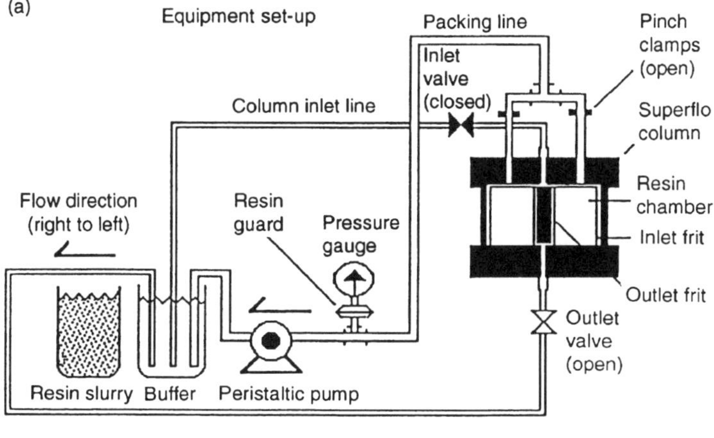

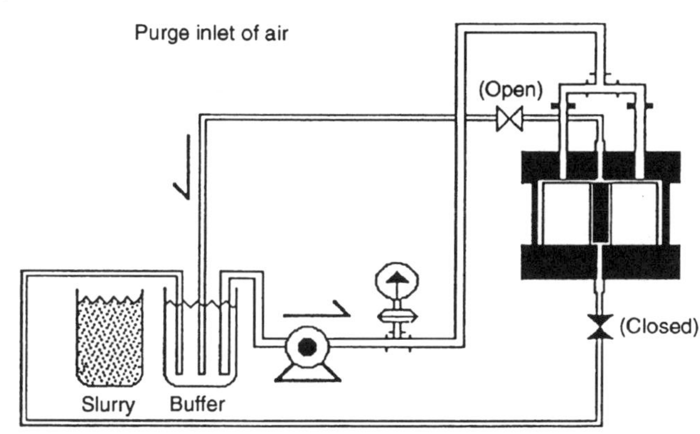

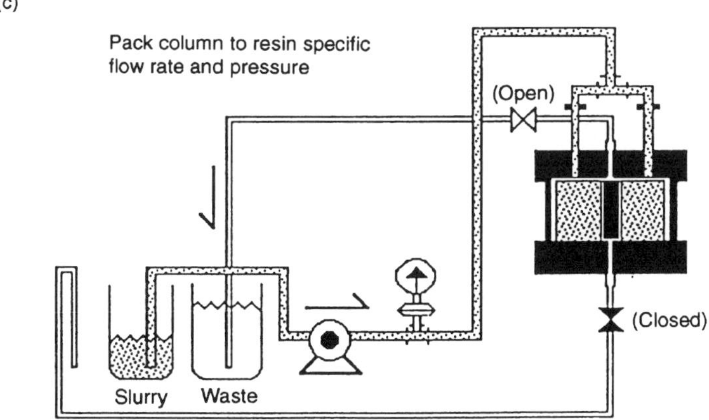

Figure 14.6 Column packing procedure for radial flow columns

Table 14.1 Pressure increases of different resins

Chromatographic resin	Column packing pressure increase (mbar)
Sepharose 4B*	34
Cellulose	70–210
Agarose	345–550
Trisacryl	345–550
Sepharose CL-4B*	345–550
Sepharose Fast Flow*	830–1035
Fractogel**, polydextran beads, Macroprep	830–1035
Silica	1725

Trade marks: *Pharmacia
                **TSK

minute. The outlet valve is closed and excess buffer exits to the buffer tank through the inlet port.

When the column is almost packed, a rapid increase in back pressure is observed. Once the pressure has reached the specified pressure shown in Table 14.1, the flow of resin slurry is stopped. These pressures have been empirically derived [9], and allow for a highly reproducible packed bed and possible automation of the column packing process. The packing manifold is removed and flushed with fresh buffer to remove all remaining resin. At the same time, packing port plugs are inserted. These press directly on to the column bed, removing any possible dead zones. In the final stage of column packing, the bed is conditioned by back-flushing the column with packing buffer. Equilibration of the column ready for the first stage of the purification process can then be carried out, generally at a flowrate of between one third to one column volume a minute.

14.5 Applications for Radial Flow Columns

Ion exchange, affinity and hydrophobic interaction techniques have all been carried out successfully on radial flow columns. Table 14.2 includes some of the types of products currently using radial flow technology for purification [10].

A recent study on factor IX purification utilized an immunoaffinity resin packed into a 50 mL laboratory-scale radial flow column and also into a 48 mm diameter glass axial column [11]. The antibody was monoclonal and was coupled to Sepharose CL4B. After equilibration with five column volumes of buffer (10 mM magnesium chloride, 100 mM sodium chloride, 20 mM phosphate, pH 7.0), the lyophilized coagulation factor IX was loaded. Antibody capacity appeared to be identical for both columns, suggesting that radial dispersion, mass transfer and intraparticular diffusion do not have a

Table 14.2 Types of products purified by radial flow chromatography

Product	Scale (where known) (L)
Monoclonal and polyclonal antibodies	
Interleukins	10
Colony stimulating factors (G-CSF, GM-CSF, EGF)	20
Interferons	
Blood factors, VII, IX, etc.	350
Recombinant protein products	100
Diagnostic enzymes, restriction enzymes	
Vaccines — Hepatitis B, etc.	5
Human growth hormones — Biotropin, Somatotropin	100
Growth factors	50

significant impact on immunoaffinity chromatography. An alternative route to the purification of factor IX from human plasma has utilized an un-crosslinked cellulose (Whatman DE52). In the existing process, the cryogenic precipitate is adsorbed on to the resin in a batch process because the amount of precipitate present in the supernatant prevents the use of normal column techniques. In another study [12], the DE52 was alternatively packed into a 100 mL radial flow column and 5 L of the cryoprecipitate loaded on to the column. Binding was achieved efficiently and the precipitate cleared the column without causing any increase in column back pressure.

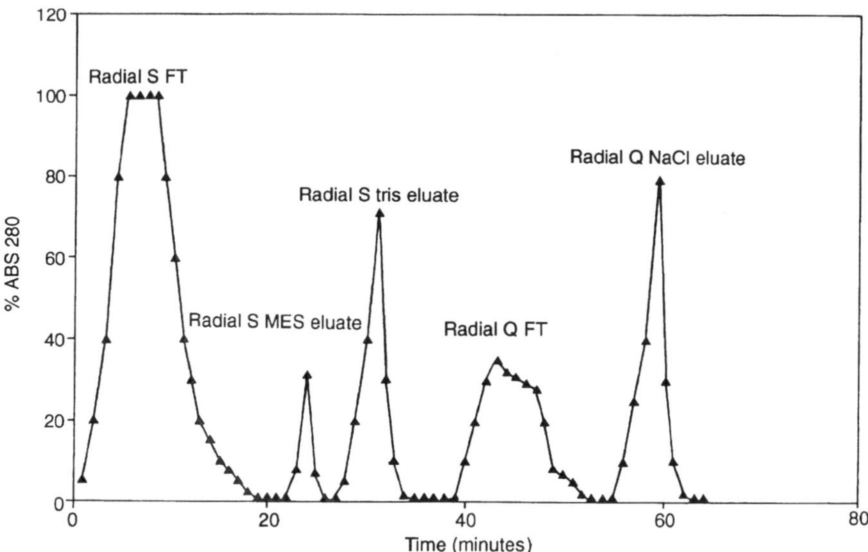

Figure 14.7 Chromatography profile of Interleukin-2 purification

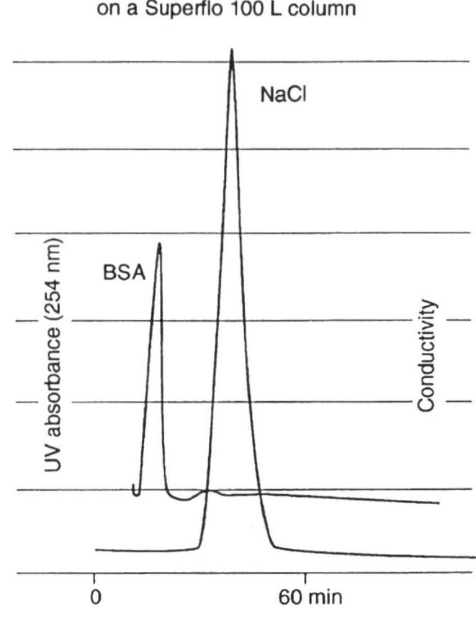

Figure 14.8   Desalting procedure for BSA on a process radial flow column

Antimelanoma antibody, produced from ascites, has been purified on a
1500 mL radial flow column packed with Protein A Sepharose [8]. The ascites
was diluted with a phosphate buffered saline, loaded (at 104 mL/min) and
washed with a pH 6 citrate buffer (170 mL/min). The monoclonal antibody was
eluted with a pH 3 citrate buffer at a flowrate of 92 mL/min. Purity was
determined to be greater than 95%, comparable with other techniques, with gel
filtration and double immunodiffusion. The total process took 3.5 hours.

A process has been developed for the purification of recombinant bovine
interleukin-2. After cell removal from the yeast fermentation mash, the product
was loaded and bound to a 500 mL radial flow cation exchange column at a
flowrate of 500 mL/min (Figure 14.7). The eluate from this was directly loaded
on to a 100 mL radial flow anion exchange column in the same buffer. Gram
quantities of interleukin-2 were rapidly and efficiently produced for clinical
trials in this manner.

14.5.1 DESALTING ON A RADIAL FLOW COLUMN

The low bed height of a radial flow column means that they can also be used
for rapid desalting, providing an alternative to diafiltration techniques. In one

study, bovine serum albumin (BSA) was desalted on a 100 L radial flow column using isocratic elution in tap water (Figure 14.8). A 40 g sample of BSA was dissolved in 20 L of 1M sodium chloride and loaded on to the radial flow column packed with BioRad P-6DG. The desalting process was completed in just under one hour.

14.6 Conclusions

Significant improvements in production rates can be achieved through the use of radial flow columns in process chromatography. Packing and operating times are also considerably reduced and their compatibility with an extensive range of resins provide the versatility required by the process chromatographer. Separations can be scaled up linearly and predictably without intermediate re-optimization steps, thus minimizing the time required for scale-up. With the exception of gel filtration, radial flow columns can provide the solution to many of the problems frequently faced by the process chromatographer.

14.7 References

1. J. C. Janson and P. S. Hedman (1982) *Adv. Biochem. Engng*, **25**, 43–99.
2. G. Sofer and C. Mason (1987) *Biotechnology*, **5**, 239–44.
3. T. C. Ransohoff, M. K. Murphy and H. L. Levine (1990) *BioPharm*, **3**, 20–6.
4. V. Saxena (1987) US Patent 4 676 898.
5. K. Weaver, D. Chen, L. Walton, L. Elwell and P. Ray (1990) *BioPharm*, **3**(7), 25–8.
6. W. C. Lee, G. J. Tsai and G. T. Tsao (1990) *ACS Symp. Ser.*, **427**, 104–17.
7. V. Saxena and A. E. Weil (1987) *BioChromatogr.*, **2**, 90–7.
8. V. Saxena, A. E. Weil, R. T. Kawahata, W. C. McGregor and M. Chandler (1988) *International Laboratory*, Jan/Feb, 50–7.
9. Sepragen Corp., San Leandro, California (1994) Internal report.
10. Sepragen Corp., San Leandro, California (1995) Internal report.
11. J. Tharakan and M. Belizaire (1995) *J. Chromatogr. A*, **702**, 191–6.
12. P. Hellman (1993) Sepragen Corp. Application Bulletin. San Leandro, California.

15 HIGH-PERFORMANCE LIQUID CHROMATOGRAPHY

Peter R. Shelley

15.1 Introduction

High-performance liquid chromatography (HPLC) is of great value in downstream processing of natural products both as an analytical tool and as a preparative procedure. Several modes of HPLC may be distinguished according to the type of stationary phase used. The principal phases used are adsorption, liquid–liquid partition, chemically bonded, size exclusion and chiral. Of these adsorption and particularly reversed phase (chemically bonded) HPLC have been of the greatest use in the last two decades in the monitoring of microbial metabolites produced on a pilot plant or production scale from fermentation origin. The ability to rapidly determine optimum fermentation harvest times and to quickly assess stage yields during isolations from up to 3000 L of broth have frequently proved crucial to the successful separation of many, sometimes unstable, novel compounds. Because of this it is recommended that there should be no departmental barriers between separation scientists and analysts to ensure the most rapid feedback of results to large-scale processes. Similarly preparative HPLC has been employed on selected process streams to exploit the diversity of metabolites produced from an organism under investigation. Stationary phases for preparative procedures were formerly distinctly different from those used for analytical purposes but now their chemistries are similar, making scale-up a more predictable process.

This chapter combines a review of the present state of HPLC in this area with illustrations drawn from the author's experience, and shows how HPLC has become an indispensable part of modern natural product separations.

Downstream Processing of Natural Products. Edited by Michael S. Verrall
©1996 John Wiley & Sons Ltd

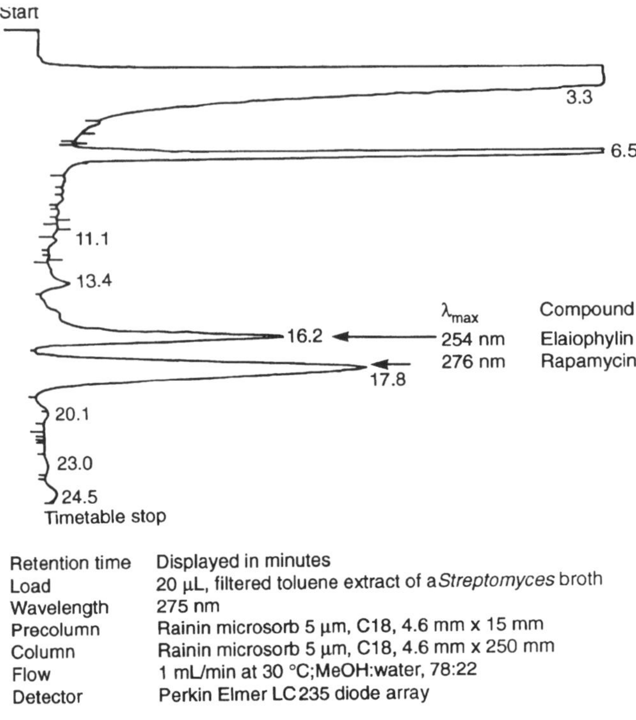

Retention time   Displayed in minutes
Load   20 µL, filtered toluene extract of a Streptomyces broth
Wavelength   275 nm
Precolumn   Rainin microsorb 5 µm, C18, 4.6 mm x 15 mm
Column   Rainin microsorb 5 µm, C18, 4.6 mm x 250 mm
Flow   1 mL/min at 30 °C;MeOH:water, 78:22
Detector   Perkin Elmer LC 235 diode array

Figure 15.1   Repamycin: methanol extract of broth

15.2 Sample Treatment

Many analytical samples require pretreatment before HPLC can be carried out.
A recent survey [1] listed 33 techniques currently in use, with liquid–liquid
extraction and solid phase extraction among the most frequently used. As well as
allowing separation of a low molecular weight entity from a complex biological
matrix, correct choice of solvent may also give a considerable degree of selectivity
to the primary extract. For example, Figure 15.1 shows a methanolic extract of a
Streptomyces broth producing the intracellular metabolite rapamycin. Comparison with standards showed the peak at 17.8 minutes was rapamycin and the peak
at 16.2 minutes an unwanted component, elaiophylin (Figure 15.2). However,
extraction with toluene still extracted rapamycin in good yield but eliminated
elaiophylin. The small peak unmasked in Figure 15.3 was a rapamycin analogue
previously hidden in Figure 15.1 by elaiophylin. In both cases ultrasonic
disruption of microbial whole broth was used to ensure intimate mixing of
solvent and biological matrix. Solvent extraction may also facilitate sample
concentration and so increase assay sensitivity. Further selectivity may be
obtained by careful choice of pH for ionized analytes. This is particularly true for
preliminary solid phase extraction techniques usually using a small cartridge with

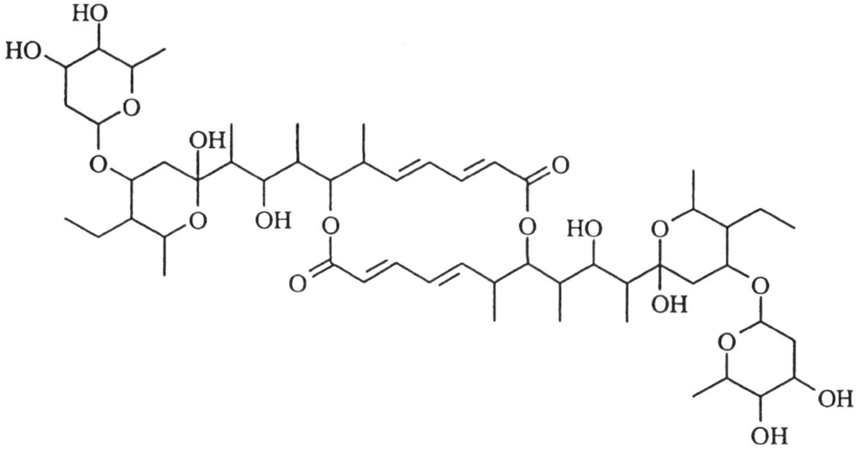

Elaiophylin

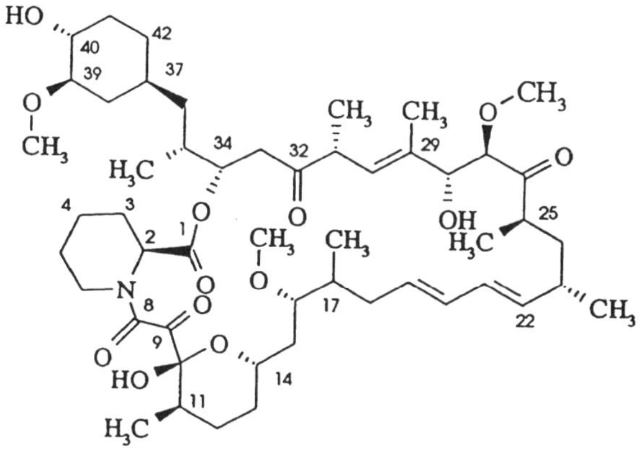

Rapamycin

Figure 15.2 Structures of elaiophylin and rapamycin

40 μm diameter packing, where buffered elution protocols are routine. Sample pretreatment may be time consuming so considerable effort has been spent in automating these procedures (see Refs. [2] and [3]).

15.3 Analytical Packings and Columns

15.3.1 TYPES OF STATIONARY PHASE

Most HPLC packings at present are based on porous silica gels. These consist of fused aggregates of colloidal silica giving usually spherical particles 3–10 μm

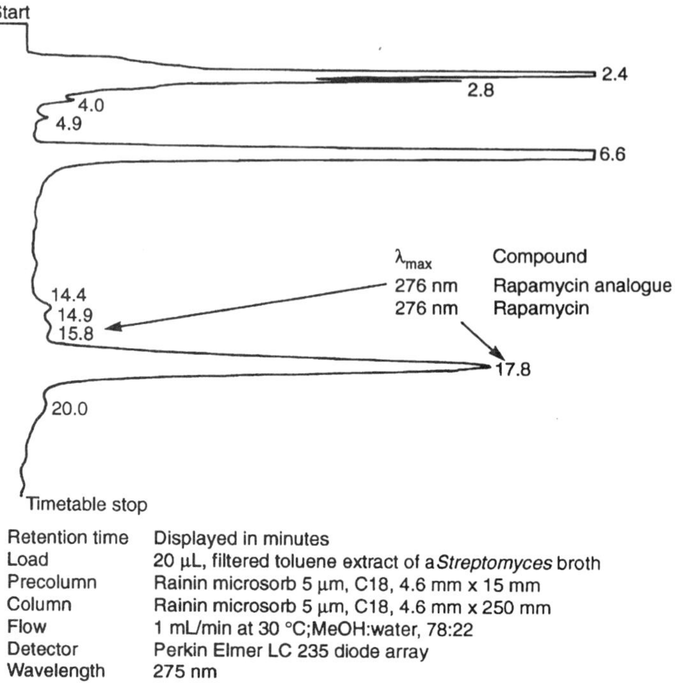

Figure 15.3 Rapamycin: toluene extraction of broth

in diameter of very high surface area (up to $400\,\mathrm{m^2/g}$). Adsorption chromatography using this type of packing has been widely used and reported for compounds soluble in organic solvents, and is particularly useful for isomer separations. However, while useful for the separation of non-polar or moderately polar materials it is of less utility for ionic or very polar compounds, which tend to suffer from poor peak shape or irreversible adsorption. Because of this a series of chemically bonded packings has been developed and is now in very widespread use for many types of solute.

To form a chemically bonded phase, silanol groups on the surface of a silica particle are reacted with reagents such as octadecyl trichlorosilane to give a hydrophobic surface and so form a reversed phase (C_{18}) packing which readily binds components with hydrophobic domains from an aqueous mobile phase. Column equilibration is faster than with 'naked' silica particles and method development more straightforward, with simple gradient systems such as water–acetonitrile or water–methanol sometimes being very effective. Aqueous buffered systems are frequently combined with such gradients to control solute ionization and to protect unstable compounds. In addition to octadecyl trichlorosilane several other reagents have been used to form a large group of derivatized silicas, many of which can be used in reversed phase mode. Part of the range available from just one manufacturer is shown in Table 15.1. They

Table 15.1 Types of reversed phase or modified silica

Type	Functional group	Notes
Silica	—	Claimed stability up to 600 bar
C8	Octyl	Not end capped
C18	Octadecyl	Hexamethyldisilazane end capping
Ph	Phenyl	Good interaction with aromatics
CN	Nitrile	Normal or reverse phase operation
NO_2–N	$(CH_2)_3$ nitrophenyl	Useful for polycyclic aromatics
NH_2	Amino	Normal or reverse phase or anion exchange
$(CH_2)_3N(CH_3)_2$	Dimethylamino	Weak base anion exchanger
SA	Sulfonic acid	Strong cation exchanger
SB	Quaternary ammonium	Strong anion exchanger

are available with $100\,\text{Å}$ pores and most in $5\,\mu m$ particle diameter, from Macherey-Nagel; similar ranges are available from other manufacturers.

Even among a single type of derivatized silica, for example C_{18} with the same particle shape and pore size, different results may be obtained from different manufacturer's packings according to the carbon load and degree and type of end capping used to remove unreacted silanol groups. This is particularly so

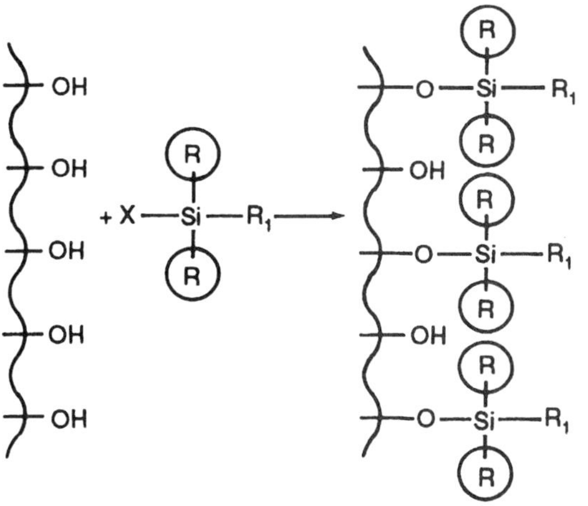

Reaction of monofunctional sterically
protected silane with hydroxylate silica
R=isobutyl side group. R_1=octadecyl group

Figure 15.4 Sterically protected bonded phase

with analytes containing basic groups, which may interact with incompletely end-capped silica producing poor resolution and trailing peaks. For this reason, some workers have ranked commercial HPLC columns in tabular form according to the quality of separation they provide for basic compounds (see Refs. [4] and [5]). Such a table tends to date quickly because of the rapid development of new column packings, so examining a number of modern packings when analysing a problem basic susbtance is worth while.

In addition, adsorption of basic substances on to active silanol sites may be controlled by adding triethylamine (25 mM) to the mobile phase. This saturates active binding sites on the column and allows good recovery of analyte. A number of base-deactivated silicas are also available from column manufacturers. These are designed to eliminate unwanted interactions, but are not always successful, as has been shown in a study of methods for the determination of tetracycline antibiotics [6].

15.3.2 STABILITY OF HPLC COLUMN PACKINGS

Standard silica columns are degraded by highly acidic or alkaline conditions so should not be used with a pH less than 2 or greater than 8. These limits also generally apply to common C_{18} stationary phases as well. However, Kirkland *et al.* [7] have reported that some of the disadvantages of conventional C_{18} bonded phases could be overcome by combining a sterically protected bonded silane phase with highly purified low acidity porous silica microspheres. The resulting bonded phase contained di(isobutyl)-n-octadecylsilane groups densely coupled to the silica support (see Figure 15.4) and provided a longer life than non-protected C_{18} phases when used under low-pH, high-temperature conditions. Similar commercial columns are now available as Zorbax SB-C18 (Rockland Technologies Inc.).

Non-silica-based packings such as porous graphitized carbon (available as 'Hypercarb' from Shandon) and polymeric styrene–divinyl benzene (available as 'PLRP-S' from Polymer Labs) offer stability at extremes of pH and an opportunity to exploit their different selectivities.

15.3.3 CHOICE OF ANALYTICAL COLUMN DIMENSIONS

Knox [8] in 1978 suggested that the minimum column diameter, in order to obtain the highest performance with 5 µm particles, was approximately 5 mm, with a typical column length of 100–250 mm. Commercial prepacked columns with these dimensions are still in use today. Our group favours columns 250 mm in length for routine analysis of secondary metabolites from fermentation broths. This gives the best chance of detecting unexpected metabolites arising from changes in fermentation parameters and enables the resolution of small concentrations of analogues from one another in a complex biological matrix.

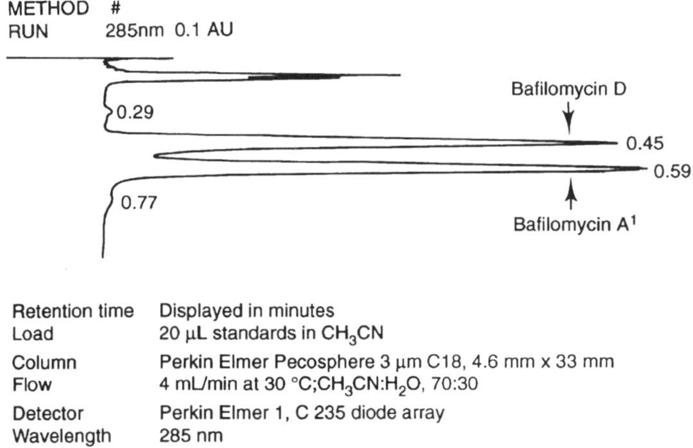

Retention time	Displayed in minutes
Load	20 µL standards in CH_3CN
Column	Perkin Elmer Pecosphere 3 µm C18, 4.6 mm x 33 mm
Flow	4 mL/min at 30 °C;CH_3CN:H_2O, 70:30
Detector	Perkin Elmer 1, C 235 diode array
Wavelength	285 nm

Figure 15.5   High-speed chromatography of bafilomycins

Once an isolation process has become established the emphasis may well switch from one of maximum resolution to that of obtaining satisfactory resolution in the minimum time, and so facilitate rapid monitoring and hence rapid optimization of plant-scale operations. For this aim shorter columns with smaller particles are ideal. For example, columns for high-speed chromatography are available with typical dimensions of 30–50 mm × 4.6 mm i.d. These columns with 3 µm diameter packing possess an optimum flowrate higher than that of columns with 5 µm packings, and typical plate counts > 5000 per column. They may also be used at higher than optimum flowrates with only a small reduction in efficiency, so that analysis may well take place with a flowrate of up to 4 mL/min. Using this technique a considerable saving in time and solvent may be achieved. For example, Figure 15.5 shows the separation of bafilomycin D, retention time 0.45 min, from bafilomycin A1, retention time 0.59 min. This represents a nearly 30-fold time saving compared with an earlier analysis developed on a longer column. In addition, smaller-diameter columns are now available for use in cases where reduced solvent consumption and increased relative peak height are a priority. Part of the range from one manufacturer (Chrompack) is shown in Table 15.2 together with a summary of chromatographic performance.

15.3.4 ADDITIONAL TYPES OF STATIONARY PHASE PARTICLE

While 3 and 5 µm spherical porous particles may be used for many natural product analyses, increasing emphasis on solvent saving or speed of analysis has led to the development of different particle types. For example, 1.5 µm diameter non-porous silica particles (such as ChromSpher UOP C18) are now available. Such columns claim advantages in terms of reduced analysis time

Table 15.2 The effect of column dimensions on chromatographic performance

	Standard columns		Minibore columns	High-efficiency columns	Microspeed columns	High-speed columns
Column dimensions $L \times$ i.d. (mm)	250×4.6	200×3.0	250×2.0	100×4.6	100×2.0	50×4.6
Particle size (μm) packing material	5	5	5	3	3	3
Plate number	18 000	16 500	18 000	11 500	11 500	6000
Flowrate (mL/min)	1.0	0.4	0.2	1.5	0.3	1.5
Retention time (min)	10	8.5	9.5	2.7	2.5	1.4
Solvent use (mL/analysis)	10	3.4	1.9	4.1	0.8	2
Relative peak height	1.0	2.8	5.3	2.0	10.6	2.9
Maximum detector cell volume (μL)[a]	15–30	5–10	2–5	7–15	1–3	5–10
Maximum injection volume (μL)[a]	30	10	5	15	3	10
Peak volume (μL)						
$k' = 1$	200	70	40	100	20	70
$k' = 5$	600	200	110	300	60	200

[a]The values given here are general recommendations only. Higher values may cause slight efficiency loss.
Data taken from Chrompack technical bulletin.

and solvent use as well as enhanced pH stability to alkaline conditions. Reduced surface area, however, means that the sample capacity of the packing is reduced. Instead of reducing the particle diameter to allow higher flowrates, which are subject to back-pressure constraints, a different approach to solving the mass transport problem is to design particles with two distinct types of pores. One class, the 'through pores', are large enough to allow some convective flow through the particle and the other, smaller, diffusive pores provide a larger adsorptive surface area. This approach, termed 'perfusion chromatography', has been taken by PerSeptive Biosystems with crosslinked polystyrene–divinyl benzene particles to produce the 'Poros' range of stationary phase which are particularly effective for macromolecules. Some examples of this material and its use to separate proteins at high flowrate are given in Ref. [9].

The desirability of analyzing low molecular weight compounds in biological fluids containing high protein levels without elaborate clean-up methods led to the development of internal surface reversed phase packings [10]. These phases contain a hydrophilic bonded external phase and a hydrophobic internal phase. Proteins are excluded by the hydrophilic surface and compounds of interest interact with the reversed phase inside the pores. Modern versions of these columns are available such as ChromSpher 5 Biomatrix, based on 5 μm small-pore (13 nm) silica. This has an internal phenyl moiety for hydrophobic interaction and an external alkanol functionality to reject proteins.

15.4 Detectors

Monitors based on UV–visible absorbance are still probably the most widely used in HPLC [11], but a definite trend can be observed towards the more frequent use of information-rich detectors in the early stages of a natural product isolation. This is particularly so for those groups investigating sources of natural products as producers of novel biologically active materials [12]. The use of photodiode array detectors [13, 14], for example, is now widespread and applications for HPLC/mass spectroscopy and HPLC/NMR (nuclear magnetic resonance) continue to expand at a rapid pace.

Diode array detectors generate data based on absorbance at multiple wavelengths and retention time. This may be stored for subsequent processing and displayed by suitable software programs. Such data manipulation may be available internally in detectors with limited post run manipulation facilities or externally by means of a suitable PC interface. Several types of graphical output are available including rotatable three-dimensional displays and colour-coded contour plots which permit all absorbance, wavelength and retention time data to be presented simultaneously. The presence of non-homogeneous chromatographic peaks can be detected by distortion of the contour symmetry. Many other routines may also be provided for indicating the presence of peak contaminants. Two of the most widely used are based on the comparison of

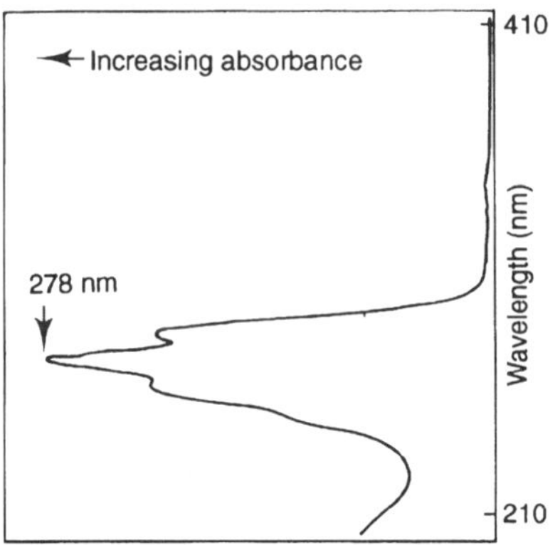

Figure 15.6 UV spectrum of rapamycin

normalized spectra at apex and inflexion points of the peak, and an absorbance ratio plot of two different wavelengths.

Secondary metabolites are often produced as groups of closely related compounds, and such structural analogues may have similar absorption spectra. If the compounds of interest possess a distinct absorption maximum, chromatographic peaks obtained from analysis of a fermentation broth, for example, can be searched for those containing similar values. This directed examination carried out early on in the examination of a culture may enable potential analogues to be quickly pinpointed and aid their subsequent purification to allow full structural determination by off-line methods such as high-resolution mass spectroscopy or NMR. For example, rapamycin, an antifungal and immunosuppressive molecule with a macrocyclic lactone structure containing a triene portion (Figure 15.2), has a specific absorption maximum (λ_{max}) of 278 nm and a characteristically shaped UV spectrum (see Figure 15.6). Searching chromatograms from high-resolution reverse phase columns obtained from methanolic extracts of several Streptomycete broths for peaks with a λ_{max} of 278 ± 3 nm allowed several different novel rapamycins to be provisionally identified. After isolation and full structural determination the close relationship of these molecules to rapamycin was confirmed (see Figure 15.7). A further example of this approach is shown in Figure 15.8. This chromatogram was produced by monitoring a reverse phase column at 286 nm and injecting a solvent extract from a known producer of bafilomycin A1, a macrolide antibiotic with λ_{max} values of 249 and 287. Searching the stored spectra of other peaks present in the chromatogram for similar values enabled the peak at 12.9 minutes to be tagged as a potentially related compound. After isolation, the structure of the compound detected at 12.9 minutes was found to be that of bafilomycin D, a compound not previously reported to be produced by this culture.

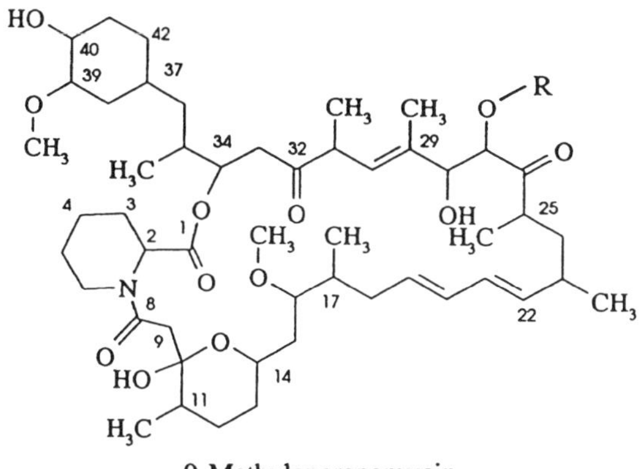

9-Methylenerapamycin

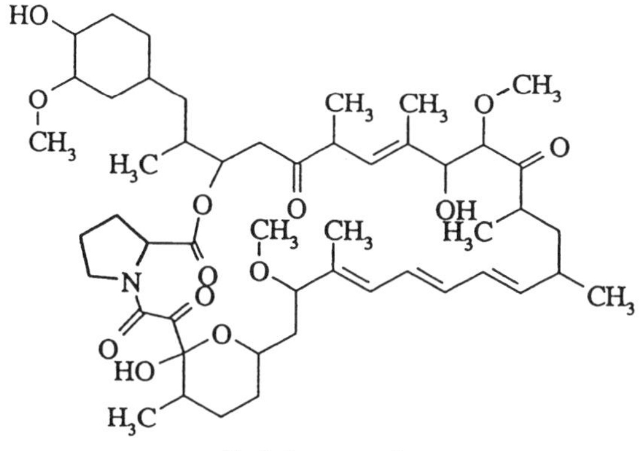

Prolylrapamycin

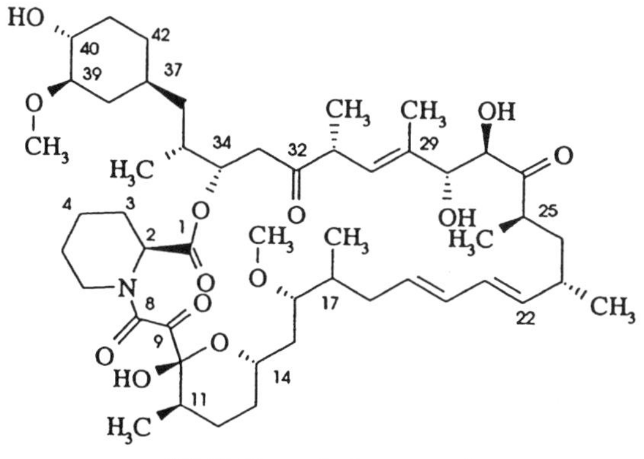

27-*O*-Demethylrapamycin

Figure 15.7 Structures of rapamycin analogues

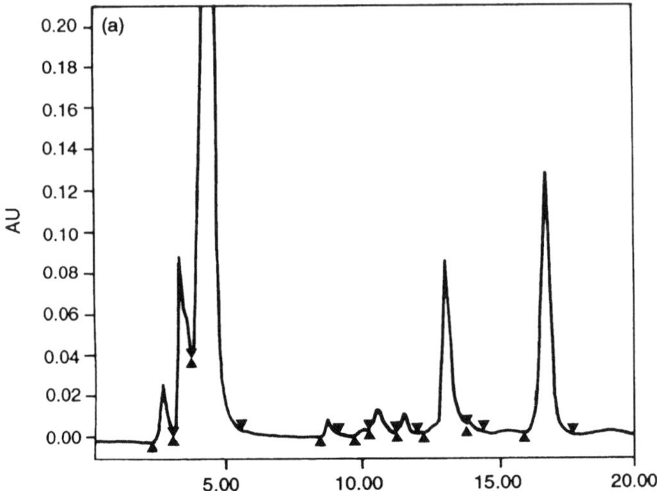

Load	20 µL filtered solvent extract from a *Streptomyces* broth
Precolumn	Rainin Microsorb 5 µm C_{18}, 4.6 mm x 15 mm
Column	Rainin Microsorb 5 µm C_{18}, 4.6 mm x 250 mm
Flow	1 mL/min at 30°C; CH_3CN: H_2O, 80:20
Detector	Waters 996 diode array

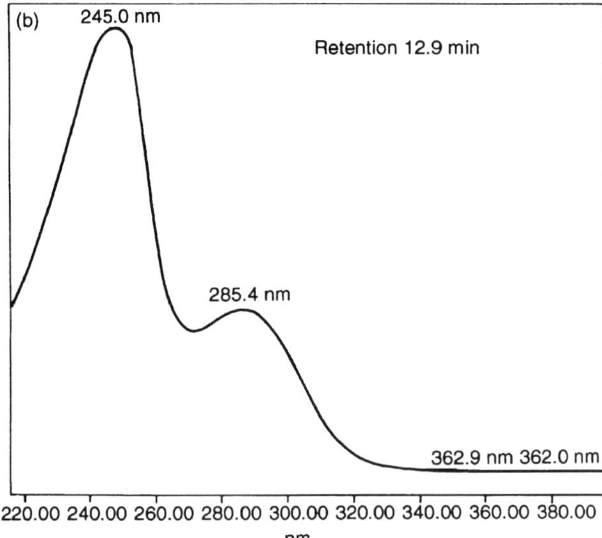

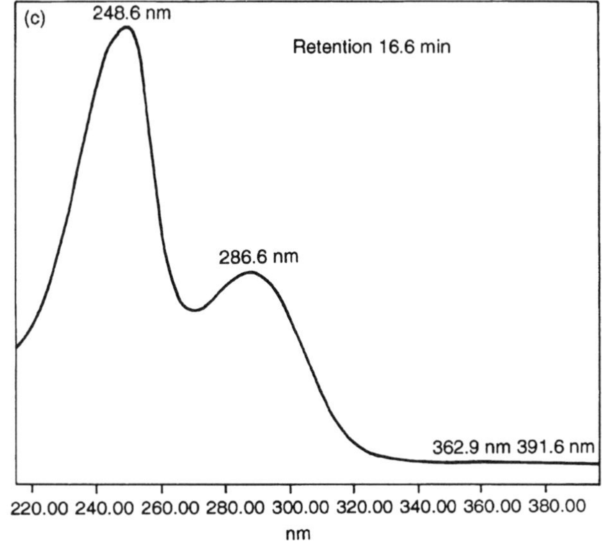

Figure 15.8 (a) Chromatogram of bafilomycins: diode array detection. (b) UV spectrum of bafilomycin D. (c) UV spectrum of bafilomycin A1. (d) Structures of bafilomycins A1 and D

15.5 Use of Databases

The screening of micro-organisms for the production of novel biologically active substances has been carried out for decades and many thousands of natural products have been reported in the literature [15]. For a new screening programme to be efficient, compounds which have already been reported need to be eliminated from further investigation at as early a stage as possible. Analytical HPLC with diode array detection and HPLC/mass spectroscopy carried out on filtered fermentation broths or cell extracts are ideal tools to carry out such a task. Preliminary correlation of biological activity with chromatographic retention and physical properties such as UV adsorption spectrum λ_{max}, molecular weight, together with solvent solubility and chemical spot test data, may frequently allow early identification of an unknown by comparison with entries in a commercial or proprietary database.

15.6 Preparative HPLC

15.6.1 INTRODUCTION

Advances in modern NMR and mass spectrometry have reduced the time needed for structural elucidation of natural products, so that the time taken to isolate and purify a new antibiotic or metabolite may be the limiting factor in operating a discovery programme. This need for the reduction in the time taken to purify a new metabolite has been one of the driving forces in the increasing use of modern instrumented preparative HPLC systems for natural product purification over the last decade or so. Others include the need for very selective separations and the importance of high-purity products used to determine biological activity.

15.6.2 TYPES OF STATIONARY PHASE

As with analytical HPLC, silica particles used for adsorption chromatography and reverse phase (C_{18}) silica particles have been found particularly useful in separating secondary microbial metabolites from microbial broths. Typical established downstream separation protocols involve low-resolution high-capacity procedures for recovery of a major product, with preparative HPLC employed near the end of a process train to recover minor components. For a compound or culture that is undergoing an investigation where proof of structure is urgently required or a sample of product is required for preliminary biological evaluation, preparative HPLC may be employed at an earlier stage.

15.6.3 CHOICE OF PARTICLE SIZE

This important decision may be reduced in essence to deciding between small (5–15 µm) diameter particles or medium (30–60 µm) diameter particles.

Reference [16] provides a suitable theoretical background based on a selectivity factor (α) and theoretical plate count (N) approach. It describes how separations with α values of > 1.3 may be achieved with columns packed with 40–60 µm particles, giving a theoretical plate count of about 250. More difficult separations with α values of about 1.1 are shown to need higher efficiency columns containing about 2000 plates, which require the use of small particle size packings, usually 10 µm diameter. Two important points are that trial separations used to maximize α should be carried out on an analytical column containing a support with identical packing to that used in the corresponding preparative column, and that efficiency rapidly decreases as load is increased for columns containing small (10 µm) packings. Further examples of the effect of flowrate and sample size on different particle size columns are contained in Refs. [17] and [18]. The implication of these findings is that small-particle preparative columns are a viable option but loading levels should be carefully controlled when they are used. Because of this, automatic repetitive injection and fraction collection is frequently employed with small-particle columns [19].

15.6.4 CHOICE OF COLUMN TYPES

Verzele [20] surveyed different column hardware available and concluded that smaller particles were preferable and column bed compression was desirable. Axial compressed columns have been available for some time [21] and currently columns are available with diameters up to 450 mm or greater with high-pressure ratings suitable for 10 µm packings [22]. While such large columns are suitable for production use on high-value products and require a large, specialized (e.g. zone 1: see Chapters 19 and 20) hall space for their location, much smaller columns may also use substantial quantities of solvent in continuous operation [23]. For example, a single preparative HPLC column of 21 mm i.d. (small particle size) operated at 25 mL/min will consume 36 litres of mobile phase every 24 hours. Clearly, even with relatively small preparative columns adequate precautions must be taken with respect to the handling and storage of large amounts of solvent.

15.6.5 APPLICATION OF PREPARATIVE HPLC TO THE PURIFICATION OF MILBEMYCINS

The foregoing discussion can be exemplified by considering an extraction process used for the isolation of two novel antihelmintic metabolites (see Figure 15.9). These metabolites, VM 48130 and VM 48633 (Figure 15.10), are both milbemycins which were isolated from a *Streptomyces hygroscopicus* broth. Although present in only minor amounts, separation of these compounds and their positive identification was desired as part of a structure-activity programme investigating antiparasitic agents.

In this example the fermentation broth was harvested after 404 hours and the milbemycins were present with a low titre. The procedure used involved

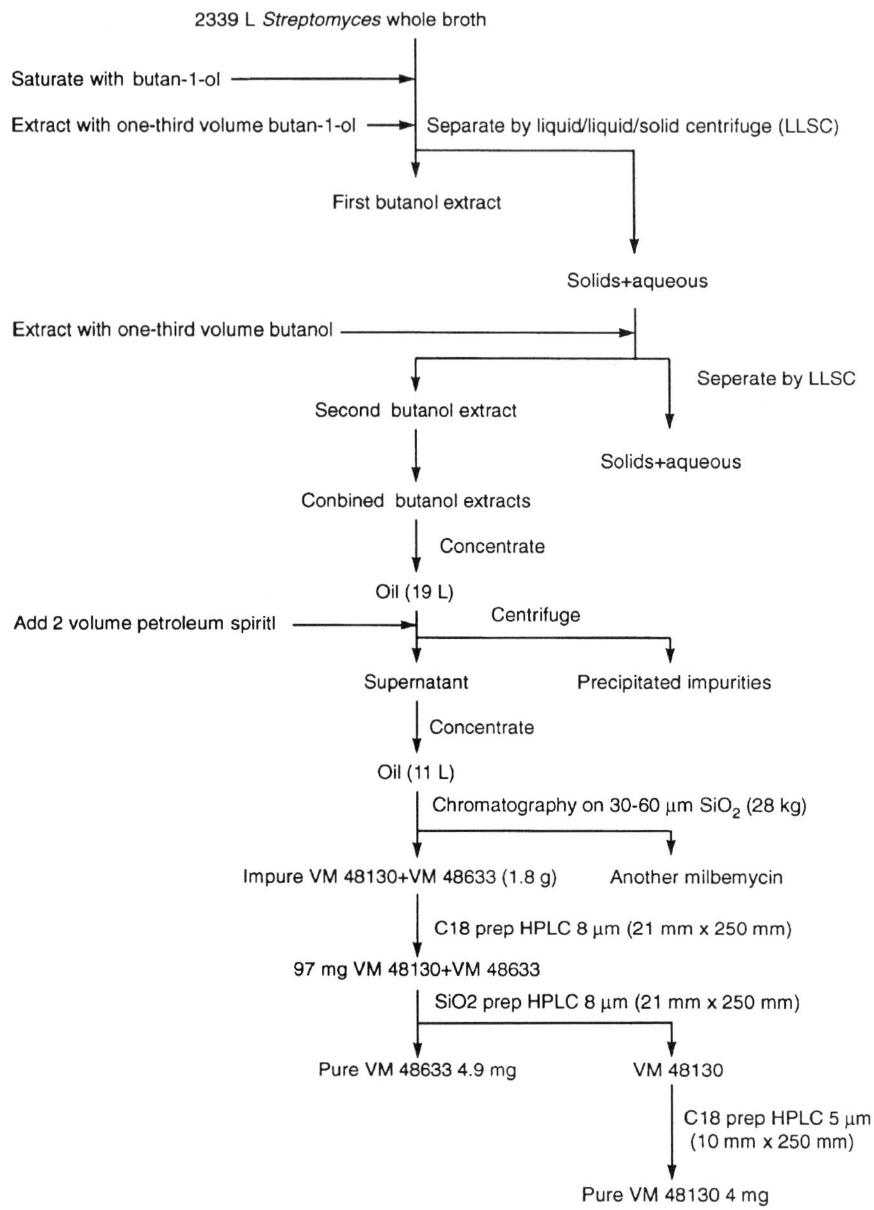

Figure 15.9   Extraction of milbemycin analogues

initial solvent extraction of whole broth followed by concentration of the rich
solvent extract. This readily reduced the volume from 2339 litres of whole
broth to 19 litres of oil and shows how effective a solvent extraction step can
be (see also Chapter 5). Petroleum spirit was then used to remove 8 kg of

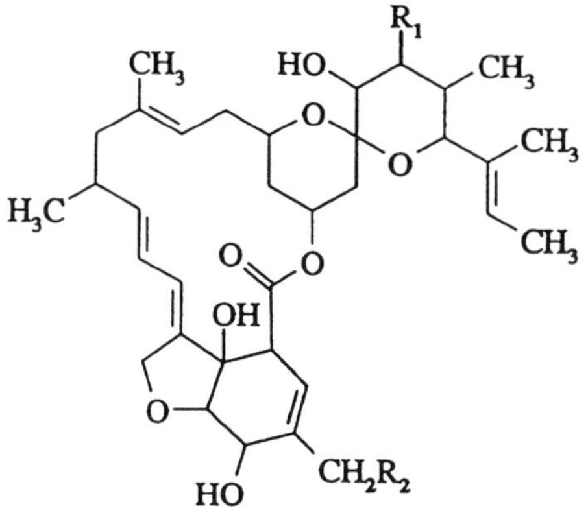

VM 48130: R$_1$ = –O–CO–CH(CH$_3$)$_2$,   R$_2$ = H

VM 48633 R$_1$ =   H   ,   R$_2$ = –O–CO = C(CH$_3$)$_2$

Figure 15.10   Structures of VM 48130 and VM 48633

impurities by precipitation. Removal of bulk impurities by changing solvent polarity in such a way can frequently aid microbial metabolite purification and reduce the solids load on subsequent chromatography stages. The remaining oil was loaded on a conventional adsorption silica column packed with 30–60 μm particles. This column was operated at low pressure with step gradient elution using increasing concentrations of ethyl acetate in petroleum spirit. This technique gave sufficient resolution to remove another milbemycin present and gross impurities from VM 48130 and VM 48633, but was unable to resolve the two minor milbemycin components in pure form.

Purification was continued by first removing further impurities using reverse phase preparative HPLC (methanol–water, with UV monitoring at 244 nm) giving > 18-fold increase in purity. Normal phase preparative HPLC (hexane–acetone gradient) gave pure VM 48633 and changing back to reverse phase preparative HPLC, but with the smaller particle size of 5 μm (methanol-water gradient) gave pure VM 48130. Alternating modes of chromatography were used to prevent the co-purification of similar polarity impurities which may occur towards the end of a purification sequence. VM 48130 and VM 48633 showed parasiticidal properties against nematodes such as *Trichostrongylus colubriformis* and they or their derivatives could be useful for the treatment of helminthiasis in mammals [24].

15.7 References

1. R. Majors (1993) 'Comparative study of European and American trends in sample preparation', *LC-GC Int.*, **6**, 130–40.
2. R. Majors (1993) 'Automation of solid phase extraction', *LC-GC Int.*, **6**, 349–50.
3. L. Jordan (1993) 'Automating a solid phase extraction method', *LC-GC Int.*, **6**, 594–9.
4. J. W. Dolan (1993) 'Amine adsorption: a case study', *LC-GC Int.*, **6**, 142–5.
5. L. R. Snyder (1988) *Practical HPLC Method Development*, John Wiley, New York, p. 62.
6. C. White, W. A. Moats and K. L. Kotula (1993) 'Comparative study of HPLC methods for the determination of tetracycline antibiotics', *J. Liq. Chrom.*, **16**, 2873–90.
7. J. J. Kirkland, C. H. Dilks and J. E. Henderson (1993) 'Technologies for an improved C_{18} stationary phase in reversed phase HPLC separations' *LC-GC Int.*, **6**, 436–40.
8. J. H. Knox (Ed.) (1978) *High Performance Liquid Chromatography*, Edinburgh University Press.
9. N. Afeyan, S. P. Fulton, N. F. Gordon, I. Mazaroff, L. Várady and F. E. Regnier (1990) 'Perfusion chromatography: an approach to purifying biomolecules', *Bio/technology*, **8**, 203–6.
10. I. H. Hagestam and T. C. Pinkerton (1985) 'Internal surface reverse reversed phase silica supports for liquid chromatography', *Anal. Chem.*, **57**, 1757–63.
11. D. B. Arkhipov and B. G. Belenkii (1993) 'Trends in the development of liquid chromatography instrumentation', *LC-GC Int.*, **6**, 370–6.
12. G. Wagman and R. Cooper (Eds.) (1989) *Natural Products Isolation*, J. Chromatography Library, Vol. 43, Elsevier, Oxford, p. 57.
13. A. F. Fell and B. J. Clark (1987) 'Computer aided multichannel detection in liquid chromatography: current developments and future perspectives', *European Chromatogr. News*, **1**(1), 16–22.
14. L. Huber and S. A. George (Eds.) (1993) *Diode Array HPLC*, Marcel Dekker, New York.
15. J. Bérdy (1985) in M. S. Verrall (Ed.), *Discovery and Isolation of Microbial Products*, Ellis Horwood, Chichester, p. 9.
16. R. D. Sitrin, P. A. DePhillips and J. J. Dingerdissen (1987) 'Preparative reversed phase HPLC of polar fermentation products', *Dev. Ind. Microbiol.*, **27**, 65–75.
17. M. Verzele and C. Dewaele (1986) *Preparative High Performance Liquid Chromatography*, TEC, Groot Brittaniëlaan, Gent, p. 115.
18. M. Verzele (1990) 'Preparative liquid chromatography', *Anal. Chem.*, **62**, 265A–269A.
19. G. Franke and F. Verillon (1988) 'Economic laboratory practice in preparative column HPLC', *J. Chromatogr.*, **450**, 81–9.
20. M. Verzele, M. de Coninck, J. Vindevogel and C. Dewaele (1988) 'Column hardware in preparative liquid chromatography with axial flow', *J. Chromatogr.*, **450**, 47–69.
21. P. Hairsine (1986) 'Preparative HPLC', *Lab. Practice*, January, 37–9.
22. G. B. Cox (1990) 'Design of large diameter columns for preparative liquid chromatography', *LC-GC Int.*, **3**(10), 10–16.
23. J. Krohn and J. Rzittky (1991) 'Preparative HPLC on a laboratory scale', *LC-GC Int.*, **4**(4), 36.
24. US Patent, 5 045 457.

16 SUPERCRITICAL FLUID EXTRACTION AND CHROMATOGRAPHY

Anthony A. Clifford

16.1 Introduction

Supercritical fluid technology is now some 15 years old. In this time it has been applied largely in the food industry, but it has been researched in other areas and it is widely believed that the most likely large area of exploitation will be in pharmaceuticals. This is for two reasons: firstly, because the techniques allow the use of environmentally friendly solvents which do not leave residues of organic solvents in the products and, secondly, because the technique involves specialized high-pressure equipment, which is less of a consideration for high-value products.

This chapter begins by explaining what a supercritical fluid is and making some general points about these media. Supercritical fluid extraction (SFE) is then described. Wide experience of this technique has been obtained from analytical sample preparation and this is readily applied to extractions of plant and microbial products on a small scale. Pilot plants are readily available for the scaling-up of SFE. The next section describes supercritical fluid chromatography (SFC), which again is well developed as an analytical technique. Preparative-scale SFC is beginning to be carried out using modified preparative HPLC (high-performance liquid chromatography) equipment, and systems are being developed for process-scale SFC on a pilot plant scale. The final section describes briefly the production of microparticles and coated microparticles from the rapid expansion of solution (RESS), which can produce drugs suitable for drug formulations at the end of an SFE or SFC process. Thus coupling of these techniques, i.e. SFE–SFC, SFE–RESS, SFC–RESS and indeed SFE–SFC–RESS processes can be envisaged.

Downstream Processing of Natural Products. Edited by Michael S. Verrall
©1996 John Wiley & Sons Ltd

16.2 Supercritical Fluid

16.2.1 THE NATURE OF A SUPERCRITICAL FLUID

When two molecules approach each other in a fluid, at a temperature where their relative speed is likely to be low, their mutually attractive forces will bring about a temporary association between them. If there is a sufficient density of molecules, there is the possibility of condensation to a liquid. On the other hand, if the temperature and the probable relative speeds are high, the attractive force will be too weak to have more than a slight effect on the molecular velocities, and condensation cannot occur however high the molecular density. It is therefore reasonable to expect, on the basis of molecular behaviour, that for every substance there is a temperature below which condensation to a liquid (and evaporation to a gas) is possible, but above which these processes cannot occur.

That there is a *critical temperature* above which a single substance can only exist as a fluid and not as either a liquid or gas was shown experimentally 170 years ago by Baron Charles Cagniard de la Tour. He heated substances, present as both liquid and vapour, in a sealed cannon which he rocked back and forth and discovered that, at a certain temperature, the splashing ceased. Later he constructed a glass apparatus in which the phenomenon could be more directly observed. These experiments can be explained by reference to Figure 16.1, which is the phase diagram of a single substance. The areas labelled S, L and G show where the substance exists as a single solid, liquid or gas phase respectively, and T is the triple point where the three phases coexist. The lines represent coexistence between two of the phases. If we move upwards along the gas–liquid coexistence curve, which is a plot of vapour pressure

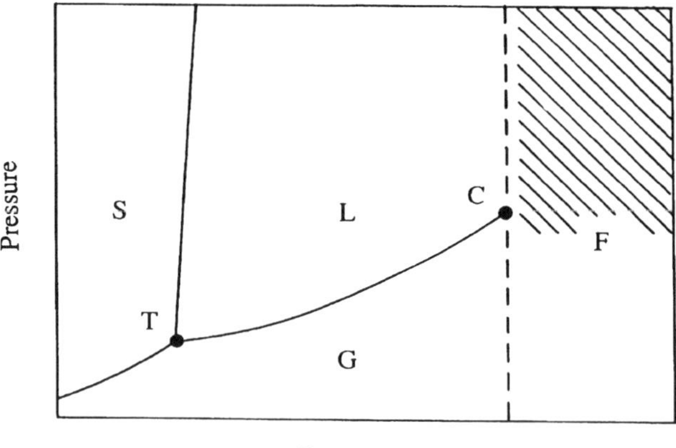

Figure 16.1 Phase diagram of a single substance: S = solid, L = liquid, G = gas, T = triple point, C = critical point, F = supercritical fluid

Table 16.1 Critical parameters of selected substances useful as supercritical fluids

	$T_c(°C)$	$p_c(atm)$	$\rho_c(10^3 kg/m^3)$
CO_2	31.3	72.9	0.47
N_2O	36.5	72.5	0.45
SF_6	45.5	37.1	0.74
NH_3	132.5	112.5	0.24
H_2O	374	227	0.34
$n\text{-}C_4H_{10}$	152	37.5	0.23
$n\text{-}C_5H_{12}$	197	33.3	0.23
Xe	16.6	58.4	1.10
CCl_2F_2	112	40.7	0.56
CHF_3	25.9	46.9	0.52

versus temperature, both temperature and pressure increase. The liquid becomes less dense because of thermal expansion and the gas becomes more dense as the pressure rises. At the critical point, C, the densities of the two phases become identical and the distinction between the gas and the liquid disappears. The substance can only be described as a fluid. The critical point, C, in Figure 16.1 has pressure and temperature coordinates on the phase diagram, for a particular substance, which are referred to as the critical temperature, T_c, and the critical pressure, p_c.

In recent years, fluids have been exploited above their critical temperatures and pressures and the term 'supercritical fluids' has been coined to describe these media. The greatest advantages of supercritical fluids are typically not too far (say 10–50 K) above their critical temperatures. Nitrogen gas in a cylinder is a fluid, but is not usually considered as a supercritical fluid, but more often described by an older term as a permanent gas. Sometimes substances are exploited a little below the critical temperature in the liquid region where the behaviour is similar to that of a supercritical fluid. For this reason some people prefer the term near-critical fluids. The term supercritical fluid has gained currency, is convenient and not a problem provided too rigid a definition is not applied. The region for supercritical fluids is thus the hatched area in Figure 16.1

Table 16.1 shows the critical parameters of some compounds useful as supercritical fluids. One compound, CO_2, has so far been the most widely used, because of its convenient critical temperature, cheapness, non-explosive character and non-toxicity, all of which are important in the pharmaceutical industry. Because the molecule is non-polar it is classified as a non-polar solvent, although it has some limited affinity with polar solutes because of its large molecular quadrupole. Thus pure CO_2 can be used for many large organic solute molecules even if they have some polar character.

16.2.2 USE OF ENTRAINERS

It has been the practice for CO_2, if insufficiently polar, to be modified with polar entrainers, such as methanol, ethanol, propanone, halocarbons and tributyl phosphate. The original motive was to increase solubilities of polar compounds, but it is now believed that equally important effects are the interaction of the modifiers with a solid matrix during extraction and with the stationary phase in chromatography. If the latter are the reasons for modifier addition, only small percentages of modifier need be added. For the pharmaceutical industry, entrainers, such as ethanol, can be chosen which are acceptable in residual amounts in a product. If RESS is being carried out, entrainers are not used because of the solubility of the particles produced in the entrainer.

If entrainers are used, it is important to be aware of the entrainer fluid phase diagram to ensure that the solvent is in one phase and obtain the advantage of supercritical conditions. For a mixture, such as ethanol–CO_2, this means that, for a given temperature, the pressure must be above a given value, depending on composition. The highest pressure that is required (for the *critical composition*) is called the 'critical pressure of the mixture' for the particular temperature. This can be understood by considering a mixture (of the critical composition) of CO_2 and ethanol above the critical temperature of CO_2. If this system is at relatively low pressure, it will consist of two phases: one liquid ethanol containing some dissolved CO_2 and the other supercritical CO_2 containing solvated ethanol. As the pressure is raised, the liquid phase will dissolve more CO_2 and the supercritical phase solvate more ethanol and will increase in density. Eventually, at the critical pressure of the mixture, the compositions and densities of the two phases will become identical and only one supercritical fluid phase will exist. Provided that the system is above the critical pressure of the mixture, only one (supercritical) phase will exist, whatever the composition. This pressure for most common entrainers in CO_2 is usually 100–150 bar at, say, 50 °C and below most pressures that are chosen for other reasons.

16.2.3 ADVANTAGES AND USES OF SUPERCRITICAL FLUIDS

Although often pursued in practice for environmental reasons, toxicological considerations and for the ease of separation of the solvent, other advantages arise from supercritical fluids because they can have properties intermediate between those of typical gases and liquids. Compared with liquids, densities and viscosities are less and diffusivities greater. These properties may be optimum for a particular process or experiment. Furthermore, properties are controllable by both pressure and temperature, and the extra degree of freedom, compared with a liquid, can mean that more than one property can be optimized. Any advantage has to be weighed against the cost and inconvenience of the higher pressures needed. Consequently, supercritical fluids are exploited in particular areas. Apart from the topics discussed in this chapter, chemical reactions are being studied, and there is a plant now in operation in Japan producing 40 000 tons of butanone per annum. Recrystallization from supercritical fluids is also being pursued in areas other than as a method of producing micronized particles. Spray-painting and dyeing are also being carried out from supercritical fluids.

For SFE and SFC advantages arise from the greater diffusivities and faster mass transfer that occur in supercritical fluids. In SFE, there is more rapid diffusion along channels and pores in a natural product matrix and extraction is faster than in a liquid. In SFC narrower peaks or bands are obtained because the more rapid diffusion of a solute between streams of different velocity keep the velocities of individual molecules closer to the average.

Pressurization under CO_2 followed by rapid depressurization is now being studied as a method of disrupting cells prior to extraction. This process can be incorporated as a preliminary stage to SFE.

16.3 Supercritical Fluid Extraction

16.3.1 SFE ON A SMALL SCALE

SFE needs to be carried out on as small scale in preliminary studies for an SFE process and some of the methods and equipment developed for analytical sample preparation are suitable for this. It is therefore described first and many of the principles are relevant to larger-scale operation. Many of the commercial analytical systems, which involve automated sample introduction and solute trapping, are, however, rather too sophisticated for the simple experiments required in this context. A simple SFE system, as shown below, is adequate and a number of versions have been assembled at little cost in our laboratory. There is also at least one low-cost simple system available commercially. Moreover, for the design of a process, it will be necessary at some stage to carry out studies of the kinetics of SFE, to produce a curve of percentage recovery versus time, and this is not always easy on some commercial instruments. It should be mentioned that a small-scale SFE apparatus can be used to obtain samples of natural product extracts, which can be screened for biological activity.

A simple but effective apparatus for carrying out SFE is shown in Figure 16.2. It consists of: a solvent supply (A), a pump (B), a cooler to cool the pump head (C), an extraction cell (D), which is mounted in a ceramic heater tube (E), a heater controller to control the temperature of the extraction cell (F), a restrictor connected to the outlet of the cell (G), a restrictor heater (H) to prevent deposition and blockage in the restrictor and a collection vial (I) containing a liquid solvent. If the pump is set to control the pressure, the restrictor controls the flowrate. If the pump is set to control the flowrate, the restrictor maintains the pressure of the extraction. Cooling of the pump head

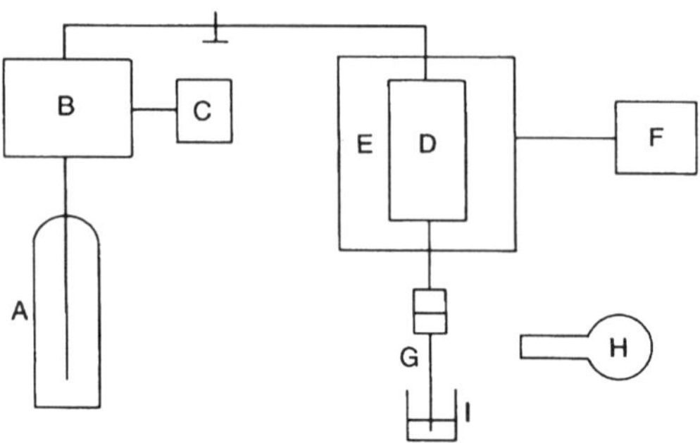

Figure 16.2 Schematic apparatus for carrying out SFE: solvent supply (A), pump (B), cooler to cool the pump head (C), extraction cell (D), ceramic heater tube (E), heater controller (F), restrictor (G), restrictor heater (H), collection vial (I)

can be carried out for fluids, like CO_2, whose critical temperatures are not much above ambient, to ensure that the fluid in the pump head is always liquid and the flow is uniform. The alternative is to use CO_2 in a cylinder with an overpressure of helium of *ca.* 100 bar. The CO_2 then contains some 3% w/w of helium, which only has a small effect on the results. Heating of the restrictor is to prevent blocking by, for example, fats and ice (the latter due to cooling because of the expansion of the fluid) and is not always necessary.

In a representative experiment, the cell would have a volume of 1 mL and be loaded with about 0.5 g of matrix to be extracted, which would be ground into particles of *ca.* 0.1 mm size. (Techniques of sample preparation are discussed in the next section.) CO_2 would be pumped at a rate of 0.5 mL per minute, measured for the liquid at the pump. The temperature would be 50 °C and the pressure 400 bar, maintained by a restrictor consisting of a silica tube of 25 μm internal diameter and length around 12 cm. The effluent would be trapped in about 3 mL of solvent in a vial, ready for analysis. A reference standard would be added to the vial, for quantitative analysis. It would probably be necessary to top up the solvent during extraction because of evaporation. The extraction would be carried out for 30 minutes. Conditions of experiments vary widely, however.

Pumps for SFE are either those used for HPLC or products specially sold for SFE application, which are derived from HPLC designs. They can be of the syringe or reciprocating type. Because typical fluid substances are more compressible at ambient temperature, pump control systems are often more sophisticated than for HPLC pumps, with advantages for the operator. Syringe pumps have the disadvantage of needing refilling, but have the advantage that they can be designed to more rapidly come up to pressure when the pump is opened to the extraction cell. Pumps may be designed to control either pressure or flowrate, but often measure both.

Modified fluids may be delivered by the following methods:

(a) Using a two-pump system with perhaps a proprietary device for mixing the two fluids.
(b) Using a single pump, fed by cylinders of premixed solvents, of which many are now available commercially, e.g. CO_2 containing 10% w/w ethanol. The entrainer concentration will vary a little as the cylinder is used up, especially towards the end.
(c) Filling a syringe pump first with the modifier and then the major fluid component. Mixing of the components relies on turbulence during the filling process. There is a chance that the mixture is not uniform when this method is used.
(d) In a well-equipped laboratory, a dual pumping system, which is part of a supercritical fluid system, can be 'borrowed' to fill several syringe pumps, when the system is not being operated, via a network of tubing.
(e) Adding the modifier/entrainer to the matrix in the cell at the beginning of the experiment and carrying out an initial static extraction, as described below.

SFE can be carried out in the dynamic mode, as described above, or alternatively in a static mode, e.g. for adding entrainer, as in method (e) above. For static use, valves are fitted on either side of the cell. The outlet valve is closed, the cell filled with fluid and the inlet valve is closed. The fluid is then left in contact with the matrix for, say, 1 hour. The valves are then opened and the fluid in the cell is swept into the analyte trap by performing a short dynamic extraction.

For a preliminary study of a possible process, kinetic studies of extraction are desirable, for which the simple system of Figure 16.2 is appropriate. During a kinetics experiment the solvent collection vial is replaced at timed intervals. Because of the form of the extraction curve, these times are not evenly spaced, but are shorter at the beginning of a run. In a representative experiment, vials would be changed at 2, 4, 8, 12, 20, 30 and 40 minutes after the start of extraction and the run continued until 60 minutes. After analysis, curves of percentage extracted versus time can be drawn, if a value of the total amount of material (100%) is obtained from exhaustive liquid extraction, perhaps with ultrasonication. The form of the curves obtained are discussed in Section 16.3.3.

16.3.2 SFE ON A PILOT PLANT SCALE

Pilot plants for SFE, with pressure vessels of a few litres volume, are now readily available and an example is given in Figure 16.3. A schematic diagram of a simplified pilot plant is given in Figure 16.4, consisting of: a fluid supply (A), a fluid reservoir (B), cooled by a cooler (C), a pump (D), a heat exchanger for further cooling (E), also cooled by (C), an extraction vessel (F), whose temperature is controlled by a hot water supply (H), and a separation vessel

Figure 16.3 The SFE and RESS pilot plant at Leeds (Separex, Champigneulles, France), showing a series of separation vessels

(G), whose temperature is controlled by another hot water supply (I). Valves for controlling pressure (J and K) are included in the system. The control of flowrate and pressures in the extraction and separation vessels can be achieved using the combination of a variety of types of valve and pump controls. In operation, the fluid is condensed from the fluid supply into the fluid reservoir

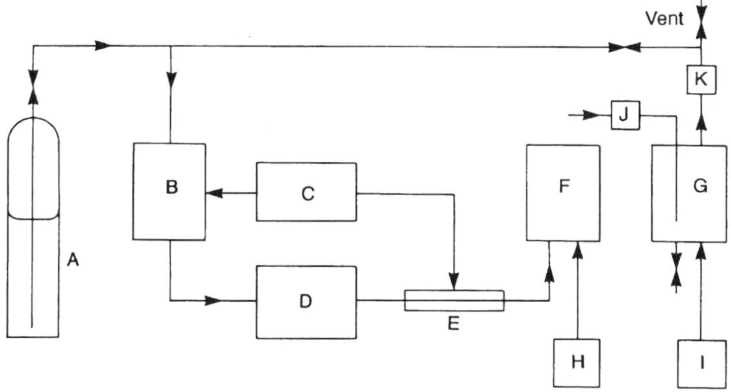

Figure 16.4 Schematic diagram of a simplified pilot plant: fluid supply (A), fluid reservoir (B), cooler (C), pump (D), heat exchanger (E), an extraction vessel (F), separation vessel (G), temperature controllers (H and I) and pressure-control valves (J and K)

and pumped, at the desired flowrate, via the cooling heat exchanger into the extraction vessel, which is maintained at the desired temperature and pressure for extraction of the solute. The pressure of the fluid then drops as it passes through valve J and into the separation vessel, which is maintained at another pressure and temperature at which the desired solute precipitates. On emerging from the separation vessel, the fluid can then either be vented or recondensed into the fluid reservoir. In a representative run, 1 kg of dried plant tissue, ground to a particle size of 0.3 mm, is loaded in the extraction cell between two metal frits. This material is extracted at 50 °C and 250 bar with CO_2 for 1 hour at a fluid flowrate of 3 kg per hour. The pressure is dropped to 80 bar in the separation vessel, which is also maintained at 50 °C and the CO_2 vented to the atmosphere. At the end of the run, the extracted material is removed from the bottom of the separation vessel. As with small-scale extraction, a static extraction, followed by a dynamic extraction to sweep out the material, can be carried out by suitable valve control.

The simple separation system given above can be made more complex to give improved separation of the extract. Firstly, the single separation vessel can be replaced by a series of vessels connected by pressure-control valves, each at a lower pressure than the previous vessel and at different temperatures, as illustrated in Figure 16.3. The extract can then be crudely fractionated during collection. Secondly, a number of separation vessels can be configured so that they can be alternately switched into line as a complex programme of conditions of pressure, temperature and entrainer concentration is operated in the extraction vessel. Finally, the separation vessel can be replaced by a vertical column, packed with inert particles, along which a temperature gradient is maintained. The highest temperature is arranged typically, though not necessarily, at the top of the column, the important point being that the solubilities of the compounds of interest should fall as they move up the column. The column is at almost constant pressure and is chosen so that the density of the fluid is close to its critical density, where solubilities change most rapidly with temperature. The eluent from the extraction vessel is introduced at the bottom of the column and a countercurrent of fluid and separated extract is set up in the column, analogous to a distillation column but at much lower temperatures due to the solvating effect of the fluid. To bring off the less soluble fractions towards the end of this process, the temperature of the column (still with a gradient) must be lowered (typically) or the pressure raised.

In the case of dried plant material loading the cell is straightforward, but in other cases preparation of the matrix to be extracted is necessary. The type of preparation may be different from what is possible in an eventual large-scale process, which may be continuous and mechanical mixing with the fluid introduced. In general material must be of such a form that particles of a fraction of a millimetre across are packed tightly enough so that fluid does not find its way along easy channels and bypass the matrix, but can still pass easily between the individual matrix particles; this situation should be maintained throughout the extraction process. Wet particles such as a paste of microbial cells, for example, tend to shrink into impenetrable hard lumps, which the fluid bypasses. Such material needs to be freeze-dried and powdered or intimately mixed with a water-absorbing support such as silica or diatomaceous earth. An alternative is to extract a concentrated slurry in water, by sparging small bubbles upwards through a frit. Oils and fats need to be supported on an inert support such as fine sand, with which they must be mixed intimately to evenly coat the support particles and leave passages for the fluid. Lipid material which might be expected to be solid at the extraction temperature is likely to go into a liquid phase as it absorbs CO_2 or whatever fluid under pressure. From the average diameter of the support particles and the relative densities of the support and oil, the thickness of the coating can be calculated and it should be one-fifth or less of the particle diameter.

As with small-scale extractions, entrainers can be introduced in a number of ways, which include adding a separate entrainer pump and adding the entrainer initially to the extraction cell and carrying out an initial static extraction. Since the function of a modifier is often its effect on the matrix, the latter method has appeal and may reduce the amount of entrainer involved. When entrainer is used, the extract produced is dissolved in at least part of the entrainer. It is advantageously recovered more easily from the separation vessel, but of course has to be separated from the entrainer.

16.3.3 PRINCIPLES OF SFE

Extraction by a supercritical (or any) fluid is never complete in finite time. It is relatively rapid initially, but there then follows a long tail in the curve of percentage extracted versus time. For a successful SFE procedure, the curve of

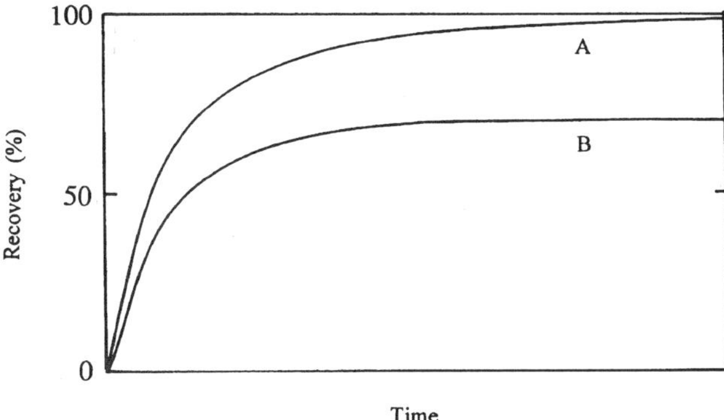

Figure 16.5 Schematic curves of percentage extracted versus time for SFE: a curve with matrix effects largely absent (A); a curve showing substantial matrix effects (B)

percentage extracted versus time has the form of curve A in Figure 16.5. In a typical situation 50% is extracted in 10 minutes, but it may be 100 minutes before *ca* 99% is extracted. In worse circumstances the kinetic behaviour may be of the form of curve B. Here there is a short slow period, followed by a rapid extraction and then a very slow extraction for a substantial fraction of the compound of interest, so that the SFE seems to have finished when only part of the compound has been recovered. Kinetic curves like Figure 16.5, obtained under different conditions, allow economic decisions to be made on the percentage of substance to be recovered in a process, which will also depend on the value of the product.

It is considered that three interrelated factors influence recovery, as shown in the SFE triangle:

For extraction to be successful, the solute must, firstly, be sufficiently soluble in the supercritical fluid. Often modest solubility only is required, as the other factors will also delay recovery. Solubility depends on the nature of the fluid, its density during SFE and the temperature. Density rather than pressure is important: raising the temperature at constant temperature will reduce density and the net effect may be a reduction in solubility. Information about solubility for compounds of interest in possible fluids may be obtained from the literature directly or inferred from reports of successful applications of SFE or SFC. If an SFC apparatus is available, experiments for the analyte/fluid systems can be carried out. The most direct method to determine whether the solubility is sufficient is to deposit the analytes of interest on to filter paper and attempt to extract them using a small-scale SFE system under various conditions. Control of solubility can allow stepwise extraction. For example, the bulk of the triglycerides may be extracted from a lipid material by CO_2 alone before compounds of interest are extracted by CO_2 containing ethanol.

Secondly, the compound of interest must be transported sufficiently rapidly by diffusion or otherwise from the interior of the matrix in which it is contained. The transport process is often diffusion of the fluid into the matrix followed by diffusion of the analyte through the fluid along pores in the matrix. The time-scale for diffusion/transport will depend on the shape and dimensions of the matrix or matrix particles. Of these the shortest dimension is of great importance, as the times depend on the inverse of the square of its value. Values for the shortest dimension of 1 mm or preferably less are usually necessary.

The third factor is that of the matrix (other than its effect on diffusion). Examples of the effect of the matrix are adsorption of compound molecules on surface sites and the need for compound molecules to penetrate cell walls in a biomatrix. All of these processes are capable of providing a rate-determining step in the later stages of an extraction process. This factor is considered responsible for the very slow final stage in some applications, as illustrated by curve B in Figure 16.5. Often this final stage is so slow that it appears that not all of a compound is 'extractable', the rest being locked into the structure of the matrix or too strongly bound to its surface. In fact the compound is still being extracted slowly. Of the three factors, that of the matrix is the least well understood at present. The key to this problem is often the nature of the supercritical fluid, which, if the use of CO_2 is desired, means the choice of entrainer, which is limited by consideration of toxic residues. Raising the temperature also improves the situation, but this may not be possible if the compound of interest is thermally labile. Pretreatment of the matrix by physical or chemical degradation is another possibility.

16.4 Supercritical Fluid Chromatography

SFC is well established as an analytical technique and a large volume of information is available in textbooks and the literature in general. Analytical SFC cannot therefore be covered in this chapter, which will be restricted to a discussion of preparative SFC. Of course, for any particular separation problem background information on analytical behaviour from the literature of experiments is very useful. The equipment described in the next section was built in-house, but commercial systems on a preparative and large scale are being developed. Preparative SFC is being used for separations that are difficult by HPLC, because of the increased efficiency of SFC over HPLC as a result of more rapid diffusion. Because peaks are narrower in SFC, and for these difficult separations very little overloading can be done to take advantage of the narrow peaks, the maximum amount of material obtained in a run is of

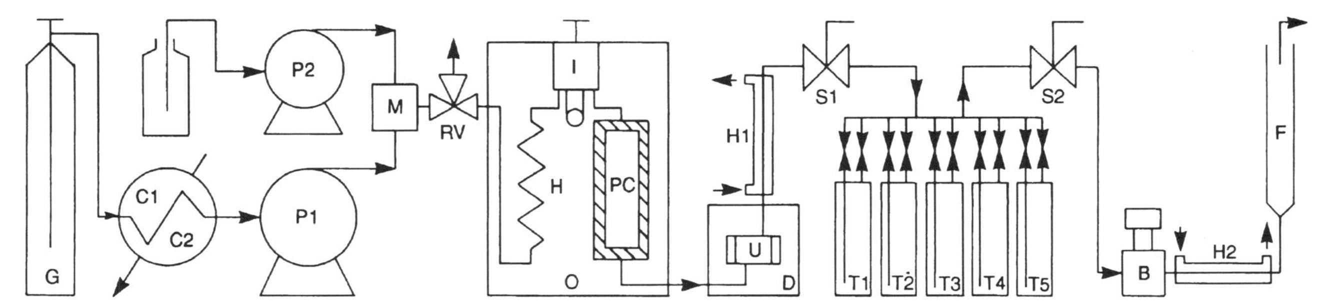

Figure 16.6 Preparative SFC system: CO_2 pump (Varex) (P1), liquid pump (Gilson) (P2), dynamic mixer and damper (Gilson) (M), pressure release valve (RV), oven (Dupont) (O), preheater (H), injection valve (Rheodyne) (I), preparative column (phase separations) (PC), UV detector (Jasco) (D), UV cell (in-house) (U), heat exchanger for cooling (in-house) (H1), seven-port valves for switching between traps (Rheodyne) (S1, S2), traps (in-house) (T1–T5), heated back-pressure regulator (Jasco or GO) (B) and bubble flowmeter (F)

the order of 100 mg in SFC compared with the 1 g amounts obtainable sometimes in HPLC.

16.4.1 A PREPARATIVE SFC SYSTEM

Figure 16.6 shows a preparative system (built at Leeds by Dr S. A. Jafar with advice from Dr C. D. Bevan of Glaxo, but representative of similar systems built elsewhere), which consists of a pumping system, a core chromatographic section, containing an injection loop, column and detector, and a trapping system followed by a back-pressure regulator to maintain the pressure of the whole system. The pumping system is capable of delivering a mixed fluid (e.g. carbon dioxide modified with ethanol) at a flowrate of 20 mL of CO_2 per minute and 5 mL per minute of modifier at a pressure of 300 bar. The CO_2 pump head is cooled by a commercial cooler, and a dynamic mixer and damper is used to provide the even flow and composition needed for efficient SFC. After passing through a preheater to bring it to supercritial conditions, the fluid then passes through an injection valve with a 1 mL sample loop. A preparative column designed for HPLC (250 mm long and 20 mm internal diameter) is enclosed in a pressure safety vessel, made in-house. It is installed in an HPLC oven controllable between ambient and 100 °C. A UV detector is used to follow the separation and the eluent is cooled in a coil in a water-bath to ambient temperature. Trapping of fractions of the peaks is thus done at room temperature where the mobile phase is liquid. The five traps are 20 mL in volume and are connected sequentially into the flow via the seven-port valves at times indicated by the chromatogram received by the detector. The traps are filled with the liquid CO_2 before the separation is carried out and during collection not more than 50% of liquid CO_2 is displaced. At the end of the separation, the traps are frozen in liquid nitrogen and opened to the atmosphere, when the mobile phase slowly evaporates. The solute is then collected from the traps using an organic solvent and analysed. The back-pressure regulator (mechanical or electromechanical) maintains the required pressure in the system and a bubble flowmeter is installed to check the flowrates indicated on the pumps.

16.4.2 OPTIMIZING PREPARATIVE SFC

The parameters to be optimized during preparative SCF are column type, temperature, pressure, modifier concentration, flowrate and loading. The initial search for conditions is carried out by observing the chromatogram obtained, without trapping. It is more economical if at least some of this work is carried out on an an analytic-scale SFC, particularly column, pressure, temperature and modifier concentration, and information may be available in the literature. Further optimization can then be done by carrying out trapping experiments in an apparatus such as that described in the last section. After analysis of the fractions and also by combining (theoretically) the fractions, a curve of purity

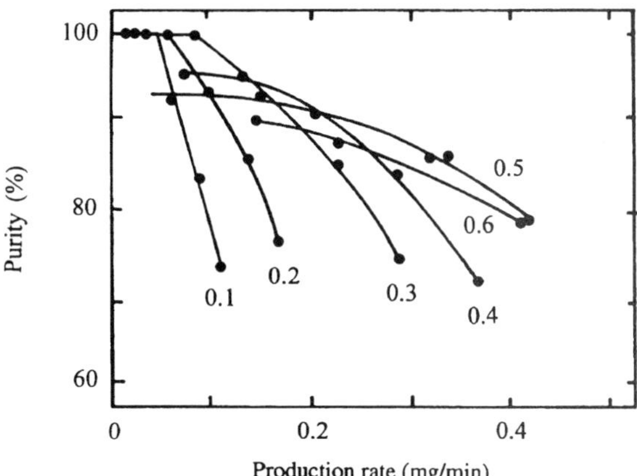

Figure 16.7 Purity versus production rate for the preparative SFC of milbemycin α_2 at various loadings of crude extract, shown in grams on the figure. Column, 250 mm × 20 mm internal diameter cyanopropyl bonded 10 μm silica particles; temperature, 47 °C; pressure, 120 bar; mobile phase, 5% methanol in CO_2 (v/v at pumps); flowrate of CO_2, 20 mL per minute

versus production rate can be drawn. An example, from the work of Ms N. Malak at Leeds with help from Mr M. S. Verrall at SmithKline Beecham, is given in Figure 16.7 for the isolation of milbemycin α_2 from a crude microbial extract at various loadings, other conditions being constant.

Below 100% purity, these curves are fitted to a quadratic and the equations obtained are used to calculate values of the production rate at a given purity for the various loadings. Curves can then be drawn of production rate versus loading for a given purity, which is shown in Figure 16.8 for the milbemycin data in Figure 16.7. As can be seen, there is an optimum loading for a certain required purity. Similar curves have been obtained for flowrate and modifier concentration. Theoretically, a multidimensional optimization could be carried out, and although we have had some success in extending it to two parameters, errors in the experimental data make this difficult. However, optimization along the lines suggested is worth while for a preparative SFC that is likely to recur and especially for the system that will be taken up to production scale.

16.5 Microparticle Formation

Solutions of a product in supercritical CO_2 or other fluids, obtained from SFE or SFC (or of course made for the express purpose), can be deposited as small particles by the process of RESS. Rapid expansion to atmospheric pressure of a solution in CO_2 produces quantities of dry ice, so this has to be avoided by heating the solution to, say, 80 °C, before expansion and/or reducing the pressure not to atmospheric but to a pressure of, say, 70 bar, where the product

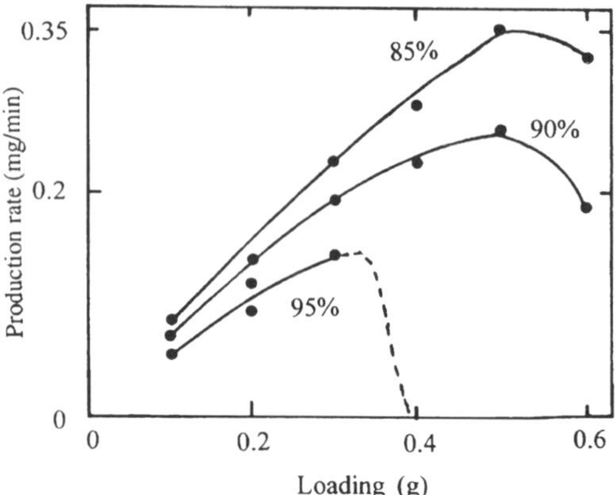

Figure 16.8   Production rate versus purity for the preparative SFC of milbemycin α_2 for various required purities shown on the figure. Column, temperature, pressure and flowrate as for Figure 16.7. A purity of 95% was not obtainable at loadings of 0.04 g and above and so a schematic dashed line is shown

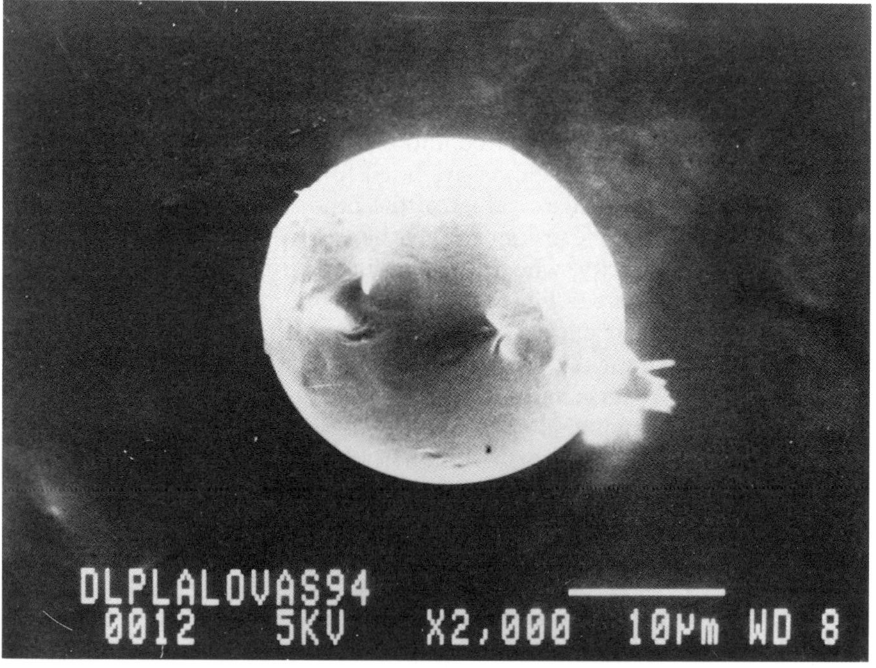

Figure 16.9   Scanning electron micrograph of a particle containing crystals of lovastatin embedded in a sphere of poly(lactic acid). (Reproduced with permission from Professor P. G. Debenedetti, Princeton University, New Jersey, USA)

is insoluble. The latter may have the advantage that unwanted, more soluble substances will not be deposited with the product. The expansion occurs through a fine nozzle, which may be a short length of stainless steel capillary or a fine hole cut by a laser in a stainless steel plate. Various shapes of nozzle have been suggested and the flow may be supersonic or subsonic in the nozzle. Using a nozzle of 25 µm, particles of around 1 µm diameter of narrow size distribution can be produced. These are suitable for formulations of drugs that are taken orally and are not very soluble in water. A variant of this method is to co-precipitate a drug with a slowly dissolving polymer to provide particles that can be injected into the bloodstream to achieve the slow release of the drug. Such particles are typical of 50 µm diameter and an example is given in Figure 16.9. Such co-precipitations are more technically involved than simple RESS and conditions for the formation of particles of the required configuration need detailed research.

16.6 Bibliography

There is now a very large amount of literature on SFE and SFC and much of this is in published proceedings of conferences and as chapters in edited books. Consequently, rather than list individual papers, a list of the main volumes in the area is given as a bibliography below.

Paulitis, M. E., Penninger, J. M. L., Gray, R. D. and P. Davidson (Eds.) (1983) *Chemical Engineering at Supercritical Conditions*, Ann Arbor Science, Michigan, Illinois.

Charpentier, B. A. and Sevenants, M. R. (Eds.) (1988) *Supercritical Fluid Extraction and Chromatography*, ACS Symposium Series 366, American Chemical Society, Washington D.C.

Johnston, K. P. and Penninger, J. M. L. (Eds.) (1988) *Supercritical Fluid Science and Technology*, ACS Symposium Series 406, American Chemical Society, Washington D.C.

Perrut, M. (Ed.) (1988) *Proceedings of the International Symposium on Supercritical Fluids*, Institut National Polytechnique de Lorraine, Nancy, France.

Vetter, G. (Ed.) (1990) *Proceedings of the 2nd International Symposium on High Pressure Chemical Engineering*, GVC, Erlangen, Germany.

Bruno, T. J. and Ely J. F. (Eds.) (1991) *Supercritical Fluid Technology*, CRC Press, Baton Rouge, Florida.

McHugh, M. A. (Ed.) (1991) *Proceedings of the 2nd International Symposium on Supercritical Fluids*, Johns Hopkins University, Baltimore, Maryland.

King, M. B. and Bott, T. R. (Eds.) (1993) *Extraction of Natural Products Using Near-Critical Solvents*, Blackie, Glasgow.

Kiran, E. and Brennecke, J. F. (Eds.) (1993) *Supercritical Fluid Engineering Science*, ACS Symposium Series 514, American Chemical Society, Washingon D.C.

McHugh, M. A. and Krukonis, V. J. (1994) *Supercritial Fluid Extraction, Principles and Practice*, 2nd edn, Butterworth-Heinemann, Stoneham, Massachusetts.

Perrut, M. (Ed.) (1994) *Proceedings of the 3rd International Symposium on Supercritical Fluids*, Institut National Polytechnique de Lorraine, Nancy, France.

Rizvi, S. S. H. (Ed.) (1994) *Supercritical Fluid Processing of Food and Biomaterials*, Blackie, Glasgow.

17 LIQUID MEMBRANES IN DOWNSTREAM PROCESSING

Julian B. Chaudhuri and Paul J. Pickering

17.1 Introduction

Liquid–liquid extraction is a well-established technology in the chemical industry. In the separation of natural biological products it has been confined to the antibiotics industry. With the developments of novel solvent extraction processes liquid–liquid extraction is beginning to be recognized as a potentially useful separation step for other classes of biochemicals and in enzyme recovery and separation.

There has been much interest in a new technique based on the principles of liquid–liquid extraction. Liquid membranes exploit the selective partition of a low molecular weight product into an organic solvent and enhance the separation by removing the equilibrium constraint for extraction. Separation is achieved by the selective transport of the product from an aqueous feed phase across a film of organic solvent into a second aqueous phase. It has significant potential for the selective separation and concentration of low molecular weight chemicals produced by fermentation.

17.2 Liquid Membrane Separations

17.2.1 SEPARATION PRINCIPLES

In membrane separation technology we usually think of the membrane as being solid. However, a liquid film which permits selective solute transport can also be considered as a membrane, a liquid membrane. The transport of a dissolved solute through this liquid membrane gives rise to the process of liquid membrane extraction (LME).

Downstream Processing of Natural Products. Edited by Michael S. Verrall
©1996 John Wiley & Sons Ltd

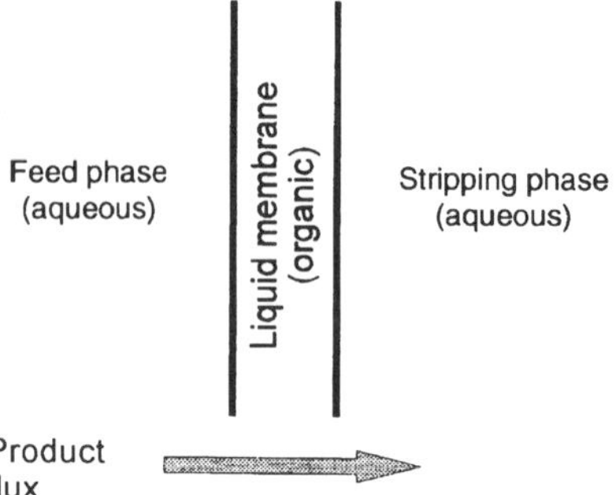

Figure 17.1   Liquid membrane system

A liquid membrane system consists of three liquid phases: an aqueous feed phase, a film of an organic solvent (the 'liquid membrane') and a receiving aqueous phase known as the 'stripping phase'. This arrangement is shown in Figure 17.1. The overall transfer process can be described as follows:

(a) The solute, originally in the feed phase, diffuses to the interface with the liquid membrane.
(b) The solute dissolves in the membrane phase and diffuses across it, either as a free solute or in a complexed form.
(c) At the interface with the stripping phase, the solute passes across this interface and into this phase, either by partition or via a chemical reaction.
(d) The extracted solute is prevented from back-transporting to the feed phase by alteration of its solubility in the organic solvent.

In liquid membrane technology the liquid film is usually an organic solvent, which usually contains a selective extractant, which is immiscible with the aqueous feed and recovery phases. The selectivity of liquid membranes towards different solutes derives from differences in the membrane phase permeabilities. This is the product of the solute solubility and rate of diffusion in the membrane solvent. For small molecular weight solutes the rates of diffusion are similar, and thus in most applications the main condition for membrane selectivity is solute solubility in the membrane phase.

Liquid membrane extraction (LME) may be thought of as an enhanced version of conventional solvent extraction, with both the extraction and stripping stages occurring simultaneously. However, it is important to note that LME is driven by a dynamic solute driving force and not by equilibrium between the two phases, unlike solvent extraction. Liquid membrane extraction has several advantages over solvent extraction. As the liquid film is very thin and high specific interfacial areas are available for mass transfer, separation is fast. As only one extraction stage is required, with respect to solvent extraction there is a reduction in equipment and solvent requirements. Also, as liquid membrane extraction involves the contacting and separating of two liquid

Table 17.1 Applications of liquid membrane processes

Industry sector	Application
Biotechnology	Recovery of fermentation products (organic acids, amino acids, penicillin)
Pharmaceuticals	Removal of toxins Controlled drug release Separation of chiral drug intermediates
Environmental	Effluent clean-up (ammonia, phenol, heavy metals)
Hydrometallurgy	Recovery of metal ions

phases, extant liquid–liquid contacting equipment may be used, e.g. centrifugal contactors, mixer-settlers, countercurrent column contactors. Examples of the practical application of liquid membranes is given in Table 17.1.

Development of a liquid membrane process involves selection of the extraction chemistry, choice of the membrane configuration and then optimization of the extraction conditions. There are two main configurations by which liquid membrane extraction can be exploited: the supported liquid membrane and the emulsion liquid membrane. Both systems will be discussed in detail below.

17.2.2 SELECTION OF CARRIER SPECIES FOR LIQUID MEMBRANE SEPARATIONS

The key to selectivity in liquid membrane extraction is the choice of a selective carrier species incorporated in the organic solvent to increase the solute solubility. The use of a carrier enhances selectivity by the formation of a reversible complex between the carrier and the solute, which is only soluble in the organic solvent. Through the carrier molecule which reacts preferentially with the solute of interest, it is possible to enhance solute mass transfer rates over the rates obtained with unfacilitated transport systems. This increased mass transfer results from the increased membrane phase permeability of the solute, conferred by the carrier. Although the rate of diffusion of the complexed solute is likely to be less than its uncomplexed form, its solubility in the membrane phase can be several orders of magnitude higher than the uncomplexed form [1] (see also Chapter 6).

Further enhancement of mass transfer in facilitated extraction systems may be achieved by the use of counter and co-transport, i.e. coupled transport systems. Both systems are able to pump a solute against its concentration difference [1]. Counter transport is illustrated in Figure 17.2:

(a) Solute A complexes the membrane soluble carrier C at the membrane external phase interface, in a thermodynamically favourable interaction.

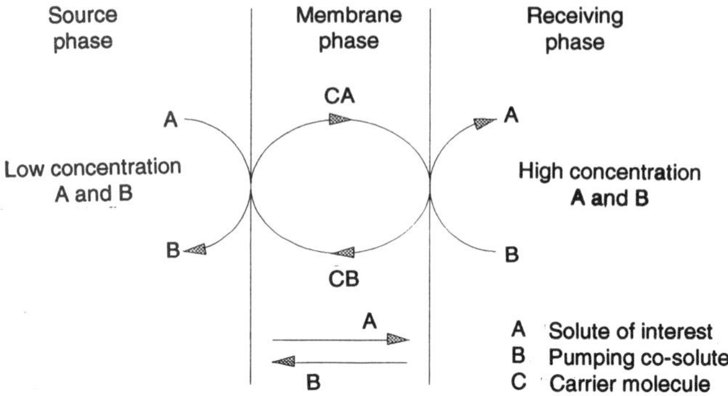

Figure 17.2　Counter transport of a solute in a liquid membrane

(b) The complex CA then diffuses down its concentration gradient to the stripping phase interface where counter solute B displaces solute A from the carrier C, again due to a thermodynamically favourable interaction. As solute A is usually at best sparingly soluble in an already saturated membrane phase, it must diffuse out into the stripping phase, regardless of its concentration in that phase.

(c) The carrier–counter solute complex CB diffuses back across the membrane to the external phase interface, again down its concentration gradient.

(d) Counter solute B is released by another thermodynamically favourable interaction with the external phase (e.g. solute hydration). The carrier is then ready for further interaction with solute A.

Thus counter solute B provides the system free energy for pumping A against its concentration difference.

The less common mode of coupled transport is illustrated in Figure 17.3. In this case, solute A and co-solute B intereact with the carrier at the external phase interface and diffuse across the membrane to be released at the stripping phase interface. By using a high concentration of B in the external phase, it is possible to drive A against its concentration difference.

A fundamental treatment of coupled transport mechanisms is detailed by Cussler [1].

17.2.2.1 How to Choose a Carrier

The most important factor in carrier selection is that it displays high selectivity towards the solute of interest. In addition, the carrier and carrier solute complex should not be lost from the membrane phase through dissolution in the aqueous phases.

A high degree of interaction is required between the solute and carrier at the membrane–external phase interface with respect to other species present, so as

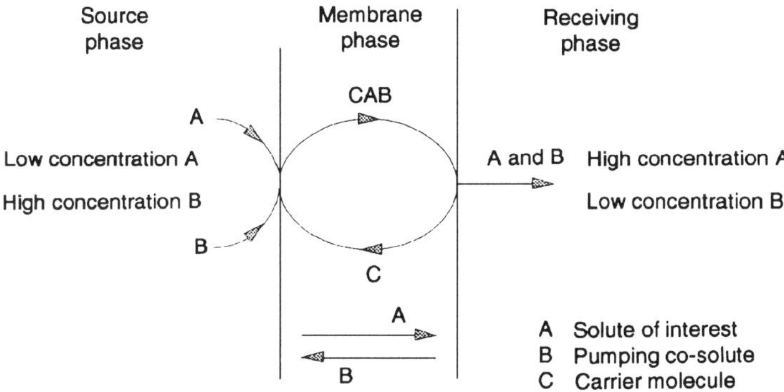

Figure 17.3 Co-transport of a solute in a liquid membrane

to engender selectivity. However, it is important that these interactions are reversible at the membrane–stripping phase interface. Pellegrino and Noble [2] have recommended interaction energies of approximately 50 kJ/mol, for both fast complexation and dissociation at the feed and stripping membrane interfaces respectively. Interaction energies of greater than 50 kJ/mol require undue energy input at the stripping membrane interface to effect solute extraction, i.e. carrier poisoning occurs.

Pellegrino and Noble [2] have also listed several other recommended carrier characteristics to aid selection:

(a) no side reactions,
(b) no irreversible or degradation reactions,
(c) no co-extraction of solvent from the feed phase.

An indication as to the efficacy of a potential carrier species can be found from a simple batch extraction experiment. The carrier should be dissolved in a suitable solvent and mixed with an aqueous solution containing the solute of interest. This should result in the formation of two phases with some transfer of the solute into the organic phase. This organic phase should then be mixed with an aqueous phase containing the intended stripping reagent. This should result in the solute–carrier complex dissociating, depositing the solute in the stripping aqueous phase. This experiment can be used as an initial screening method for suitable carriers.

This screening procedure is an equilibrium process as characterized by the distribution coefficient of the system. If the distribution coefficient is very low only small amounts of the product may partition into the organic phase. It should be remembered that liquid membrane extraction is a rate process and so the low distribution coefficient may not hinder the transport rate in the liquid membrane system. Similarly, the liquid membrane system can efficiently utilize

Table 17.2 Carrier/solvent combinations used in liquid membrane extraction

Carrier	Solvent	Extracted solute	Reference
Di-2-ethylhexyl phosphoric acid (D2EHPA)	Telura 619 (paraffin)	L-Phenylalanine	30
Alamine 336	70% n-heptane/30% paraffin	Lactic acid	3
Dicyclo-hexano 18-crown-6	Toluene	Potassium	29
D2EHPA	n-Heptane	Zinc	31
Aliquat 336	100 Neural (paraffin)	L-Phenylalanine	28
Aliquat 336	95% kerosene/5% decanol	Chromium (VI)	32
Aliquat 336	Shellsol A (paraffin)	Citric acid	15

low concentrations of the carrier species because of the cyclic nature of facilitated transport. Table 17.2 lists some of the common carriers that have been used for the recovery of biological products.

Where charged carriers are involved in an ion exchange or ion-pairing reaction, the pH at which extraction is carried out is an important parameter and is related to the dissociation constant of the solute. An example of this is extraction of lactic acid by the tertiary amine Alamine 336 [3]. It is important to operate at a low pH to ensure that the H^+ ion concentration is high. The free H^+ ions protonate the amine, enabling the lactate ion to complex with the protonated amine which can then be transferred across the organic phase [4].

Ion exchange type mechanisms for enhancing selectivity may not be as selective as required in 'real' feedstocks. The tertiary amine Alamine 336 described above was used to recover lactic acid from a fermentation broth. Extraction yields from the fermentation broth were found to be low when compared to the model system, which showed fast extraction with significant product concentration [5]. This was a result of competition for the carrier by ionic 'impurities' in the broth, e.g. metabolic products, media salts and pH controlling reagents.

Highly selective separations are possible with liquid membranes, in particular the use in chiral separations. Various attempts have been made to utilize liquid membranes for enantioselective transport of amino acids and drug intermediates. Most attempts have involved the use of a chiral carrier molecule which interacts more favourably with one enantiomer than the other. Chiral transport of sodium mandelate has been based on the optically active ·carrier N-(1-naphthyl)methyl-7a 1-methyl-benzylamine dissolved in chloroform. The choice of anion was found to influence chiral selectivity. Separation of racemic N-(3,5-dinitrobenzoyl) 7a 1-amino acid derivatives was achieved using (S)-N-(1-naphthyl)-leucine octadecyl ester as the carrier in dodecane [2]. A bulk liquid membrane was used for this application. Enantiomeric resolution of phenylalanine using the carrier N-decyl-(L)-hydroxyproline in an emulsion liquid membrane has also been demonstrated [6].

Pellegrino and Noble [2] and Noble *et al.* [7] describe in detail the selection of species for use as carrier molecules.

17.3 Liquid Membrane Configurations

17.3.1 EMULSION LIQUID MEMBRANES

An emulsion liquid membrane (ELM) is formed by creating, under high shear, a dispersion of the stripping phase within the organic solvent which forms a non-porous film around the stripping phase droplets. The emulsion thus formed (stabilized by a surfactant) is dispersed into the feed phase containing the solute (Figure 17.4). Depending on the dispersion conditions, the globule diameter will be of the order of 1–2 mm and the stripping phase droplets will be micrometre sized. The two aqueous phases cannot physically contact each other and the solute is transported into the stripping phase droplets by facilitated diffusion through the stabilized solvent film.

An ELM consists of three major components: the carrier, an oil-in-water emulsion stabilizing surfactant and the membrane solvent. The function of the membrane is to allow selective transport of the solute from the product stream into the stripping phase. In order to make this process valid it must also prevent any physical contact between the feed and stripping phases.

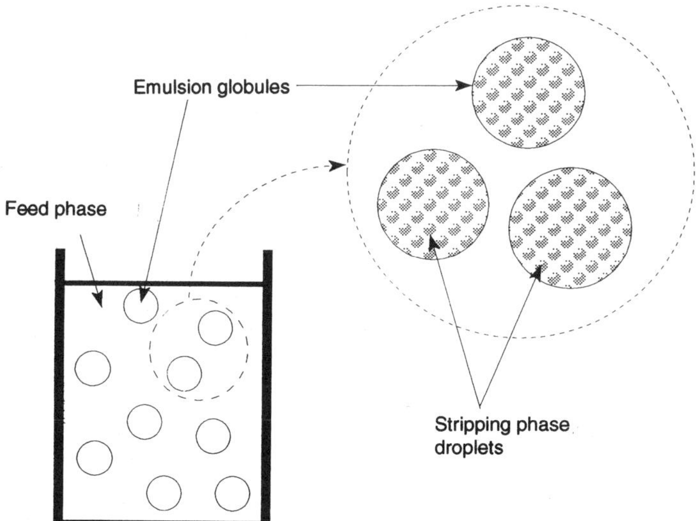

Figure 17.4 An emulsion liquid membrane

17.3.1.1 Membrane Solvent

The most important role of the solvent is to provide maximum solvation of the carrier and the carrier–solute complex, so that neither are lost to the aqueous phases. This is especially important when the carrier is expensive.

Aliphatic diluents are generally preferred as the solvent because of their lower solubility in water. In an ideal situation the solvent should have no solubility in water to ensure that there is no aqueous phase contamination by trace organics. The solvent viscosity should be kept as low as practically possible to maximize the solute diffusivity through the membrane phase. However, several workers have reported that a reduction in solvent viscosity leads to increased membrane instability and globule breakage [8–10]. Finally, it is important that the solubility of the membrane solvent in water is minimized, as organic contamination of both aqueous phases will require subsequent removal from both effluent and product streams. Two widely used solvents are S100N (Exxon) and low-odour paraffin solvent (LOPS) (Exxon).

17.3.1.2 Emulsion-stabilizing Surfactant

In an emulsion liquid membrane system a surfactant is required in order to stabilize the water-in-oil emulsion against coalescence. Surfactants are characterized on the basis of the hydrophilic/lipophilic balance of the molecule: the HLB scale. On this scale, species with high lipophilic character and which are good water-in-oil emulsifiers are assigned low HLB values: the optimum HLB value for a water-in-oil emulsifier is 5.0. The non-ionic surfactant Span 80 (sorbitan monoleate) has an HLB value of 4.3 (close to the optimum) and has been widely used in liquid membrane research (Table 17.3).

The concentration of surfactant used should be the minimum required to maintain stability for the duration of separation. Too much surfactant will increase the mass transfer resistance at the interface with the stripping and external phases and will also lead to a more stable emulsion which will be harder to break to enable product recovery.

Table 17.3 Surfactants used to stabilize emulsion liquid
membranes

Surfactant	Reference
Paranox 100	30
Span 80	3
Span 80	29
L-Glutamic dioleyl esters	31
Paranox 100	28
Paranox 100	32
Span 80	15

17.3.1.3 Carrier

The rationale behind the choice of the carrier has been discussed above. It is worth mentioning, however, that the carrier species is often a surface active molecule, or has some surface active nature. This can lead to problems when formulating the emulsion. The surface active natures of the surfactant and the carrier may give rise to emulsion instabilities and formulation problems.

17.3.1.4 Stripping Phase Reagent

The stripping phase conditions are designed to prevent back-extraction of the extracted solute. This is achieved by ensuring that the extracted solute is insoluble in the membrane phase by including a reagent in the stripping phase which ionizes the solute, e.g. for the extraction of an organic acid a base would be included to form an ionic salt. When choosing the reagent for the stripping phase the chemical compatibility of the surfactant with the reagent must be considered. For example, the surfactant Span 80 is an ester which is susceptible to alkaline hydrolysis by sodium hydroxide, which results in an unstable emulsion [11].

In discussing these parameters, it should be noted that Marr and Kopp [12] have identified the following guidelines for the formation of stable water-in-oil ELMs:

(a) organic phase soluble surfactant, 0.1–5 wt%,
(b) organic phase viscosity, 30–1000 mPa s,
(c) volume of stripping phase to membrane phase, 0.2–2.0,
(d) volume ratio of stripping phase to external feed phase, 0.05–0.2,
(e) volume ratio of feed phase to emulsion phase, 1–40,
(f) surfactant HLB value, 6–8.

Figure 17.5 shows a schematic flowsheet for an emulsion liquid membrane process. The individual operations are discussed below.

17.3.1.5 Emulsion Formation

Emulsion liquid membranes are formed by subjecting the stripping phase and the membrane phase components to high shear rates to form a dispersion of stripping phase droplets in the membrane phase. Emulsification has been carried out using several methods: e.g. using moderate stirring, 350–600 rev/min [13,14], high-speed stirrers, speeds greater than 6000 rev/min [15] or an homogenizer [16]. It should be noted that this operation requires a high specific energy input to form a stable emulsion.

17.3.1.6 Phase Contacting

Extraction of the solute is carried out by dispersing the emulsion into the product stream. The strategies used for contacting the emulsion and the feed

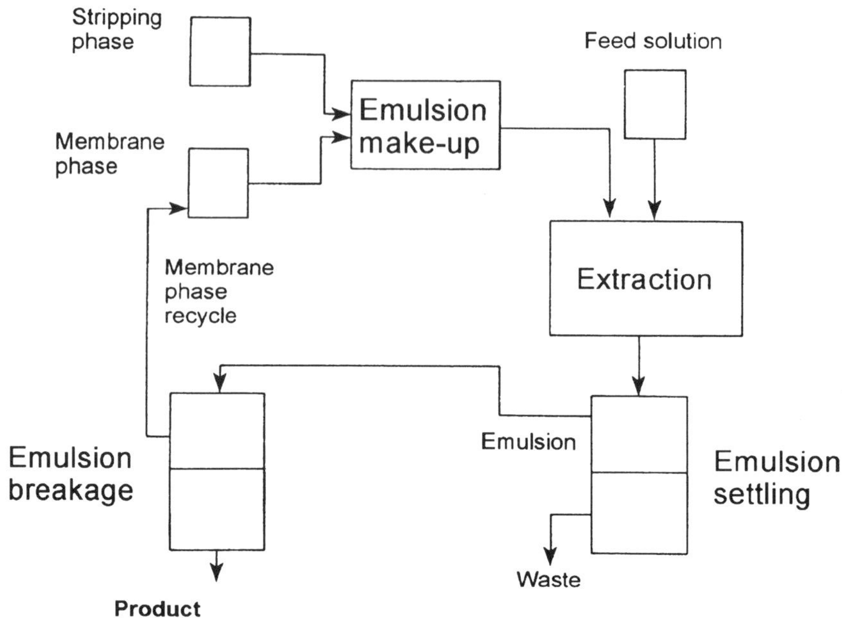

Figure 17.5 A schematic flowsheet for an emulsion liquid membrane process

solution have arisen from solvent extraction technology. It is not surprising therefore that the most frequently reported method of contacting the emulsion and solute solution is by dispersing the emulsion in a stirred batch reactor [11,17–19].

Other contacting configurations which have been reported include a countercurrent column for zinc extraction [20], a spray column [19], a turbine-stirred countercurrent column [21] and a novel mixer utilizing internal recycle [22]. The criteria that should be applied to choosing the contacting method include low shear mixing so the emulsion is not broken, creation of a large interfacial area for mass transfer and good phase separation with low entrainment.

17.3.1.7 *Emulsion Recovery*

After extraction the product is inside the emulsion in the stripping phase. It is necessary to recover the emulsion from the depleted feed phase. Because of the density difference between the emulsion and the aqueous feed phase this is readily achieved by settling under gravity. Mixer-settlers as used in liquid–liquid extraction have been successfully used.

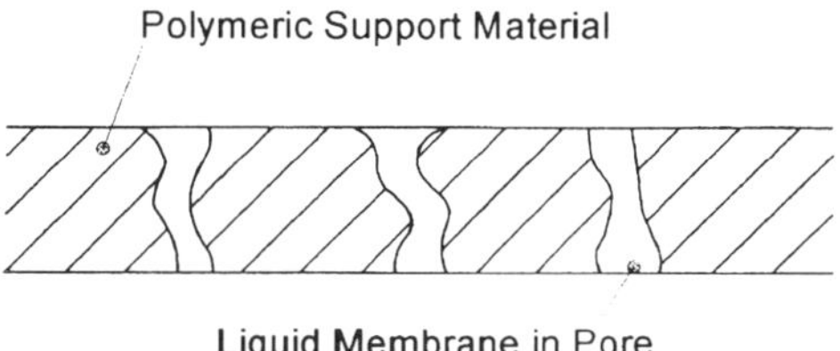

Figure 17.6 A supported liquid membrane

17.3.1.8 Emulsion Breakage

The emulsion must be broken to release the stripping phase and hence recover the extracted solute. There are several methods available for emulsion breakage, such as thermal demulsification, chemical methods and centrifugation. The various demulsification methods work by enhancing the slow, natural coalescence of the emulsion globules.

In order to make ELM separations economically attractive, a restraint imposed upon demulsification is that the membrane phase components can be recovered and re-used. With this in mind, it is now generally recognized that electrostatically enhanced coalescence is the best method available, giving good separation of the stripping and membrane phases. Liquid membrane systems are designed to be stable throughout the time required for extraction so a high voltage is required to break down the emulsion in order to recover the solute [11]. Yan *et al.* [23] reported that demulsification using a high-pressure pulse wave generator is superior to that obtained using an a.c. generator. Although a high voltage is required to break emulsions, Hsu and Li [24] found that application of electric fields directly to emulsions leads to sparking and emulsion component deterioration. To remedy this Hsu and Li [24] found that by applying an alternating electric field via an insulating material to the emulsion, sparking and sponge formation were largely avoided during demulsification.

17.3.2 SUPPORTED LIQUID MEMBRANES

A supported liquid membrane (SLM) is the configuration where the solvent is impregnated in a porous polymeric membrane. This membrane separates the two aqueous phases as shown in Figure 17.6. The solvent is held in place by capillary forces that exist within the pores. Transport occurs only across these solvent-impregnated pores.

Three geometries are used: planar, spiral wound and hollow fibre. The planar membrane is used in laboratory work but is not suitable for larger-scale use because of the low surface area per unit volume characteristics. On the other hand, the spiral wound and hollow-fibre systems have high surface areas of 100 and 10 000 m^2 (support)/m^3 (equipment) respectively [25].

A supported liquid membrane consists of the solid membrane, the organic solvent and the product-specific carrier species. The solid supports used are generally microporous polymeric films, e.g. PTFE (polytetrafluoroethylene), polypropylene, polysulfone or other hydrophobic materials. The membrane material should be chemically stable to the solvent and the aqueous phases. Typical dimensions are a membrane thickness of 25–50 µm, with pore sizes between 0.02 and 1.0 µm.

The choice of liquid membrane support material is based on several factors including surface properties, reactivity, pore size, porosity, tortuosity and thickness [26]. In practice pore size, tortuosity and membrane thickness are minimized and porosity is maximized for optimum mass transfer and membrane performance.

As with ELMs, aliphatic solvents are generally used. In order for the membrane pores to be effectively wetted, the surface tension of the solvent must be less than the critical surface tension of the membrane polymer.

The driving force for the extraction of solutes by liquid membranes is the concentration difference of the solute between the stripping and external phases. It is possible to sustain this solute concentration difference beyond normal equilibrium conditions by reacting the solute in the stripping phase with a reagent to form a different reaction product which may be a non-transportable species, typically an ion in inorganic liquid membrane systems [18]. This reduces the concentration of the transported species in the stripping phase to zero, thereby sustaining a finite solute concentration difference across the membrane until all the external phase transported solute is extracted or the stripping phase reagent is exhausted.

A schematic flowsheet for a supported liquid membrane extraction system is shown in Figure 17.7. The membrane would be soaked in the solvent/carrier solution to enable take-up into the fibres. Impregnation may be enhanced through the use of vacuum or positive pressure conditions. The feed and stripping phases are recirculated around the membrane. It is important to maintain the trans-membrane pressure so that the solvent is not forced out of the membrane pores. Provided that membrane stability is maintained there is no mixing of the two aqueous phases. Conditions for transport may be maintained quite easily as compared to the ELM system.

17.3.3 COMPARISON OF CONFIGURATIONS

The main advantages of emulsion liquid membranes are very fast transfer rates that arise from the small droplet sizes and high specific surface area of the emulsion and the fact that the solute can be simultaneously separated and concentrated by making the stripping phase volume smaller than that of the

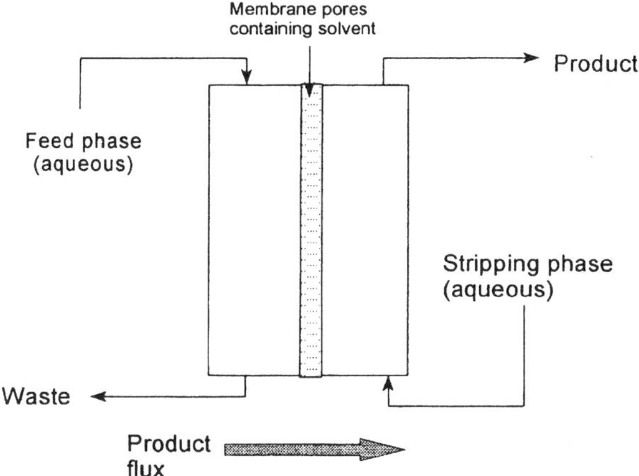

Figure 17.7.   A schematic flowsheet for a supported liquid membrane extraction system

feed phase. In addition, the highly efficient membrane phase enables very expensive carriers to be used at low concentrations.

The need for emulsion formation and breakage are drawbacks of this technique. However, two specific process problems are membrane stability and swelling. During solute extraction some of the extracted solute can leak back into the external phase. Usually this is accompanied by leakage of the stripping phase reagent which can then transform the solute into a non-extractable form. This is primarily an emulsion formulation problem. The emulsion is designed so that it is stable under process conditions, but is also easy to break to recover the extracted solute. The degree of emulsion breakage is small, less than 2%, and occurs during the initial stages of extraction [27,28].

As a result of osmotic pressure differences across the membrane phase, water can be transported from the external phase into the stripping phase. This results in swelling of the emulsion and subsequent dilution of the stripping phase contents. In some cases it is possible that the transported water carries dissolved solute and impurities across the membrane, hence reducing the selectivity of the process. Additionally, the apparent viscosity of the emulsion increases and can lead to emulsion breakage. Swelling can be minimized by judicious choice of organic solvent and surfactant; e.g. the surfactant Span 80 leads to high levels of swelling.

The main problem associated with SLM extraction is loss of membrane integrity and thus reduced membrane life. This loss is usually the result of several factors such as mechanical faults, instability of the membrane to system chemicals, and loss of carrier from the membrane pores. Carrier losses can occur because of the solvent or carrier dissolving in the feed or stripping solutions, ingress of water into the membrane pores or carrier loss from high-pressure differences across the membrane.

In conclusion, ELM extraction displays several advantages over SLM extraction:

(a) *Sustainable extraction efficiency.* The emulsion globules of ELMs may be thought of as very large numbers of small-volume SLMs extracting in parallel. Therefore should the membrane of one of these globules fail, a small volume of stripping phase would be dispersed in the external phase, leading to a small loss in extraction efficiency. However, a single pore failure in an SLM leads to immediate cross-contamination of the bulk separated phases, leading to rapid loss of extraction efficiency.

(b) *Ability to concentrate.* As the stripping phase is much smaller than the external extraction phase, the potential for simultaneous concentration with separation is much higher than for SLMs.

(c) *Reduced solvent requirement.* Far larger membrane phase volumes are required for similar transport rates in SLM systems compared to analogous ELM systems, mainly due to the inherently high interfacial area of ELM systems [29].

(d) *Simplicity of unit operation.* The requirements for contacting the emulsion with the external phase are identical to those of conventional solvent extraction; thus the same techniques and equipment may be used.

References

1. E. L. Cussler (1984) *Diffusion Mass Transfer in Fluid Systems*, Cambridge University Press, Cambridge.
2. J. J. Pellegrino and R.-D. Noble (1990) *TIBTECH*, **8**, 216–24.
3. J. B. Chaudhuri and D. L. Pyle (1992) *Chem. Engng Sci.*, **47**(1) 49–56.
4. J. B. Chaudhuri and D. L. Pyle (1992) *Chem. Engng Sci.*, **47**(1), 41–8.
5. C. Schöller, J. B. Chaudhuri and D. L. Pyle (1993) *Biotechnol. Bioengng.*, **42**, 50–8.
6. P. J. Pickering, J. B. Chaudhuri and K. J. Carpenter (1994) *The 1994 IChemE Research Event*, Institution of Chemical Engineers, Rugby, pp. 253–5.
7. R. D. Noble, C. A. Koval and J. J. Pellegrino (1989) *Chem. Engng Prog.*, **1989**, 58–70.
8. P. T. Gadekar, A. V. Mukkolath and K. K. Tiware (1992) *Sep. Sci. Technol.*, **27**(4), 427–45.
9. T. Kinugasa, K. Watanabe and H. Takeguchi (1989) *J. Chem. Engng Japan*, **22**(6), 593–7.
10. T. Kinugasa, K. Watanabe and H. Takeguchi (1992) *J. Chem. Engng Japan*, **25**(2), 128–33.
11. S. C. Boey, M. C. Garcia del Cerro and D. L. Pyle (1987) *Chem. Engng Res. Des.*, **65**, 218–23.
12. R. Marr and A. Kopp (1982) *Int. Chem. Engng.*, **1**, 44–60.
13. E. L. Cussler and D. F. Evans (1974) *Sep. Purif. Meth.*, **3.**, 399–421.
14. S. W. May and N. N. Li (1972) *Biochem. Biophys. Res. Commun.*, **47**(5) 1179–85.
15. C. del Cerro and D. Boey (1988) *Chemistry and Industry.*, **1988**, 681–7.
16. M. P. Thien, T. A. Hatton and D. I. C. Wang (1986) in J. A. Asenjo and J. Hong (Eds.), *Separation, Recovery, and Purification in Biotechnology*, American Chemical Society Symposium Series 314, pp. 67–77.

17. R. E. Terry, N. N. Li and W. S. Ho (1982) *J. Membrane Sci.* **10**, 305–23.
18. M. P. Thien and T. A. Hatton (1988) *Sep. Sci. Technol.*, **23**(8–9), 819–53.
19. I. Ortiz Uribe, S. Wongswan and E. S. Perez de Ortiz (1988) *Ind. Engng Chem. Res.*, **27**, 1696–701.
20. J. Draxler and R. Marr (1986) *Chem. Engng Process*, **20**, 319–29.
21. E. Rautenbach and O. Machhammer (1988) *J. Membrane Sci.*, **36**, 425–44.
22. T. A. Hatton, E. N. Lightfoot, R. P. Cahn and N. N. Li (1983) *Ind. Engng Chem. Fund,.* **22**, 27–35.
23. Z. Yan, S. Li, Y. Yu and X. Zheng (1987) *Desalin*, **62**, 323–8.
24. E. C. Hsu and N. N. Li (1985) *Sep. Sci. Technol.*, **20**(2–3), 115–30.
25. R. D. Noble and J. D. Way (1987) in R. D. Noble and J. D. Way (Eds.), *Liquid Membranes. Theory and Applications*, ACS Symposium Series 347, pp. 1–27.
26. J. D. Way, R. D. Noble and B. R. Bateman (1985) in D. R. Lloyd (Ed.), *Materials Science of Synthetic Membranes*, ACS Symposium Series 269, pp. 119–28.
27. A. Shere and H. M. Cheung (1988) *Chem. Engng Commun.*, **68**, 143–64.
28. M. P. Thien, T. A. Hatton and D. I. C. Wang (1988) *Biotechnol. Bioengng.*, **32**, 604–15.
29. R. M. Izatt, R. D. Lamb and R. L. Bruening (1988) *Sep. Sci. Technol.*, **23**(12–13), 1645–58.
30. H. Itoh, M. P. Thien, T. A. Hatton and D. I. C. Wang (1990) *J. Membrane Sci.*, **51**, 309–22.
31. M. Goto, H. Yamamoto, K. Kondo and F. Nakashio (1991) *J. Membrane Sci.*, **57**,(2–3), 161–74.
32. E. Salazar, M. I. Ortiz, A. M. Urtiaga and J. A. Irabien (1992) *Ind. Engng Chem. Res.*, **31**, 1523–9.

18 LYOPHILIZATION

John W. Snowman

18.1 Introduction

Many labile, heat-sensitive products in dilute aqueous solution require to be stabilized by drying. Lyophilization, also called freeze drying, avoids most of the problems of liquid phase drying. A simple definition of lyophilization is that it is 'a means of drying, achieved by freezing the wet substance and causing the ice to sublime directly to vapour by exposing it to a low partial pressure of water vapour'. In this context, 'low' means below about 6 mb (4.6 mmHg), the triple point of water. Below this pressure adding heat energy can turn ice directly into water vapour (sublimation) without it first melting to liquid. Because the subliming ice crystals leave cavities, lyophilized dry material contains a myriad of interstices into which water can penetrate to give rapid and complete re-solution when needed. The name lyophilization is derived from this property which comes from the Greek 'to make solvent loving'. Because it is frozen, constituents of the material are kept immobilized during sublimation drying by the ice crystals. The shape of the dry substance is broadly the same as that of the frozen wet substance and migration to the surface to form a skin is reduced. Because drying takes place at a low temperature, damage is minimized and volatile components are retained.

Unfortunately the apparent simplicity of the process masks the fact that it causes complex and potentially damaging effects in the material being dried [1]. The formulation of the wet starting material exerts a major influence on the way it can by lyophilized. Thus formulating a product for lyophilization and developing an optimum drying cycle are mutually interacting activities which will often require many iterations before a satisfactory product and process are found. A noted authority in the field with extensive resources has commented that it usually takes at least 6 months of work to optimize a drying cycle. Familiarity with process, product and equipment does not guarantee success but clearly are a great help.

Downstream Processing of Natural Products. Edited by Michael S. Verrall
©1996 John Wiley & Sons Ltd

The principal advantages of lyophilization can be summarized as:

(a) minimum damage to, and loss of activity in, delicate heat labile materials,
(b) creation of a porous, friable structure,
(c) possibility of accurate, clean dosing into final product containers and
(d) speed and completeness of rehydration.

The principal disadvantages are:

(a) high capital cost of equipment (about three times more than other
    methods),
(b) high energy costs (two to three times more than other methods),
(c) lengthy process time (typically 24–48 hours per drying cycle but often
    longer),
(d) possible damage to products due to changes in pH and tonicity as the
    solute concentrates during the freezing process.

Lyophilization is useful when the product meets one or more of the following
criteria:

(a) It is unstable.
(b) It is heat labile.
(c) Minimum particulates are required.
(d) Accurate dosing is needed.
(e) Quick and complete rehydration is required.
(f) The product is of high value.

Lyophilization is a method of choice for parenteral products because the wet
material can be sterile filtered just before it is filled into final containers,
reducing particulate and bacterial contamination.

18.2 Description of the Freeze Drying Process

A lyophilization process consists of three phases as shown in Figure 18.1:

(a) *freezing* to solidify the material,
(b) *sublimation* drying to reduce moisture to below 20% w/w of the dry
    product and to leave an apparently dry cake of solid residue which is
    substantially the same size and shape as the frozen mass,
(c) *desorption* or secondary drying to reduce bound moisture to the required
    final value (often below 1% w/w).

The three phases usually overlap to some extent in different parts of the cycle.
For example, if a low melting point material such as ethanol is present, it will
not freeze completely at the temperatures available in practical lyophilizers. At
the commencement of primary drying the ethanol will evaporate from the
liquid phase before true freeze drying of the product begins. Towards the end

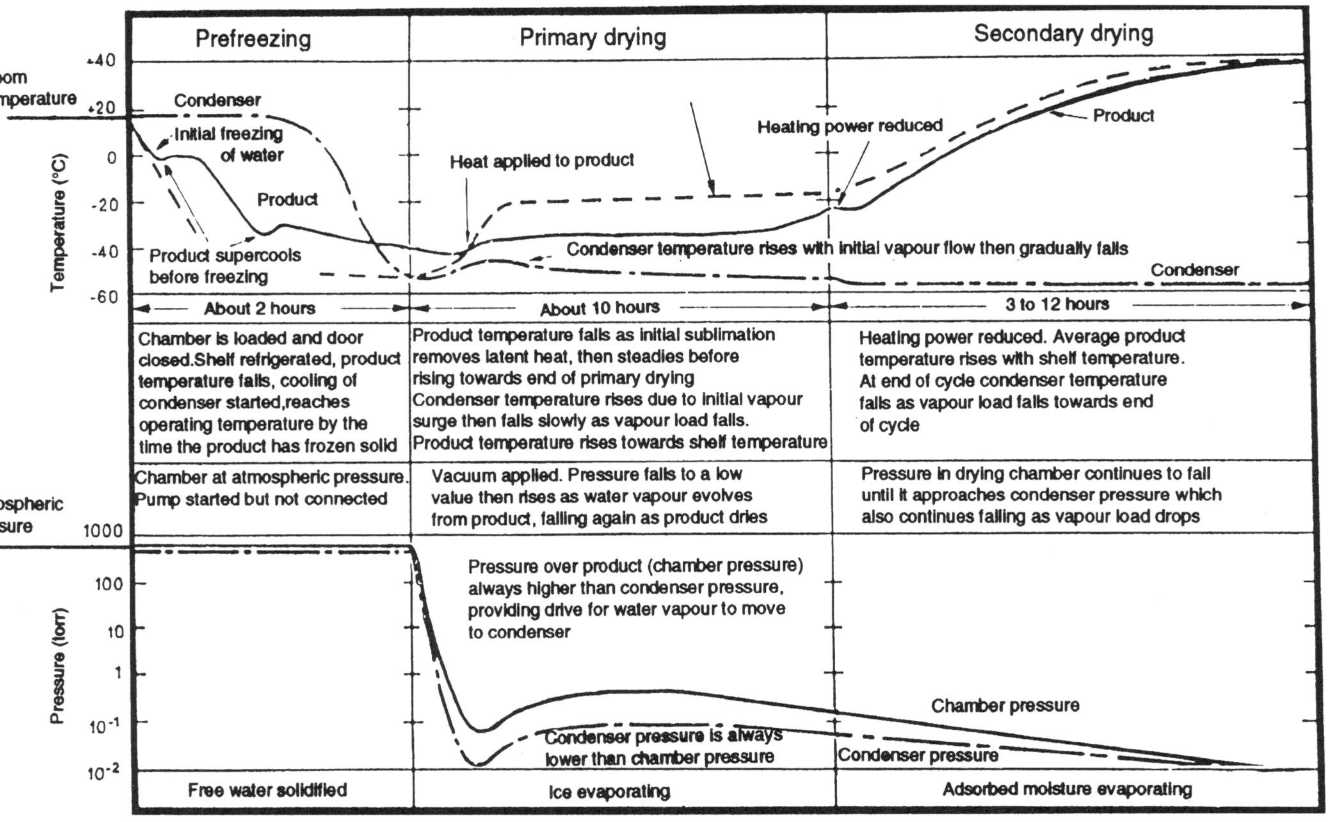

Figure 18.1   Chart of complete freeze drying (FD) cycle

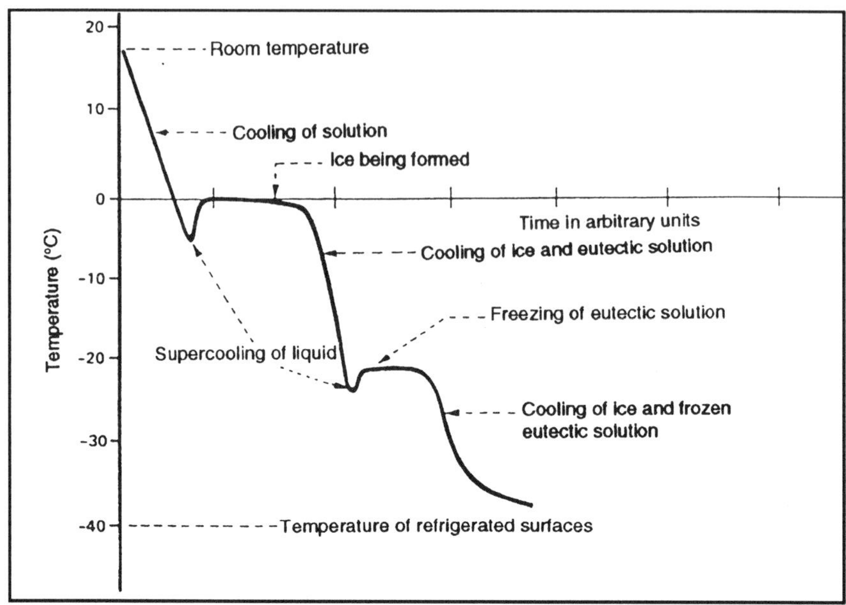

Figure 18.2　Typical freezing curve

of the cycle some desorption drying will begin in the drier parts of the cake before sublimation drying is completed.

During sublimation, vapour is generated from a distinct interface or freeze drying front which moves from the outer surface of the cake. Since the subliming vapour must pass through the channels left from the ice crystals that formed on freezing, their shape affects the speed of drying. If the crystals were small and discontinuous, the escape path for the vapour is limited; if large crystals were formed, escape is easy and the product can be dried more quickly. The method and rate of freezing are thus critical to the course of sublimation. Quick freezing methods such as the use of liquid nitrogen must be viewed with caution because of the formation of very small ice crystals.

18.2.1 FREEZING

In ideal two-component systems water and a solute may both crystallize. In this case freezing proceeds as follows. Cooling from the initial temperature to below 0 °C eventually results in the nucleation of ice. The release of heat of crystallization raises the temperature towards 0 °C. Crystallization of water then proceeds at a progressively falling temperature related to the equilibrium melting point of the concentrating solution as shown in Figure 18.2. As the temperature falls the unfrozen solution approaches saturation and crystals of solute are precipitated. Eventually a eutectic point is reached where the

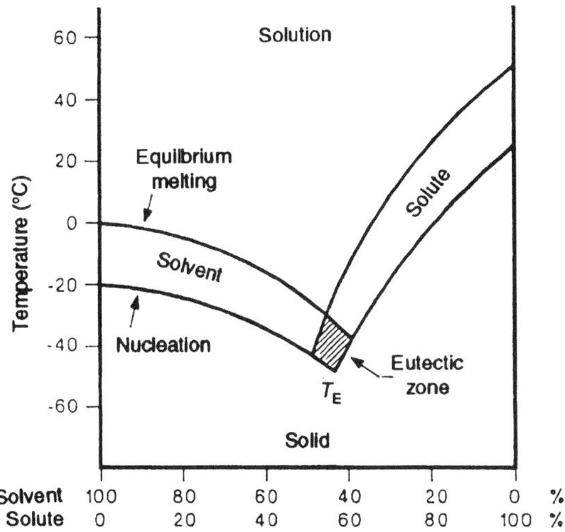

Figure 18.3 State diagram — two crystallizable components

material become wholly crystalline. When the material contains more than one crystallizable solute, a similar situation exists but the eutectic point is lower than that of any two components combined. Figure 18.3 is a state diagram of a system with two crystallizable components.

Unfortunately, in the majority of practical cases, the solute does not actually crystallize. The solution continues to concentrate as water is frozen out and its viscosity increases rapidly. Eventually it becomes syrupy, and then rubbery, as

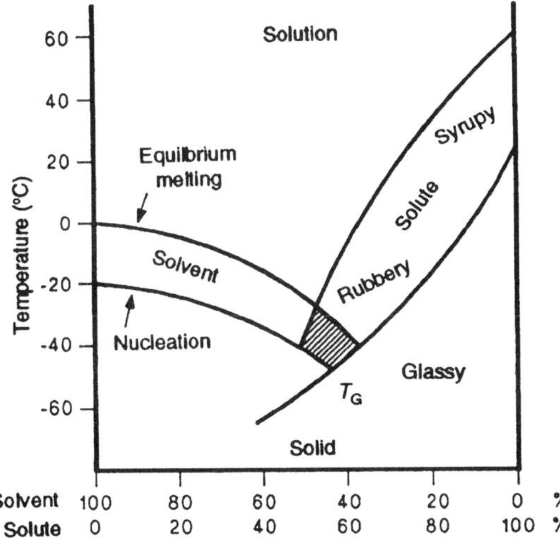

Figure 18.4 State diagram — one crystallizable and one glass-forming component

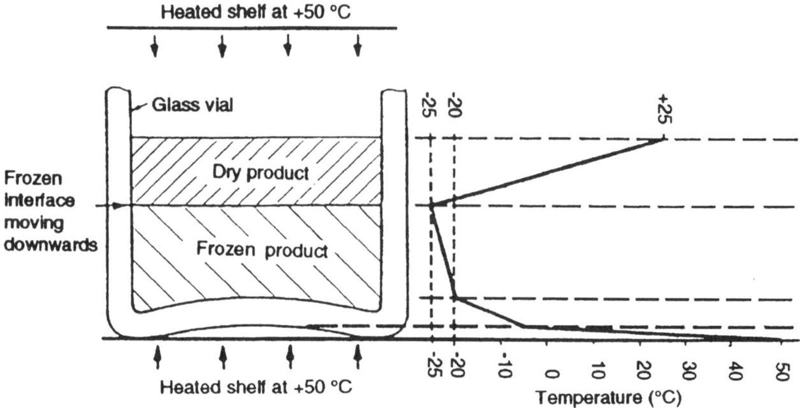

Figure 18.5 Heat transfer to vial

shown in Figure 18.4. Further lowering of the temperature finally causes the formation of an amorphous glass containing some bound water. This occurs at the glass transition temperature T_G. In such cases, there is no eutectic temperature as such, but there is a characteristic maximum product temperature for freeze drying called the collapse temperature, T_C. This has been extensively researched by Mackenzie [2], who has published the collapse temperatures for a number of substances.

18.2.2 SUBLIMATION (PRIMARY DRYING)

After freezing, the partial pressure of water vapour must be reduced below the triple point pressure to allow sublimation to take place.

If the temperature T_E of the lowest eutectic exhibited by the material is exceeded during drying, then melting can occur. This gives rise to gross material faults such as melting, shrinking and puffing. These are sometimes confused with collapse. At temperatures below the 'collapse' temperature T_c the solute will remain rigid during primary drying and a good cake will be produced. However, if drying is carried out above the collapse temperature T_c the solute will not be rigid. While it is supported by the pure water ice crystals it appears to be rigid, but as the pure water ice crystals sublime away, the support is removed and apparently dry material collapses to form an impermeable mass. The collapse temperatures of a number of commonly used substances are listed in Section 18.4.3.

Most of the heat for sublimation must be transferred to the freeze drying front through the frozen material, which usually has poor thermal conductivity. The need to avoid melting means that the thermal gradient is low and thus the heat conducted is small. Figure 18.5 illustrates the problem of transferring heat from a shelf at $+50\,°C$ through the insulating effect of a

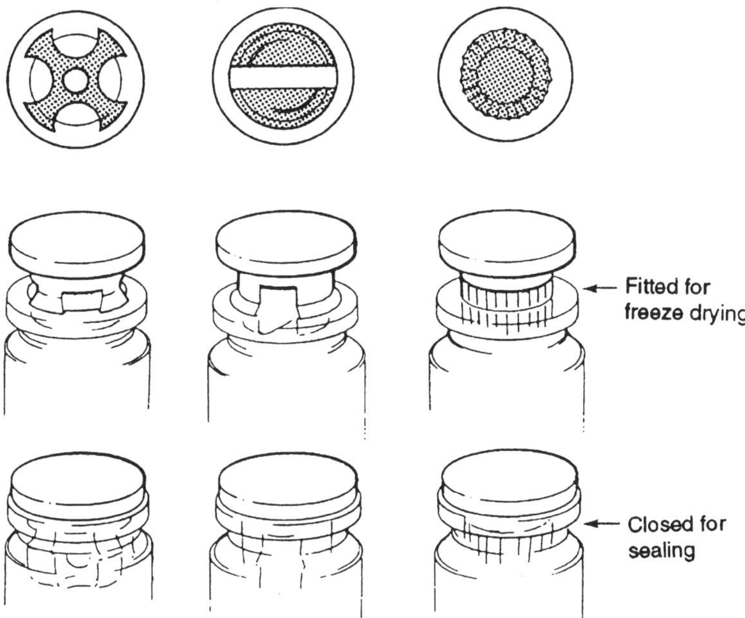

Figure 18.6   Bungs of various types

vacuum to a glass vial with minimal contact to the shelf, and from there through the glass and frozen product to the freeze drying front without overheating the product in the outer bottom part of the vial. Later in the process the vapour flow from the freeze drying front and hence the sublimation rate is restricted by the increasing depth of the dry part of the product cake.

Vials with special freeze drying bungs or stoppers are the preferred containers for aseptic freeze drying in final containers. Figure 18.6 shows various types of freeze drying bungs which are designed to be half inserted in the vial before lyophilization and then closed at the completion of the cycle before removal from the lyophilizer. It is normal to back-fill the chamber and vials with dry nitrogen gas before the stoppers are closed. The chamber is not filled to atmospheric pressure to prevent the small amount of compression that occurs when the bung is driven home from causing an overpressure which could lift it.

The bungs are usually of butyl rubber for low permeability to moisture. They are often lubricated so that they can be pushed home easily after freeze drying. The lubricant is chemically and biologically inert and can be steam sterilized after application. Stoppers should be vacuum dried under heat for at least 7 hours after sterilization.

Further factors are introduced when bulk material is to be dried in trays. Trays larger than $300 \times 300\,\text{mm}$ will distort during sublimation drying due to the thermal gradient between the bottom and top surfaces of the tray bottom,

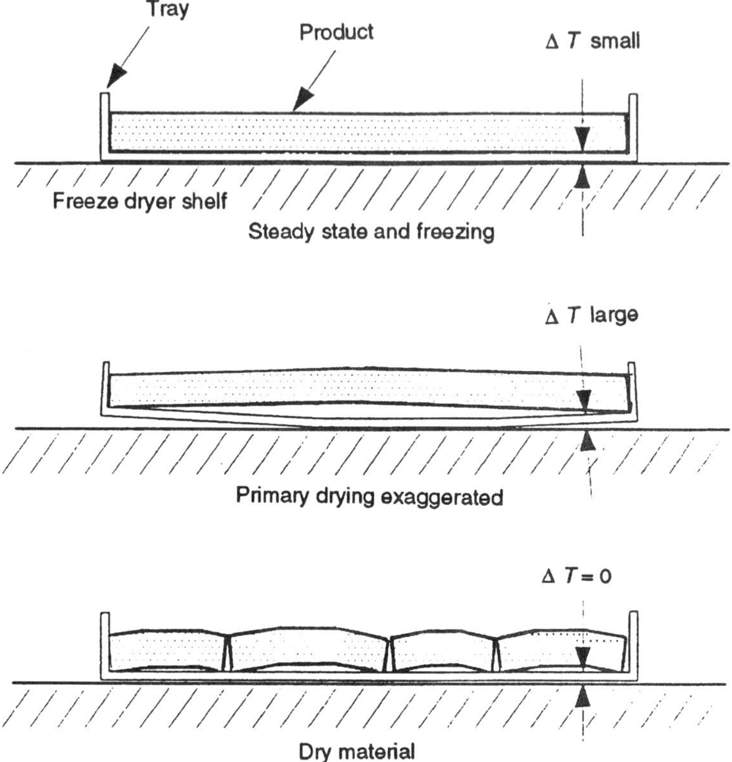

Figure 18.7   Tray distortion during drying

as shown in Figure 18.7. With large trays this effect can be reduced by putting loose dividers into the wet product before freezing. If aseptic conditions must be maintained in bulk products some form of lid must be fitted to the trays.

Heat transfer from shelf to product container and subsequently to the freeze drying front during sublimation can be increased by spoiling the vacuum [3]. In one product studied the vapour flux with a pressure of 65×10^{-3} mbar was around $0.09\,\text{g/h cm}^2$; when the pressure was increased to 350×10^{-3} mbar the vapour flux increased to $0.16\,\text{g/h cm}^2$. Spoiling the vacuum is typically done by bleeding air or dry nitrogen into the chamber of throttling the vacuum pump to maintain pressure around 250×10^{-3} mbar. This improves convective heat transfer from the heating shelf to the bottom of the product container and reduces the temperature difference required from shelf to product for a given drying rate. For sterile products, air or nitrogen is bled into the chamber through a filter, usually a $0.2\,\mu\text{m}$ steam sterilizable model.

Typical primary drying time for a 1 cm thickness cake of a 'simple' material is around 10 hours with favourable ice crystal structure and optimized temperature and pressure conditions in the drying chamber. Drying time varies with thickness approximately by the relationship

$$\text{Time} = K \, (\text{thickness})^{1.5} \qquad \text{where } K \text{ is a constant}$$

18.2.3 DESORPTION (SECONDARY DRYING)

This is the removal of bound moisture which may be water of crystallization, randomly dispersed water in a glassy material, intracellular water or absorbed water. Primary or sublimation drying accompanied by desorption removes the free water. The remaining bound water usually amounts to below 20% w/w. It can be removed by heating the product under vacuum, typically to between 20 and 40 °C.

The residual moisture content of dry material is a critical factor in storage life. Many biological materials can be damaged by overdrying and retain their best titre when dried to between 1.5 and 2% final moisture content. Examples are vaccines such as BCG: 1.5%, and live rubella, measles and others: 2%. Other materials, such as chemotherapeutics and antibiotics, must be dried to as low as 0.1% residual moisture for best results. Most freeze dried materials are hygroscopic and must be stored in sealed containers. In the early days of lyophilization, products were stored under vacuum. Current practice is to store them under dry inert gas, usually pure dry nitrogen. Other gases such as argon and helium have also been used. Factors affecting storage life have been reported [4].

Formulation and cycle development are critical factors in freezing and sublimation. Secondary drying is limited only by the maximum temperature allowed for the material.

18.3 Equipment

18.3.1 GENERAL DESCRIPTION

Aqueous solutions can be frozen by placing them in a vacuum. The more energetic molecules escape and the temperature of the solution falls by evaporative cooling. Eventually it freezes. About 15% of the water is lost.

The simplest lyophilizer would consist of a vacuum chamber in which solutions could be placed together with a means of removing water vapour so as to freeze them by evaporative cooling and then maintain the water vapour pressure below the triple point pressure. The temperature of the material would continue to fall below the freezing point and sublimation would slow down until the rate of heat gain by conduction, convection and radiation to the material was equal to the rate of heat loss as the more energetic molecules sublimed away and were removed. This simple approach presents other difficulties. When a solution is frozen by evaporative cooling it froths as it boils. This can be suppressed by low-speed centrifugation, but a more common

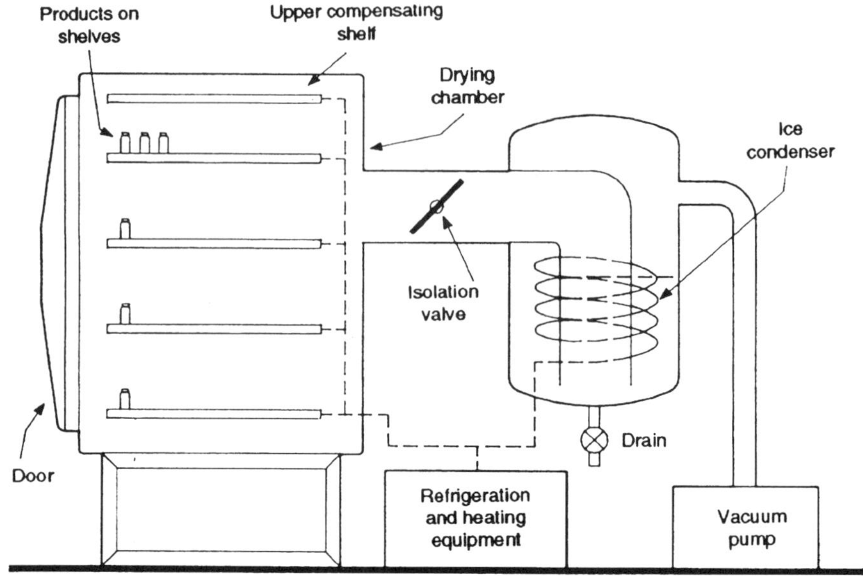

Figure 18.8   Parts of industrial FD diagrammatic

alternative is to freeze the material before it is placed under vacuum. With small laboratory lyophilizers material is usually prefrozen inside a flask by a laboratory refrigerator or a special rotary freezing bath. The flask is then attached to a manifold connected to the lyophilizer, which comprises an ice condenser and a vacuum pump.

For larger-scale equipment, it is usual to place the material on shelves inside the drying chamber which can be cooled so that the material is frozen by conduction and convection at atmospheric pressure before the vacuum is created. Without a controlled heat input to the material its temperature would fall during sublimation until drying was virtually at a standstill. A heat supply is connected to the product support shelves so that, after their initial use for freezing, they can be used to provide heat to maintain the product at a constant low temperature.

At a typical lyophilization cycle pressure, 1 mL of ice produces more than 1 000 000 mL of water vapour. The vacuum pumps cannot handle this, so a refrigerated trap (called the ice condenser) is fitted between the lyophilization chamber and the vacuum pump. Figure 18.8 shows the major parts of an industrial freeze dryer. Some important modern refinements are shown in Figure 18.9 and include:

(a) Separation of drying chamber and ice condenser to reduce cross-contamination with an isolation valve between chamber and ice condenser to allow for end-point determination and simultaneous loading and defrosting.

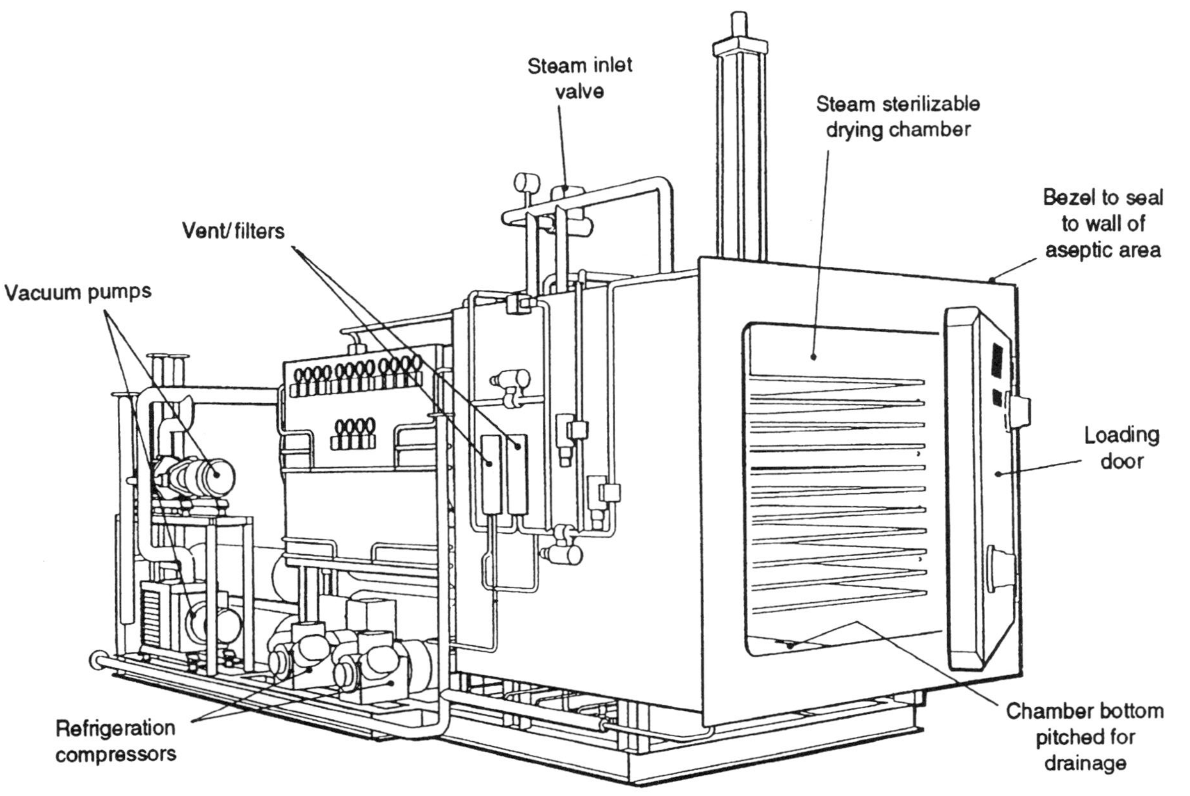

Figure 18.9 Practical industrial FD diagrammatic

(b) Pressure vessel rating for chamber and ice condenser to allow for sterilization by pressurized steam at 121 °C or higher.

(c) Use of a circulating intermediate heat exchange fluid to give even and accurate control of shelf temperature.

(d) Movable product support shelves which at the end of freeze drying close the slotted bungs used in vials. Movable shelves can also facilitate cleaning and loading.

(e) Duplicated vacuum pumps, refrigeration systems and other moving parts to enable drying to continue without risking melt-back or collapse of the product in the event of breakdown.

(f) Automatic control system with safety interlocks and alarms.

(g) Additional instruments to control, monitor and record process variables.

For low-value products where safety and possible cross-contamination are not as important the ice condenser is sometimes placed inside the drying chamber to reduce equipment costs. This is the normal method of construction for food freeze dryers.

18.3.2 FORMULATION AND CYCLE OPTIMIZATION

One batch of a pharmaceutical material in a freeze dryer can often be worth millions of dollars. Deciding the optimum conditions for freeze drying a complex product requires knowledge of its eutectic or collapse temperatures, and yield of active product depends critically on the methodology adopted.

It is essential to have access to a pilot size freeze dryer, typically having a product shelf area between 0.4 and $2.0\,m^2$ with similar layout and characteristics (shelf and ice condenser areas, temperatures and pressures) to the anticipated production equipment. If the quantity of sample product is small compared with the freeze dryer capacity shelves should be loaded with a placebo or an easily dried bulk material like skimmed milk to fill the available area.

For formulation and process development it is not necessary to use sterilizable validated equipment. However, when clinical trial batches are being dried it is essential that the freeze dryer is validated and cGMPs (current good manufacturing practices) are observed.

Effective formulation and cycle development require a knowledge of the glass transition and collapse temperatures of the product or the lowest eutectic point for the few products where the solute crystallizes. This requires use of differential scanning calorimetry (DSC) apparatus. A resistivity monitor is also useful but not essential. A freeze drying microscope helps to determine the physical condition. A sample thief is of limited application, mostly for establishing the end-point of secondary drying.

An approach to formulating a product for lyophilization and subsequent cycle development has been outlined [5]. For scale-up work, if the refrigeration and vacuum systems are proportional, it is possible to scale-up by a ratio of

Figure 18.10 Industrial freeze dryer of 20 m² shelf area before installation

20:1. For larger scale-ups the chamber geometry must be taken into account due to the variations in drying rate that occur across large shelves [6].

Lyophilizers with shelf areas between 2 and $20\,m^2$ are considered 'medium scale'. They are often custom built to suit the product with features such as special refrigeration and vacuum pumping performance, standby machinery and sophisticated control and data acquisition systems to meet regulatory authority requirements for quality control.

Large-scale installations are considered to be from $20\,m^2$ shelf area upwards in size. In the pharmaceutical industry the largest units in current use reach about $100\,m^2$ total shelf area, whereas food processing equipment is frequently built with a shelf area over $200\,m^2$. Figure 18.10 shows a modern lyophilizer of $20\,m^2$ shelf area before installation in the wall of an aseptic area. It is large enough to process some 37 000 bottles of 23 mm diameter or 400 kg of bulk material.

Large capacities require special attention to loading and unloading procedures. Operator handling is the major source of contamination of sterile products and much attention is now focused on reducing or eliminating their presence by automation. Its high cost precludes the automation of small freeze dryers under about $15\,m^2$ of shelf area although an exception occurs with highly potent or toxic materials when operators must be protected from contact.

18.4 Formulation

The final dry product

Must	Should
Retain acceptable activity	Not change significantly from its
Remain clean and sterile if required	original composition and activity
Be easy to reconstitute	Be pharmaceutically elegant
Justify the high cost of lyophilization	Be uniform

May

Need isotonicity
Need pH adjustment

18.4.1 DOSAGE

This will normally be determined after clinical trials and stated as a product requirement. However, the effects of freezing and drying on potency must be allowed for and is thus the first loop of iteration. As starting guesses assume that bacteria will lose 60% of activity and other biomolecules 20% of activity.

If the dose is very small the main task is to find the right filler after which formulation should be easy.

18.4.2 FILL DEPTH

$$\text{Fill volume} = \frac{\text{dose}}{\text{concentration}}$$

Aim for a fill depth of 5–10 mm. If typical 10 mL tube vials with an outside diameter of 23 mm are used, the inside diameter will be about 21 mm. Thus at a depth of 10 mm the fill volume will be 3.46 mL.

For tray drying the depth should be limited to about 15 mm if possible. A good starting concentration is 10–15% solids. Above 25% solids drying is slowed due to restriction of vapour escape by the layers of dry product. Below 5% solid the cake becomes very friable and may break up during drying. Below 2% solids the fluffy cake may well fly around the chamber. The concentration used may be constrained by solubility limitations. Bear in mind that, for a given dose, lower concentration means larger fill volume and hence longer drying time. The drying time increases faster than fill depth; thus increasing the fill depth 2 times may increase drying time by 2.2 times. Clearly, using larger vials will reduce fill depth but a given number of vials would then require more shelf area and so a bigger freeze dryer. Choice can be guided by the rule of thumb:

$$\text{Primary drying time (hours)} = K \, (\text{depth in mm})^{1.5}$$

18.4.3 EXCIPIENTS

Excipients are used as:

(a) *Fillers* (bulking agents) to provide a matrix for the active ingredients, especially if the dose is small or concentration is less than 1%. Commonest is mannitol because it crystallizes and it has a higher eutectic point. Glycine has similar properties and is an excellent material for freeze drying. Unfortunately it is unpopular with primary production biochemists because it encourages the growth of organisms in separation columns. It is being replaced in many formulations by Tris, which is difficult to freeze dry, especially at concentrations above about 15 millimolar, but has the advantage of being slightly bactericidal. Sorbitol is a poor choice as it does not crystallize and has a low collapse temperature. If the amount of drug in the formulation is large, bulking agents may be unnecessary.

(b) *Buffers* to control pH. Usually a phosphate buffer is used. Note that crystallization of either the acid or base component of the buffer will change the pH drastically.

(c) *Tonicity modifiers* used for osmotic pressure adjustment. Usually NaCl is chosen. It should be noted that NaCl is likely to denature proteins as the concentration increases during freezing of the water ice.

(d) *Structure modifiers* [7]. The main purpose is either:

  (i) To provide a matrix on to which material can collapse without serious effect on rehydration time. Mannitol is often used.

  (ii) To improve freeze drying speed when sugars are restricting vapour escape by forming a skin. *t*-Butanol is useful because it crystallizes quickly then sublimes at $+12\,°C$.

(e) *Stabilizers.* Used to protect the active molecule. These may be cryoprotectants, drying protectants or 'water buffers', to prevent over-drying. Sugars are particularly useful when drying proteins because they bind water. Again, mannitol is useful but sucrose, glucose or dextran are frequently employed. Dextran is especially useful due to its high collapse temperature $(-12\,°C)$. Sucrose is useful as a protectant but tends to form a skin which inhibits drying.

(f) *Collapse inhibitors.* To protect protein molecules it may be necessary to use an excipient with an undesirably low T_c. An example is lactose. A 2% solution has a collapse temperature of $-32\,°C$ but gives good protection as well as a good cake. However, the drying cycle is long. 2% mannitol $(T_c = 1.4\,°C)$ gives a good cake but poor protection. 2% glucose $(T_c = 41\,°C)$ gives a poor cake but good protection. A mixture of 2% mannitol with 1% glucose $(T_c = 16\,°C)$ gives good protein protection and a good cake with a fast cycle. T_c varies with concentration. If materials with two different T_c's are used the combined T_c will lie between the two. An example is shown in Figure 18.11.

There are many potential excipients and some are noted in the references. It is best to choose one that crystallizes and has a high collapse temperature. Some possible materials are shown in Table 18.1. However, probably 80% of formulations are based on a combination of the excipients marked * and they are recommended for initial investigation.

18.4.4 NON-AQUEOUS SOLVENTS

Sometimes low boiling points solvents are present. If there are traces left from a previous stage of production concentration is usually under 1 or 2%. In this case they will have little effect on processing. However, if they are present in larger amounts, say to increase solubility or promote crystallization, there can be a marked effect on the frozen product. There is often formation of a surface skin of the product which must be ruptured before freeze drying [8]. At

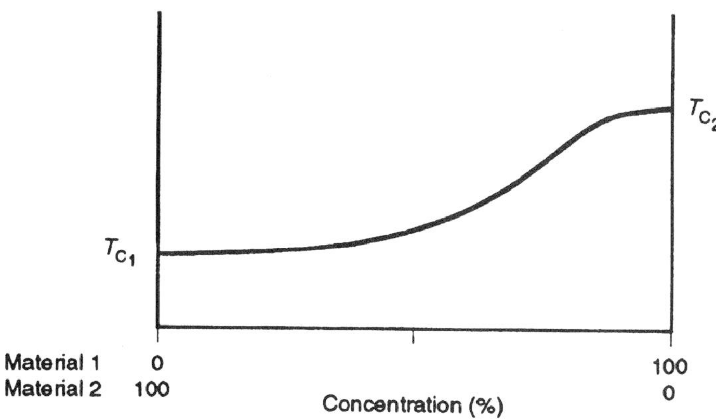

Figure 18.11   Variation of T_c with concentration

concentrations above 8% an auxiliary low-temperature trap for the evaporated solvent is often fitted to protect oil-sealed rotary pumps. A better alternative is to fit oil-free dry pumps, ideally a vertical version to permit drainage.

Some commonly used non-aqueous solvents are listed in Table 18.2. Figure 18.12 shows the vapour pressures of water and some non-aqueous solvents.

18.5 Cycle Development

The typical conditions which follow are designed as a starting point from which one parameter at a time can be changed as the cycle is developed.

18.5.1 FREEZING

Load the product and placebo to fill the shelves and allow equilibration of the temperature. Normally, this temperature would be about 5 °C.

If the fill depth of the solution is less than about 0.5 cm, a uniform freeze can be obtained by simply decreasing the shelf temperature as fast as the equipment will allow to a suitable low temperature. Usually, -40 °C is sufficient, but it must be at least 5 °C lower than the collapse temperature. With a product that crystallizes, the temperature should be 10–20 °C lower than the eutectic temperature as many eutectic systems supercool.

After the product has reached the low temperature, a 1 hour soak time will ensure that all vials are competely frozen, bottom to top. Small fill depths below 10 mm may not require a soak time.

18.5.1.1 'Tempering'

When a crystalline product is desired, but not obtained in the initial freezing, a tempering cycle will sometimes induce crystallization of bound water and reduce skin formation. The product is frozen as usual, but after the required

Table 18.1 Collapse temperatures

Substance	Collapse temperature (°C)
*Dextran	−9
Ficoll	−19.5
Fructose	−48
Gelatin	−8
*Glucose	−40
*Glycine	
Inositol	−27
Lactose	−31
Mannitol	−30
Methocel	−9
Monosodium glutamate	−50
Na$_3$CIT/citric acid (1:1)	−40
Ovalbumin	−10
Polyethylene glycol	−13
Polyvinyl pyrrolidone	−23
Raffinose	−26
Sorbitol	−45
*Sucrose	−32

low temperature is reached, the temperature is increased above the collapse temperature and held there for a time. This allows the solute to crystallize. The annealing temperature and time can be established by trial and error. If a differential scanning calorimeter is available the annealing temperature can be located using the exothermic peak. A good starting point is about 10 °C above the collapse temperature for 2–4 hours. After the tempering process, the product temperature is decreased to about 5 °C below the collapse or eutectic temperature before drying is started (Figure 18.13).

18.5.2 PRIMARY DRYING

It is desirable to maintain the product at the highest possible temperature in order to speed the process. This will be a constant temperature depending on the product. As a rule of thumb there is about a 13% increase in drying time for every 1 °C reduction in the product temperature. After freezing, the shelves

Table 18.2 Some common non-aqueous solvents

Solvent	MP	T for VP of 1 mb (°C)
Methanol	−97.9	−49
Ethanol	−114.7	−37
Propanone	−96.5	−66
Benzene	6.5	−42
Chloroform	−6.5	−63

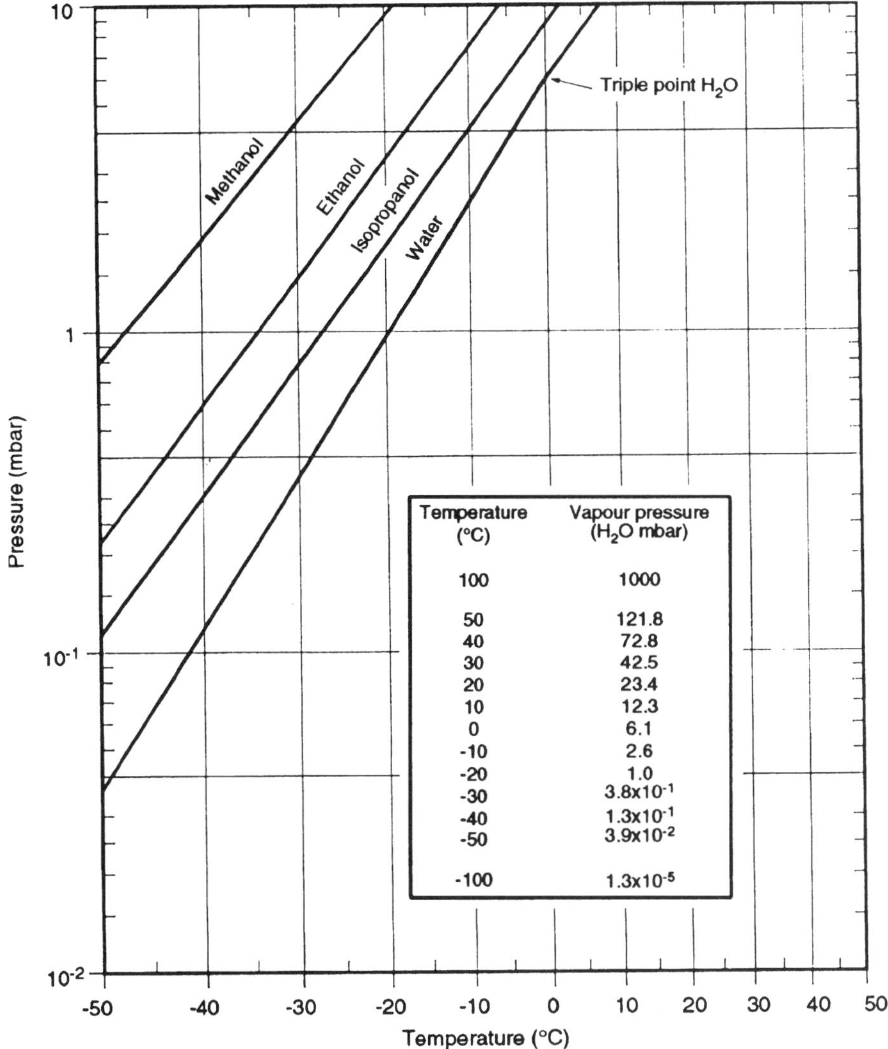

Temperature (°C)	Vapour pressure (H₂O mbar)
100	1000
50	121.8
40	72.8
30	42.5
20	23.4
10	12.3
0	6.1
-10	2.6
-20	1.0
-30	3.8×10^{-1}
-40	1.3×10^{-1}
-50	3.9×10^{-2}
-100	1.3×10^{-5}

Figure 18.12 Vapour pressure of H_2O and some non-aqueous solvents

of the freeze dryer will normally be about 5 °C below the temperature planned for the initial stage of primary drying.

The resistance of the dried product thickness increases with time, and the loss of heat due to sublimation decreases with time. Since the objective is to maintain constant product temperature, the heat input must also decrease with time. A decreasing heat input with time means that either the shelf temperature or the chamber pressure (or both) must decrease during the primary drying cycle.

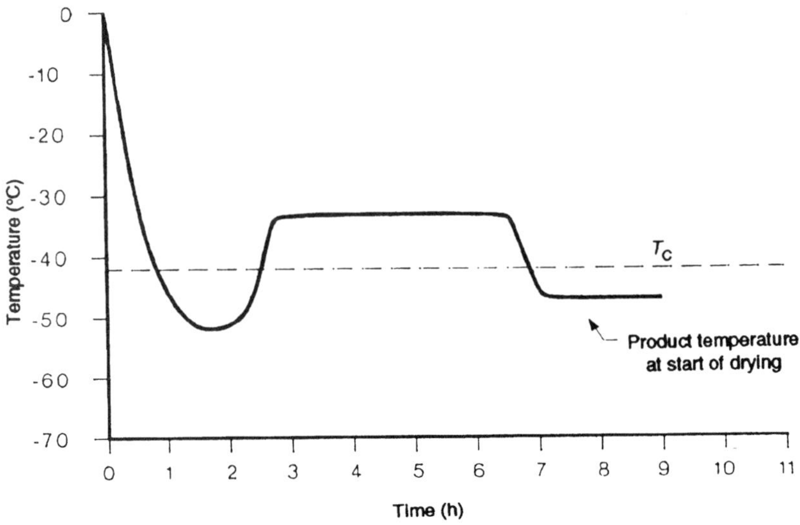

Figure 18.13   Tempering a product with T_c of $-42\,°C$

The important factors in primary drying are chamber pressure and shelf temperature. Chamber pressure should be significantly less than the vapour pressure of ice at the target product temperature. High chamber pressures decrease the driving force for sublimation. However, uniform heat input from sample to sample is usually easier to achieve at chamber pressures around $10^{-1}\,mbar$ and heat input to the product can be controlled by the chamber pressure, which can be varied more quickly than the shelf temperature. A good start is to select a chamber pressure of around 20% of the vapour pressure of ice at the target product temperature. With a low collapse temperature, a chamber pressure near the upper limit would allow operation at a pressure close to the optimum for uniformity of heat transfer. Thus, if the target temperature is $-30\,°C$ for $T_c = -25\,°C$ the vapour pressure of ice is about $3.8 \times 10^{-1}\,mbar$ and the chamber pressure would be around $1.1 \times 10^{-1}\,mbar$.

Once the decision is made on the chamber pressure, the next decision is what shelf temperature to use. In the absence of other information, simply start the primary drying run at an initial guess for the shelf temperature and adjust the shelf temperature until the product temperature is at the target. The time for equilibration after a shelf temperature change is typically 30 min to 1 hour, and since the correct shelf temperature will change with time, it will probably require several tests before the required shelf temperature/time profile is developed. A rough rule of thumb is that the initial temperature difference between the shelf and vials should be around $100 \times$ vapour pressure of ice in mbar at the target temperature.

A combination of variable shelf temperature and variable chamber pressure is often used. If T_c is low, the chamber pressure must be low and the variation in heat input is obtained by decreases in shelf temperature to maintain a constant product temperature. If the product has a high T_c, the chamber pressure can be quite high without seriously slowing the sublimation and a constant shelf temperature could be used while achieving the reduction in heat input by decreases in chamber pressure.

18.5.3 PRIMARY TO SECONDARY DRYING TRANSITION

This can be found by noting the time when the product temperature is within 2–3 °C of the shelf temperature. Take the total time in drying up to this point as the primary drying time. A delay time of roughly 20% of the primary drying time should be allowed before the shelf temperature is increased for secondary drying. This allows an atypical vial to 'catch up'.

18.5.4 SECONDARY DRYING

The optimum residual moisture for storage stability should be studied during formulation. Usually, the lower the residual moisture, the more stable is the product. However, many biological materials have better stability with higher moisture contents. The drying rate is independent of pressure up to about 2.5×10^{-1} mbar. A good pressure for secondary drying is about 1.5×10^{-1} mbar. The temperature and time of secondary drying must be established by experiment.

Assuming that the product stability is best at a low water content, say, below 1%, the secondary drying process is straightforward and is carried out for an hour or two longer than the minimum necessary to achieve the desired result. A secondary drying temperature of 24–30 °C for 3–4 hours is often good for proteins; for non-proteins the shelf temperature can usually be increased to 40 °C.

Many biological materials can be damaged by over-drying. If a higher moisture content is needed for maximum stability, the process design is more difficult since over-drying must be avoided. In this case the shelf temperature should be kept at around 15 °C for good uniformity and samples tested for moisture content at different time intervals during the secondary drying. The approach of rehumidifying with steam to a controlled level after drying has worked for products requiring very accurate residual moisture content.

Sampling for residual water measurement should include likely extremes such as shelf corners, top and bottom shelf, vials exposed to the condenser chamber and vials near a window.

18.6 Good Manufacturing Practice

International regulatory agencies, requiring users to observe good manufacturing practices (GMPs) and calling for validation of processes and equipment before and during use, are responsible for many improvements in the equipment. These can be grouped loosely into three categories: design and fabrication of equipment, calibration or instruments and registration of control software (see also Chapter 21).

Equipment must be oriented towards the elimination of potential dirt traps and the avoidance of rubbing contacts which could generate particles. Trapped volumes can generate virtual leaks. Screw threads should be eliminated where possible. Figure 18.14 indicates some good and bad construction features.

Good cleaning access must be provided or clean-in-place (CIP) systems should be fitted. Effective CIP systems require a large number of nozzles. Because of the large flowrate of water for injection (WFI) required to operate the nozzles, the chamber is often divided into zones which are cleaned in sequence. The effectiveness of the CIP system can be determined by spraying a fluorescent product such as vitamin B2 around the chamber, allowing it to dry and then running the CIP system. An ultraviolet lamp can then be used to look for remaining traces.

Surface finishes should be 240 grit or better to facilitate cleaning and avoid dirt traps of significant size. Where simple 'wipe-down' sterilization is employed, it is difficult to establish sterility, thus requiring media tests to be made. Gas sterilization using formaldehyde, ethylene oxide or mixtures of hydrogen peroxide is also less popular for safety reasons, as well as the difficulty of penetrating pipe dead legs. An additional problem with gas sterilization is the hazard presented to operators.

Steam sterilization is not a mandatory requirement but its use is becoming almost universal for parenteral production. The use of clean steam at a temperature of 121 °C or more is therefore the preferred method. Once a steam sterilization cycle has been validated it is not necessary to carry out sterility testing each time it is performed.

18.7 Leakage

The significance of leaks in a freeze dryer is frequently not well understood, since during the drying process there will always be vapour evolution from the product. Contaminants leaking into the freeze dryer have a very high probability of being swept through the system to the ice condenser or vacuum pump. The only known cases of product failure due to contamination though leaks have occurred when major mechanical breakdown has caused a loss of product due to melt-back.

Real leaks from the atmosphere into an evacuated system may be by hydrodynamic flow through channels or gaps, or by diffusion through porous materials. Mass rates of flow through leaks are not significantly affected by variations in the system pressure when it is below some tens of millibar and it is hydrodynamic.

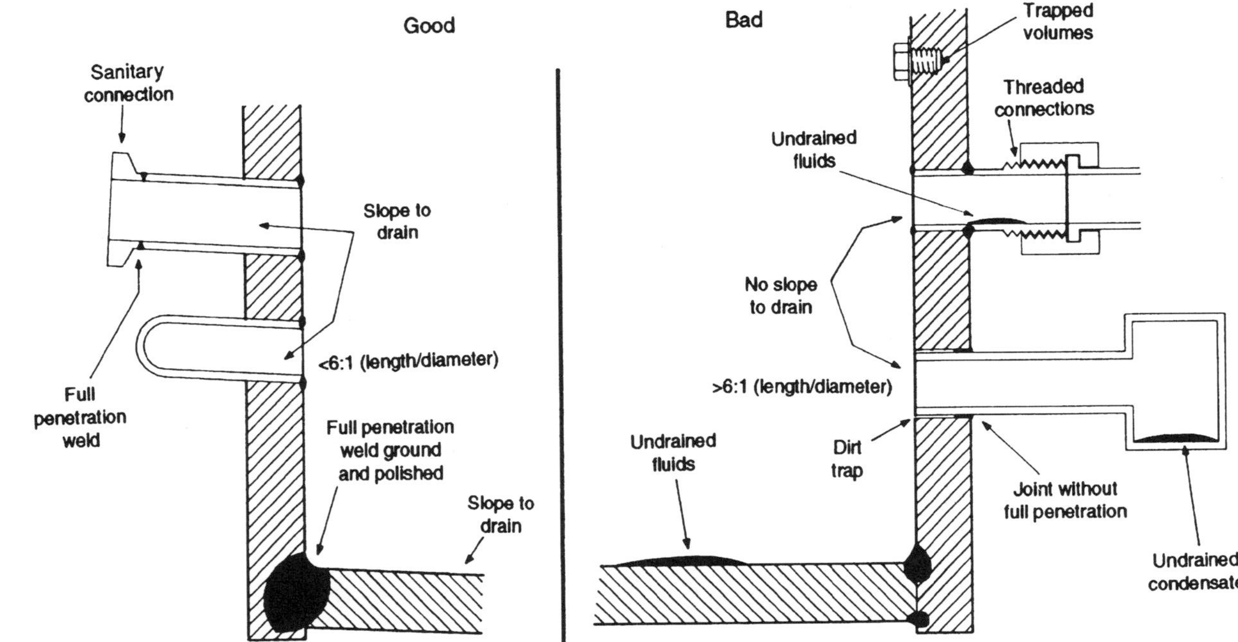

Figure 18.14 Good and bad construction features

Real leaks are distinguished from virtual leaks which are caused by desorption from surfaces, or from trapped volumes of porous materials within the system or by the generation of vapours within the system. The last is important in freeze dryers because it is difficult to dry them completely, and water vapour must be eliminated in order to obtain meaningful leak test results. Fortunately, this can be achieved by cooling the ice condenser during leak testing.

Real leaks, as opposed to virtual leaks, are the cause of concern about possible product contamination. Vacuum leaks are expressed in millibar litres per second. For a constant mass of a gas, the product or its pressure and its volume remains constant at a constant temperature. Therefore, this measure holds for the rates of passage of gas entering and leaving a leakage pathway. Frequently, tests are based on 'leak-up' rates which are pressure rise tests with the chamber blanked off and which depend on the chamber volume. In this case, the units will be mbars per second.

For example, a user of vacuum equipment first obtained a pressure of 40×10^{-3} mbar in an empty dry chamber of 180 litres volume. When the chamber was isolated the pressure increased to 120×10^{-3} mbar over a period of an hour. It may be assumed that the pressure rise of 80×10^{-3} mbar was at a constant rate over the isolation period of an hour. Working from the volume of the chamber, this represents a leak rate of $(80 \times 180 \times 10^{-3})\,3600 = 4 \times 10^{-3}$ mbar litres per second.

Typical leak rates required will be less than 0.001 mbar litres per second on a clean, dry chamber with all its nozzles carefully blanked off and less than 0.01 mbar litres per second on chambers in their final condition with all valves and connections in place.

18.8 The Cost of Lyophilization

Conventional drying from the liquid phase requires the addition of just over 2000 kJ of energy per kg water evaporated. In freeze drying, the energy flows are much greater. Allowing for typical efficiency of machinery, the energy consumed will be approximately as follows:

Freezing product	1670 kJ/kg
Drying product	3760 kJ/kg
Condensing vapour	3180 kJ/kg
Total	8610 kJ/kg

This energy cost, while high compared to the energy cost of conventional drying, is a minor factor in the overall cost of lyophilization. By far the greatest cost is the capital costs of equipment. Assume that a medium-sized industrial unit with a shelf area of 20 m^2 costing \$700 000 installed and commissioned will be amortized over 10 years with average interest rates. Assume that 180 freeze drying cycles are completed each year and that 300 kg are sublimed in each cycle. Then the cost of sublimation of each kilogram of water is about \$2.00. A more precise calculation would need to take into account the cost of cooling water for the refrigeration compressors, clean water for CIP, steam for sterilization and any other services. It should also consider the cost of loading and unloading operations, maintenance and record keeping. Capital costs of freeze dryers vary roughly in proportion to their capacity raised to the power 0.75.

18.9 References

1. F. Franks, R. H. M. Hatley and S. F. Mathias (1991) 'Materials science and the production of shelf stable biologicals', *BioPharm*, October, 38–55.
2. A. P. Mackenzie (1977) 'The physico-chemical basis for the freeze drying process', in *International Symposium on Freeze Drying of Biological Products, Washington D. C.' 1976*, Develop Biological Standard, Vol. 36, (S. Karger, Basel,) pp. 51–67.
3. M. J. Pikal (1990) 'Freeze drying of proteins. Part I: process design', *BioPharm*, **3**(8), 18–27.
4. C. C. Freyrichs and C. N. Herbert (1974) 'Long term stability studies on the International Reference Preparation of Newcastle Disease Vaccine', *J. Biolog. Standards*, **2**(1), 59.
5. M. J. Pikal (1990) 'Freeze drying of proteins. Part II: formulation selection', *BioPharm*, **3**(9), 26–30.
6. T. W. G. Rowe, D. Greiff and J. Monroe (1979) *Proceedings of XV International Congress of Refrigeration*, Venice, 23–29 September 1979.
7. D. J. Korey and J. B. Schwartz (1990) 'Effects of excipients on the crystallisation of pharmaceutical compounds during lyophilisation', *J. Parenteral Sci. and Technol.*, **43** 80–3.
8. H. Seager (1985) 'Structure of products produced by freeze drying solutions containing organic solvents', *J. Parenteral Sci. and Technol.*, **39**, 161–79.

19 INSTRUMENTATION AND PROCESS CONTROL

John Noble and Keith G. Robins

19.1 Introduction

The complexity and sensitivity of all natural product manufacturing operations
has limited the extent to which conventional process control and instrumentation
can be applied to process operations. In many cases designers must compromise,
monitoring and controlling performance by proxy rather than by direct
measurement of the desired parameter. Typically a unit operation will be
designed around specific parameters such as flowrate, pH, conductivity and
temperature which can be readily measured *in situ*. However, none of these
provide any concrete evidence of product recovery or contaminant removal. As a
result of this, actual process performance can often only be ratified by off-line
analysis of in-process samples. This tends to fragment process operations,
increasing the level of operator involvement and reducing the extent to which
integrated control strategies can be implemented. The issue is further heightened
by the value of many natural products and the belief that the extensive
involvement of highly skilled process operators in intuitive monitoring and
control may substantially reduce the potential for product corruption.

The following chapter consists of two main sections. The first covers the
general criteria applicable to instrumentation for hygenic or aseptic operation,
as is usually required for the isolation of macromolecular products. The second
section deals with a practical approach to process control strategies in a
multipurpose experimental pilot plant. A brief concluding comment covers
possible future trends.

19.2 Design of Instruments

The parameters monitored in natural product manufacturing differ little from
those in conventional chemical production. The nature of the equipment
utilized can, however, be substantially different.

Downstream Processing of Natural Products. Edited by Michael S. Verrall
©1996 John Wiley & Sons Ltd

The type of monitoring equipment installed in any process is heavily dependent upon the nature of the product. If the product has to be manufactured to the standards of good manufacturing practice (GMP) then it is likely that considerable demands of reliability and hygiene will be placed upon any equipment that comes in contact with the process stream. These measures are to ensure that the process can be operated both with repeatable quality and the minimum potential for corruption or degradation of the product by foreign matter. Where natural products are concerned their liability induces a hygienic requirement that can set the process equipment aside from other conventional products. In addition, for certain products stringent demands of containment can be set for either the producing organism or the final product.

The following sections cover the basic rules that must be applied during the design of instrumentation for natural product recovery and specific information relating to the selection of individual instruments. The safety aspects of hazardous area operation must also be considered during instrument selection.

19.2.1 DESIGN OF INSTRUMENTATION FOR NATURAL PRODUCT RECOVERY

An exhaustive list of the factors raised by the requirements of cleanliness and containment is impossible to draw up. However, most points can be covered under four basic headings:

Design for ease of cleaning
Design for sterlizability
Design for maintenance of asepsis
Design for containment

These topics and the selection of construction materials are covered in detail in the following sections. Although the regulations are always open to interpretation, the points raised below are considered to be generally accepted throughout the biotechnology industries.

19.2.2 EASE OF CLEANING

The level of cleanliness required by a process is often very difficult to quantify. However, in qualitative terms three broad areas of operation can be defined: hygienic operation such as in food and beverage manufacture where considerable efforts are made to minimize the level of contamination; aseptic operation where all possible efforts are made to operate the process free of microbial contamination but the involvement of operators and the nature of the process feed stock prevents sterile operation; finally, sterile operation in which the product is free from microbial contamination, which is typical of the final filling of injectable pharmaceutical products. Where natural products are concerned, excluding bulk foodstuffs, the majority are manufactured under aseptic conditions with considerable attention paid to the elimination of microbial contamination and also the removal of other pyrogens such as endotoxins.

If all the points listed below can be readily satisfied then there should be few problems in assuring ease of cleaning (all cleaned surfaces should be stable in the cleaning medium):

(a) There should be no crevices including exposed threads in contact with the product.
(b) Any pockets should be large and shaped so that they can be cleaned thoroughly.
(c) The equipment should be self-draining so that cleaning fluid cannot be retained and thus allowed to contaminate the process. Pockets should also be self-draining.
(d) A good surface finish should be provided to aid cleaning and to allow soiling to be visible.
(e) The equipment should be easily and quickly dismantleable, preferably with a minimum number of tools. All complex parts should be capable of being removed in their entirety.
(f) Where clean-in-place (CIP) is to be used, the flow of fluid through the equipment should be smooth. There should be no backwaters that might retain debris. In large equipment cleaned with spray balls, internal fittings should not mask other areas.
(g) As steam sterilization effectively eliminates all viable micro-organisms it may be possible to relax some of the above points when heat sterilization is to be used. This is also true of processes such as solvent extraction where the manufacturing environment is not conducive to biological growth.

19.2.3 STERILIZABILITY AND SANITIZATION

Any process plant that is to operate aseptically will generally be sterilized or sanitized before use. This minimizes the potential for microbial, viral and other foreign body contamination of the process stream. Sanitization fluids include caustic soda, hypochlorites and dilute acids. Sterilization usually infers the use of a set temperature for a set period of time. Most commonly employed is dry saturated steam at temperatures between 120 and 140 °C for times of up to four hours. The decision between utilizing sanitization is highly process dependent. In a fermentation where microbes could penetrate or escape from the process sterilization will most probably be employed. If the feedstock is a natural product, milk for example, sanitization may be sufficient to maintain a non-hazardous bioburden in the process. The points listed in the previous section cover the requirements for sanitization whilst those listed below deal specifically with sterilization.

The construction materials should not be distorted or damaged by the temperatures or pressures involved. Steam should freely enter all parts of the equipment. The possibility of air pocket formation should be avoided. The design should ensure free draining condensate.

19.2.4 MAINTENANCE OF ASEPSIS

The type of process will determine the length of time for which asepsis must be maintained. In a typical batch plant the process may only be in operation for a few hours each day and will have differing requirements to a continuous process operation. A flanged joint, once sterilized, may retain its integrity for some days but will be less likely to do so for six months.

Factors worthy of consideration are:

(a) Continuous butt welds are preferable to flanges or fittings for static joints. For pipeline fittings, orbital welding may be used.
(b) Rotating and reciprocating joints may be flushed with steam or other suitable sterilizing fluid to act as a barrier to the penetration of microbes.
(c) Some static joints may be steam flushed. The design requirements for the maintenance of containment are similar to those put forward for preventing the entry of microbes.
(d) Replaceable and disposable equipment parts should be capable of being removed with a minimum of exposure to the environment.

19.2.5 MATERIALS OF CONSTRUCTION

In spite of the wide range of materials available to equipment manufacturers few have found lasting favour with end users in the biotechnology industries. End users have tended to stick with a range of well-tried and tested materials, including austenitic stainless steels (never lower than AISI standard 304 and more often 316), borosilicate glass, PTFE (polytetrafluoroethylene), silicone, butyl and nitrile rubbers and other elastometers such as viton.

Certain broad principles can be stated as a guide to the nature and finish of materials if these are to be acceptable:

(a) Construction materials must be inert to process materials, cleaning fluids and any sterilizing agents under the conditions encountered. This includes being corrosion free, having no catalytic activity or biological toxicity.
(b) Materials must be non-porous.
(c) Materials must be normally resistant to repeated temperature cycles between ambient and 150 °C, without creep, deformation or embrittlement.
(d) Copper, zinc and alloys containing these metals are unacceptable in nearly all cases.

(e) Material surfaces should be crevice free. A polished metal surface will make soiling more easily visible, but the relationship between surface finish and cleanability is not clearly defined. The depth of surface irregularities may be less important than whether they are rounded or jagged. In this respect, mechanical polishing does not always produce a surface that is cleaned more easily than one produced through pickling or de-scaling. The highest standard of surface finish would be that produced by electro-polishing.

19.3 Selection of Instrumentation for Natural Product Recovery

The selection of instrumentation for natural product recovery can often be complicated by the enormous range of off-the-shelf equipment available and a lack of vendor understanding of the duty in question. The previous section goes some way to defining the general design and selection basis for instrumentation whilst the specific requirements of each instrument type are briefly covered below. Where possible, typical examples of installation and operational problems have been highlighted. It should be noted that the examples below represent best-case scenarios and in certain instances the designer may well have to compromise on the factors raised in Section 19.2 to ensure adequate process monitoring.

19.3.1 FLOW MEASUREMENT

In the recovery of natural products, flow measurement, especially in the pharmaceutical fields, has tended to concentrate on non-invasive methods of flow measurement. Historically techniques such as electromagnetic inductance and time of flight have been employed, with the former proving most flexible over a very wide range of operations. More recently Coriolis systems for the measurement of mass flowrates have received increasing attention, offering advantages of volumetric flowrate, fluid density, fluid viscosity and temperature measurement within a single probe. The cost of mass flow systems is significantly greater than that for electromagnetic systems and they are generally only used in areas where electromagnetic flow systems are unsuitable. One specific example is in the metering of fluids with conductivities less than 0.5–1 mS, such as pyrogen-free water where electromagnetic flowmeters are at the limit of their operating range.

Electromagnetic flowmeters consist of a flow tube housing a set of electrodes which are isolated from each other by an insulating sleeve which fits inside the tube. Only the sleeve and electrodes come in contact with the process fluid and both must be selected to ensure compatibility with process operation, especially temperature and cleaning solution components and flowrates.

In certain fields such as food and beverage manufacture the use of positive displacement, variable area and vortex shedding meters may well be acceptable. However, the intrusive and complex internal design of these systems, and the associated problems of cleaning validation, precludes most from use in aseptic processing unless there is no suitable alternative.

Where possible gas flowrates are measured prior to the clean area boundary with sterile filtration following the meter. As a result less stringent demands are placed on instrument selection. Hygienic designs are available for both vortex and variable-area gas flowmeters should the design necessitate. If absolute cleanliness is required then the use of mass flowmeters would be recommended.

19.3.2 PRESSURE MEASUREMENT

Hygienic pressure measurement is generally accomplished using diaphragm seals which isolate the pressure detection system from the process fluid. The pressure measuring element can be either a standard Bourdon-type gauge or a piezoelectric device. As the seal contacts the process fluid its design is often of greater importance than that of the measuring element. Flush-fitting tank and pipe wall-mounted. systems are available which offer excellent cleaning characteristics. In addition, where cleaning requirements are less stringent devices can be attached to minimum dead-leg 'T' pieces.

Problems can occur with the ingress of solvents to the diaphragm seal and care must be taken in matching materials compatibility. The integrity of many seals can often be compromised by over-pressure and it is essential that equipment specification accounts for all operating scenarios. For example, the maximum pressure experienced by a seal may not be during process operation but could be during CIP or steam sterilization.

19.3.3 TEMPERATURE MEASUREMENT

As with pressure measurement the temperature probe is usually isolated from the process fluid by a thermowell which is designed to accommodate all the sanitary and safety aspects of the process. The majority of natural product processes rely upon platinum-resistance temperature detectors (RTD).

When mounted in process vessels thermowells are generally positioned near the base to maximize the operating range of the vessel and so that cleaning fluid in the heel of the vessel can access the underside of the probe during CIP. The small pipe diameters encountered in natural product recoveries favour the use of RTDs externally mounted on the pipe wall. Where the accuracy or speed of measurement necessitates it may be necessary to insert a thermowell in the line via a minimum dead-leg 'T' piece. The sanitary implications of such an action should be considered before it is implemented.

19.3.4 pH MEASUREMENT

The measurement of pH is based upon a potentiometric operating principle and utilizes an electrode system to assess the potential developed between a glass and reference electrode which is proportional to the hydrogen ion concentration in solution. The two electrodes can either be in a combined form or housed separately. As the latter requires two process mountings the system is undesirable and does not have widespread application in the process industries.

The glass electrode is enclosed by a pH-sensitive glass membrane which is permeable to hydrogen ions. Inside this membrane a conductor is immersed in a buffer solution. The electrical circuit is completed by the reference electrolyte in which the reference electrode is mounted. In the combination electrode the reference system concentrically surrounds the glass electrode. The reference electrolyte must be in contact with the measuring solution.

With conventional liquid reference electrolytes external pressurization provides a constant electrolyte flux across the junction which maintains good contact with the measuring solution and ensures that the electrolyte can be kept at a constant pressure. Both of these factors are critical to the maintenance of reliable repeatable readings. As a result of this the process fluid becomes contaminated with very small volumes of electrolyte. Solid gel reference electrolytes have been developed to eliminate or greatly reduce the level of process fluid contamination. Gel electrodes are currently available in both internally pressurized and non-pressurized forms, with the latter giving zero ingress of electrolyte to the process solution.

The choice of electrode is a subject of much debate and is likely to be very process specific. Key parameters will be the tendency of the electrode to foul and the accuracy of measurement required. In general the highest accuracy can be obtained from pressurized liquid electrodes and the tendency to foul can be counteracted by electrode pressurization.

Probes can be mounted in a full range of line sizes using minimum dead-leg 'T' pieces and also mounted on vessel walls. When mounted, a pH probe must be at an angle of at least $15°$ to the horizontal to ensure electrical integrity. When mounted in vessels care must be taken to ensure that the probe can be readily cleaned either in place or out of place. Typically probes are mounted just above the dished end of the vessel to ensure that fluid in the heel can clean the underside of the probe. Finally, as probes are quite fragile the design must ensure that maintenance access is readily available.

19.3.5 CONDUCTIVITY MEASUREMENT

Typically the conductivity of a solution can be monitored using two techniques: a conductometric analysis which utilizes two electrodes and an inductance analysis which measures the potential between two inductance coils. The latter is particularly suitable for highly conductive complex solutions and solutions containing solids. As a result it has found wide acceptance in natural product extraction processes.

The analysis of conductivity requires the immersion of a measuring head in the process fluid. A flow of fluid must be maintained around this system whether the element is vessel or pipe mounted. This has implications for the mounting of the measuring system in vessels and care must be taken to ensure that not only is the system flushed during normal agitation but also readily cleaned.

19.3.6 LEVEL, VOLUME AND MASS MEASUREMENT

The measurement of level, volume and mass are interconnected by three basic measuring principles: the measurement of pressure, the measurement of displacement and the direct measurement of mass. Using one of these techniques it is often possible to directly measure one process parameter and then calculate the others using predetermined algorithms.

Using pressure measurement systems the hydrostatic head in a vessel can be measured and converted using density in to a height of liquid above the measuring head. The hydrostatic head can also be estimated from changes in fluid capacitance. This figure can then be linearized, via calibration, to yield a fill volume for the vessel. The fill volume can then, if necessary, be converted into a mass. The principal problem here is the need to determine the density of the fluid. Where this is highly variable the technique cannot be considered reliable.

Buoyancy can be used to raise a float up a measuring element as a vessel fills. The vertical movement of the float can be measured using resistive techniques and this can be linearized into a volume measurement. The calculation of the fluid volume is independent of the fluid characteristics and as a result the technique can have certain advantages over the measurement of hydrostatic head. Potential problems lie in the nature of the measuring systems which must be mounted inside the process vessels and may cause problems with cleaning in place. These problems can often be overcome by careful spray ball location.

The mass of fluid flowing through a pipework system has been covered in the section on flow measurement. Where a direct measurement of fluid mass in a vessel is required the vessel in question can be mounted on load cells. From the mass measurement and a knowledge of the fluid density both the fill volume and liquid level in the vessel can be calculated. Where the mounting of hydrostatic probes is impracticable, e.g. large vessels where the pressure range is high or small vessels where space constraints prevail, load cells have a particular attraction. However, design must take into account the movement of the vessel and ensure any fixed pipework can withstand the strains imposed.

Overall float systems and hydrostatic head are the cheapest and most commonly employed systems in small to medium sized vessels. In larger vessels load cells are preferred for mass measurement, especially if the fluid density is variable, whilst float systems are more common for volume measurement.

Other systems such as ultrasound and radiation scattering have found applications in certain areas and may be worthy of consideration. However, the techniques are invasive and consideration must be given to the cleanability of the systems.

It is also possible to obtain point measurement of level using capacitance and vibrating rod type sensors. Such systems are frequently employed as detectors to trigger control systems or alarms and they are substantially cheaper than measuring systems.

19.3.7 ANALYSIS

In liquid systems optical density can be used to monitor chemical compositions utilizing various wavelengths to assess the concentration of certain known species. Hygienically designed industrial probes are available to monitor optical density at a range of wavelengths including 280 and 620 nm, the former being particularly useful for proteinaceous products.

Light scattering can also be utilized to monitor fluid turbidity and the presence of air in a fluid stream and such systems are frequently employed on centrifugation and adsorption systems as alarms and switches. Electrical systems such as capacitance and time of flight can also be used to detect the presence of air in a process stream.

19.4 Process Control in Product Development and Pilot Plant Processes

19.4.1 CONTROL MODULE CONCEPT

Most of the facilities used for the development and scale-up of downstream processes for natural products are of a multipurpose nature. The cost of retrofitting primary instrumentation to a typical downstream processing plant handling up to 3000 litre batches of fermentation broth is estimated to be in the order of £300 000. The organization of such a pilot plant usually involves the use of a limited range of equipment at any one time. Thus much of the investment in full instrumentation would not be realizing its potential whilst certain unit operations were not in current use. Analysis of downstream operations shows that the parameters which need to be controlled are similar for many unit operations, e.g. flow, pH, temperature and level, etc.

There are three main benefits from employing automated control and monitoring of processes under development: (a) improved data quality, allowing more rapid optimization, (b) better use of available manpower and (c) enhanced consistency in control leading to better product quality. The multipurpose plant will almost certainly be used to address the cost/benefit question as well as process optimization in terms of both yield and purity.

Figure 19.1   Example of a scheme using modules: 1, feed vessel temperature controller; 2, variable-area flowmeter and flow control valve; 3, pH control module; 4, pressurized titrant reservoir for pH control module

Unit operations in these plants are frequently performed using a 'clean' but not aseptic, regime wherein equipment is assembled into a process scheme using flexible connections. Insertion of a control element into such a system therefore becomes an easy adaptation. By adopting a modular approach to sensing and control it has been demonstrated [1] that the benefits listed above may be accrued without excessive cost. Investment in items such as sensors and transmitters is minimized and the number of modules required is defined by the most complex simultaneous processing task proposed. For example, investment of approximately 10% of the sum for full instrumentation in the plant analysed above allows the control and monitoring of an in-line solvent extraction with continuous back extraction and solvent recycle. Furthermore, costs, particularly for field instrumentation, may be spread over a period of time as modules need only to be added to the system as required.

The recommended scheme uses modules consisting wherever possible of a complete control package, i.e. sensor, transmitter and control element. These items may be contained within a standard portable framework, which may be stacked one upon another to minimize the equipment footprint (see Figure 19.1). The sensors employed in these modules should be selected with reference to those used in full-scale production plant to facilitate the transfer of development processes to commercial operation. Furthermore, equipment selected from that used for production duty will have proven reliability and safety and will have undergone extensive evaluation in this rôle for any cGMP (current good manufacturing practices) operations proposed.

Using the recommended modular instrument packages, all common process parameters have been successfully monitored and controlled, i.e. liquid flow and level, conductivity, pH, pressure and mass. Although the above statement regarding instrument selection from the range in use in production is broadly true, some care must be exercised to ensure that the sensor is compatible with a wide variety of situations encountered. For example, ultrasound sensors are now popular for level sensing; however, their accuracy is dependent on the vapour density in the vessel head space. Recalibration would be required for each process solution involving a change of solvent. For this application, i.e. level sensing, resistance chain sensors have been successfully used in a wide variety of liquids. However, high-viscosity liquids may cause difficulties with these devices.

The greatest benefit from process instrumentation and control facilities can be realized by connecting the modules to either a suitable sequence controller (PLC) or stand-alone PID (proportional integral derivative) loop controllers. This arrangement provides greatest flexibility of the system. Stand-alone controllers encourage the use of process control for even simple operations as they do not require programming, although reconfiguration of process parameters, e.g. input scaling, is relatively straightforward on modern programmable controllers. More complex processes can be controlled by the sequence controller which will supervise and schedule the operation of the PID controllers, collect data for the generation of batch records and down-load process parameters, e.g. set points, at appropriate points in the process. Many sequence controllers will also have the capability to control certain processes by using a mixture of digital and analogue inputs and outputs within the PLC itself. To maintain flexibility within the system it will be necessary to be able to vary the way in which field modules are connected to PLC or PID controllers.

To create this flexible network of connections in a limited process facility a 'telephone exchange' patching system is the best solution. For larger plants, fieldbus systems are suggested as the best option with the modules connected to fieldbus outstations and signals decoded from the network by a PLC at the controller end.

19.4.2 FLAMEPROOF AND INTRINSICALLY SAFE INSTRUMENTS

Downstream processes for natural products such as secondary microbial metabolites often entail a solvent extraction step or other unit operations involving the use of flammable solvents. When scaled up this procedure will need to be performed in a designated hazardous area with due attention given to the specification of appropriate electrical equipment. The normal criterion for designation of such an area is one in which greater than 50 litres of flammable solvent is present. Hazardous areas are classified according to the nature, ignition temperature and probability of the hazard. Within the European Union the probability of the hazard existing is defined by zoning:

Zone 0. An area in which an explosive gas/air mixture is continuously present, or present for long periods.

Zone 1. An area in which an explosive gas/air mixture is likely to occur in normal operation.

Zone 2. An area in which an explosive gas/air mixture is not likely to occur in normal operation and if it occurs will exist only for a short time.

Thus vessel head spaces become zone 0 and areas around pump glands, where there is potential for leakage, become zone 1. A small pilot facility will have a number of zones overlapping from various items of equipment, and in practice the whole facility will normally warrant classification as zone 1. Electrical equipment suitable for use in such a zone will be of two types, either high power for pumps, agitators and other process equipment or low voltage for instrumentation. Both types must be designed in such a way as to prevent the ignition of a flammable atmosphere and suitably certified. For solvents commonly used in downstream processing certification to the CENELEC standard for high-voltage electrical equipment:

EEx d IIB T4

will more than cover most hazards. For instrumentation, the concept of 'intrinsic safety' is most convenient. In essence this concept specifies the design of the instrumentation and includes a device within the electrical circuit in the safe area which limits the current which is allowed to flow within the circuit in the hazardous area. Thus the energy necessary to cause ignition of a flammable vapour is prevented from reaching the zoned area. Two types of device are available, shunt-diode safety barriers or galvanic isolation barriers. Intrinsically safe equipment will bear the CENELEC certification:

$$EEx\ ia\ IIB\ T4 \qquad or \qquad EEx\ ib\ IIB\ T4$$

depending on whether the device is safe with one (ib) or two (ia) faults within the instrument. (Note that the gas group (IIB) and temperature (T4) parts of the certification may be varied according to local requirements.) An explanation of the above symbols is given in the Glossary. A clear and comprehensive discussion of the whole electrical safety issue is provided by Garside [2, 3].

All instrumentation used in the modules described by Robins [1] is certified as intrinsically safe (IS), making the modules usable in both safe and hazardous areas. The IS concept is most convenient for this modular approach as the plug and socket combinations used to connect transmitters into control loops are considered to be 'simple apparatus' and thus have no energy storage capability. Because of the high specification of flameproof or intrinsically safe electrical equipment it is often more economical to use air-driven mechanical equipment and pneumatic controls. This concept also makes for easy control of the process equipment. The arrangement coupled to the centrifuge illustrated in Figure 19.2 consists of inexpensive components and has been found to operate reliably in a variety of process configurations.

Application of this modular approach to control and monitoring of downstream processes at an early stage of development leads to more rapid optimization of that process. Batch records are improved with more information available to track the product history, giving rise to improved mass balances. Batch-to-batch variations are also reduced due to the consistency introduced by automatic controllers. Pay-back times are difficult to estimate for such a system, but with product costs at early stages of development being typically hundreds of pounds per gram, reducing losses at this stage has an obvious benefit. Additionally there is pressure to reduce development cycle times; again good data obtained at an early stage are of great benefit.

By applying a modular and documented life cycle approach to the preparation of software for sequence control, and also obtaining the appropriate documentation from instrument manufacturers (data sheets, manuals, calibration and safety certificates), the approach should be found acceptable to the regulatory authorities for work under cGMP standards.

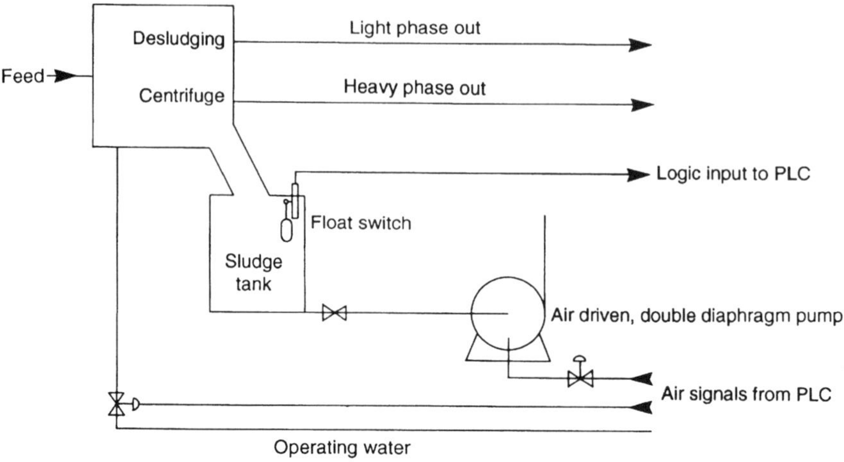

Figure 19.2 Diagram of the arrangement of the process equipment coupled to the centrifuge

Glossary (after Garside [3])

EEx d IIC Tg

E   European (CENELEC) certified
Ex Explosion protected by:
d   Protection method: d = flameproof, i = intrinsically safe, ia = with two faults, ib = with one fault
IIC Most onerous gas group in which apparatus may be used; IIC = all hazards including hydrogen
T6 Maximum surface temperature of apparatus; T6 = 85 °C

19.5 Future Trends

With the majority of natural product manufacture taking place at small scales in batch process plant the need for complex distributed control systems has been limited. In general, process control has been restricted to the sequencing of simple operations, simple feedback control and the collecting and logging of batch records.

The major problem encountered by the control engineer is the integration of the many equipment packages utilized in natural product processing and the various utility systems. Each package will have its own validated software which suppliers will be very reluctant to alter, and this must communicate with overall management systems and also process utilities, especially clean-in-place. The task is by no means simple and must be carefully shepherded through design and commissioning, especially if biopharmaceuticals are to be manufactured.

For most small-scale processes it is unlikely that the above trend will change much as technology develops. However, as the field of biotechnology develops and begins to encompass very-large-scale production of natural products the financial advantages of fully automated processes may well begin to heavily influence design. Areas of particular interest will be the make-up of large volumes of process solutions and the automation of systems for continuous or semi-continuous purification.

Finally, the use of instrumentation in natural product recovery is restricted by the availability of measurement devices. Once active probes have been developed to determine the concentration of specific biochemical species it is likely that the level of, and reliance upon, instrumentation will begin to increase, with a shift towards the use of expert systems to monitor and control extraction processes.

19.6 References

1. K. G. Robins (1992) *Advances in Process Control*, Vol. III, Institution of Chemical Engineers, Rugby.
2. R. Garside (1988) *Intrinsically Safe Instrumentation: A Guide*, Hexagon Technology, Aylesbury.
3. R. Garside (1990) *Electrical Apparatus and Hazardous Areas*, Hexagon Technology, Aylesbury.

20 SCALE-UP CONSIDERATIONS

John Noble and Robert Davies

20.1 Introduction

The life cycle of any given product will generally involve moves from research and development, through pilot plant and finally into manufacturing. As this life cycle progresses, the amount of product required continually increases, and the process must be scaled up to meet this demand. Natural product manufacturing processes are unique in the range of scales encountered. For certain very high value products a few grams may well satisfy the global market demand, whereas for a typical bulk biological polymer many tonnes may be produced in a single batch. As a result the classical concept of laboratory scale, pilot scale and manufacturing scale are somewhat redundant and it is better to consider the volume of material flows than the end market for the product. This is reflected in the three basic scales of operation given in Table 20.1.

Although consideration must be given to scale-up issues at all phases of product development, the greatest attention is required during the switch from small scale to intermediate scale. Due to pressures of time, and where pharmaceuticals are concerned the requirements of regulatory bodies, the pilot plant process will often end up as the production process with only simple linear scale-up. From this it is clear that insufficient attention to scale-up requirements could well result in significant increases in both capital and operating costs and could ultimately severely impair product market performance.

Using a few simple guidelines the implications of scale-up can be assessed in a logical manner during process development to ensure that the progression from the laboratory bench to the production plant is as smooth as possible. The following sections present a step-by-step approach to scale-up, supported by practical experience of the problems encountered.

Downstream Processing of Natural Products. Edited by Michael S. Verrall
©1996 John Wiley & Sons Ltd

Table 20.1 Scales of natural product recovery

Scale	Processing volume	End market for product
Small scale	0–25 litres	Initial evaluation Product development Clinical trials Final product
Intermediate scale	25–250 litres	Product development Clinical trials Final product
Large scale	Above 250 litres	Clinical trials Final product

A step-by-step approach to scale-up is presented in Figure 20.1. Below, each step is discussed in some detail with emphasis on the generalities of the process rather than specific consideration of equipment or unit operations. The latter is covered to some extent in the final section of this chapter which discusses practical scale-up experience.

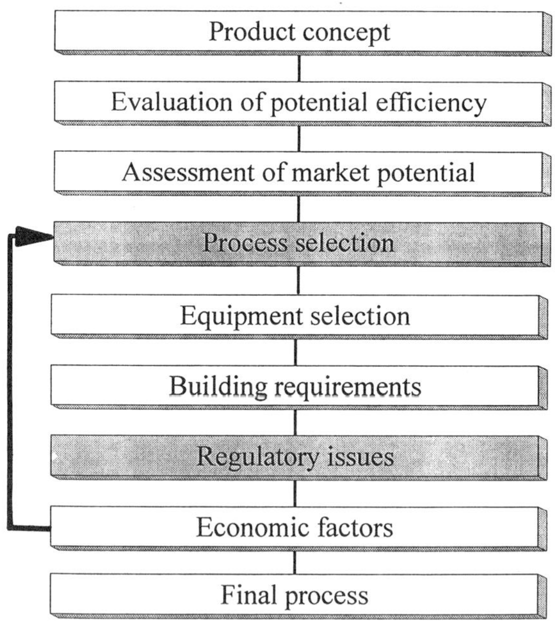

Figure 20.1 Stages in scale-up

20.2 Considerations for Scale-up

20.2.1 PRODUCT EVALUATION

During initial laboratory trials it will become clear that particular products can satisfy certain market demands. Once this has been confirmed it is necessary to assess the size of the market that the product can penetrate. From this market analysis it will be possible to estimate the volumes of the product required during market trials and for final production. Armed with this information it is possible to begin the assessment of the manufacturing options.

20.2.2 PROCESS SELECTION

From a knowledge of the volumes of product required it is possible to begin finalizing the basic manufacturing operations. Although, in many cases, the fundamentals will be reasonably defined during the testing of product efficacy the process must be revisited to ensure that the unit operations are both technically and economically feasible at larger scales. The selection process is usually iterative in nature and results in a range of process options for further assessment in the scale-up study. A schematic for the process selection operations is given in Figure 20.2.

From the product starting point, be it a fermentation broth or an animal or plant product, a range of options will be available for consideration as

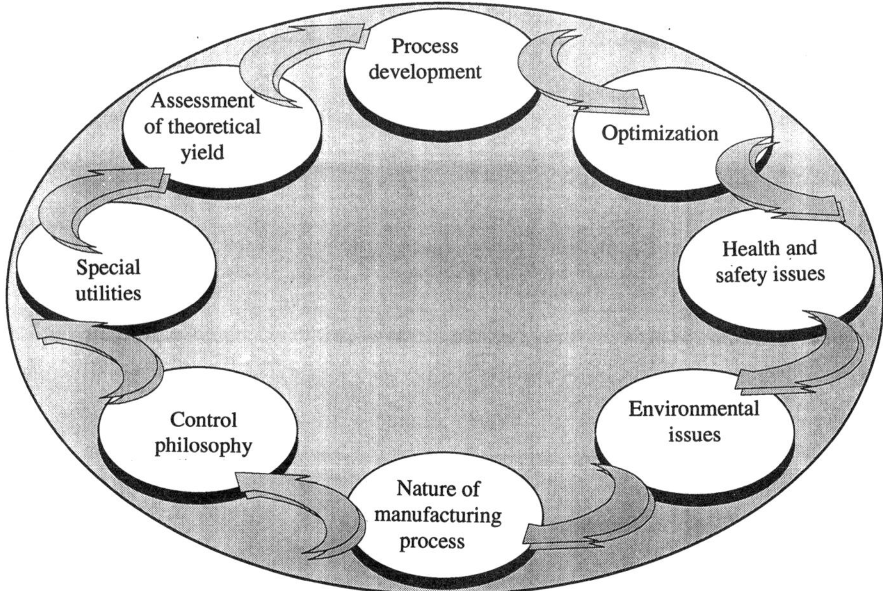

Figure 20.2 Process selection

production-scale operations. Each of these options must be optimized and the maximum theoretical product yield assessed to allow the ranking of process options in terms of efficiency. At this stage it may well be possible to eliminate certain options through poor product yields or obvious financial/technical limitations. Having generated a list of potential process routes it is then necessary to consider the safety, health and environmental (SHE) issues connected with the process.

Typically in natural product extractions health and safety issues will be related to the containment of dangerous pathogens and genetically manipulated organisms [1, 2], the control of substances hazardous to health (COSHH) [3] and the containment of potentially explosive substances [4]. COSHH issues are most often related to the handling of powders and volatile solvents whilst safety issues are strongest during the use of volatile solvents in liquid extraction and evaporation operations. All of these issues can add significant cost to a project as the containment of the hazardous agents must be designed into the processing equipment/building. To this end it is advisable to minimize the areas in which such issues influence the design. In addition to health and safety issues environmental matters must form an integral part of process development, ensuring that the principles of waste minimization and integrated pollution control are incorporated in all production steps [5]. What appears as a small speck of detritus in a beaker during development can, during production, develop into many tonnes of hazardous waste requiring cost-intensive disposal with possibly adverse public relations for the operating company.

It is likely that certain of the process alternatives will fall by the wayside during analysis of the SHE issues, presenting a further reduced set of options for the final process. Having reached this stage in the process development it is possible to consider various operational aspects of the production process. The foremost of these issues is the nature of the manufacturing process, whether it will be batch, continuous or semi-continuous or whether the plant will be dedicated, fixed or mobile.

Concerning the choice between batch and continuous operation significant economies can be made by utilizing continuous processing lines, especially at larger production scales, as equipment sizes can often be significantly reduced. A move towards continous production can also act to significantly reduce process effluents, especially those associated with washing. Many natural product unit operations are batch operations, but can be made semi-continuous by running several parallel production lines. The advantage of this is to reduce equipment duty from a daily production volume in, say, one hour to the same volume in 24 hours. It is of course essential to assess potential product losses through denaturation which may be induced by the increased processing time and balance these against potential savings. The scale of the process (see Table 20.2) will also influence the nature of the manufacturing process as small-scale batch operation utilizing mobile/flexible equipment will be most practicable; at the intermediate scale natural products can be best recovered using either mobile or fixed batch plant whilst at above the 250 litre scale fixed plant either batch or continuous is desirable.

The final stage of operability assessment is to develop options for control and utility requirements. The control aspects are discussed in detail in Chapter 19. Overall the operator must decide the level of automation in both process operation and data logging. These decisions will influence heavily the number of operators and the manner in which personnel and materials flow throughout the process. As a result, they can have a significant impact upon the overall process economics. Consideration of utilities is also of vital importance as special utilities such as water for injection (WI), clean steam, pyrogen-free water and clean-in-place (CIP) can contribute around 15% of the project capital costs. It is essential that the use of high-purity systems be optimized to ensure that such utilities are only utilized when needed and not used without question throughout the entire process. This assessment has specific relevance to the energy efficiency and waste minimization aspects of plant design. Cleaning operations can use up to 90% of process water and it is thus essential that the water quality is not over-specified.

20.2.3 EQUIPMENT SELECTION

Having defined the options available for production it is necessary to select the equipment that will be used both in the pilot- and production-scale

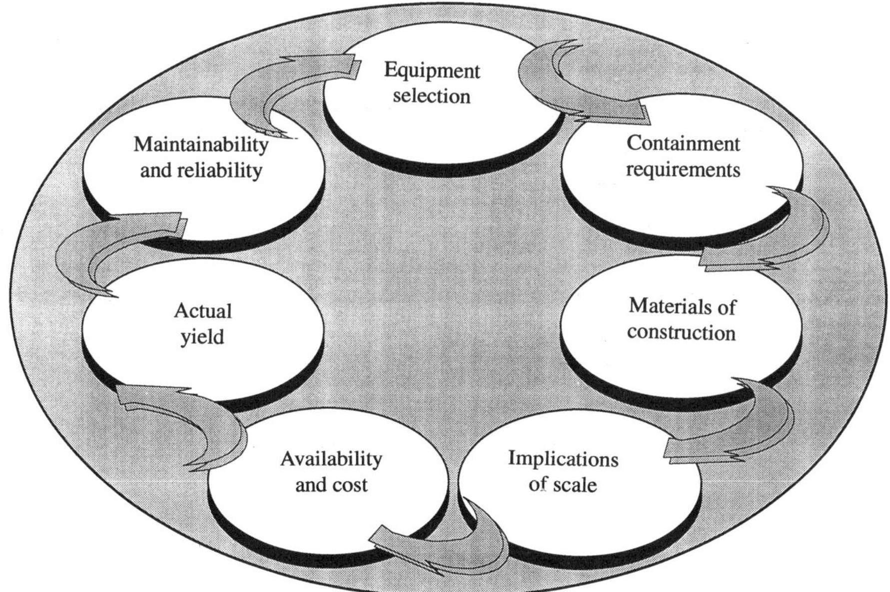

Figure 20.3 Equipment selection

manufacture. Again this is an iterative process considering a wide range of factors (see Figure 20.3). A central theme is the need to use pilot-scale systems to evaluate the best practicable approach to full-scale manufacture. It is thus important to mimic the full-scale facility at the pilot scale. Inherent in this approach is the concept of scale-down in which the equipment selection is made with reference to the full-scale duty and then the equipment item is scaled down to suit pilot-scale manufacture. Obviously this is not always possible but it is an approach which, if implemented, can significantly reduce problems associated with full-scale operation and validation.

In selecting equipment items the SHE implications raised in process development must be carried forward to ensure that suitable containment, material specification and waste minimization measures are designed into the equipment.

Shortages of high-value starting materials such as fermentation broths can often cause considerable problems during scale-up. For example, it may not be possible to source an equipment range which covers both pilot and production scales. If operating principles cannot be matched then efforts should be made to match design parameters such as residence time, rotational speed, pressure drop and mixing regimes. The latter point is of key importance at scales beyond 250 litres as heat and mass transfer become increasingly dependent upon vessel, agitator and column design. This has specific economic and design implications in the development of precipitation, extraction and adsorption processes [6]. Having considered these points it will be possible to assess the availability and cost of the potential equipment items and then select those with most potential. From this information, options can be developed for process configuration. These can then be considered against the background of projected yield and equipment maintainability, operability and reliability.

The complexity of many downstream processing operations for natural products has a significant effect on product yields with recoveries of between 10 and 40% of the norm. With this in mind it is essential that during scale-up the production equipment should be designed to minimize product loss. At pilot scales this factor can be especially important and will strongly influence the choice between fixed and mobile equipment. A particular problem is associated with the dead volume of centrifuge, filtration and pipework systems. The maintainability and reliability of equipment is also an important factor, especially if manufacture takes place in a clean room environment or hazardous area. In such circumstances the cost of downtime and the preparation required to recommission the contained areas can have significant implications for production economics.

Where special utilities are concerned consideration must be given to the supply of these systems to the process. For example, the level of automation and the flexibility required by operators will heavily influence the cost of the services. In addition, as scale-up moves from mobile to fixed plant it is imporant to ensure that equipment is readily cleaned in place and that cleaning flowrates can be accommodated by the process system.

20.2.4 BUILDING REQUIREMENTS

The hazardous nature of many of the chemical and biological agents used in the manufacture and the exceptionally high purity required of many natural products places specific constraints on building design for production facilities. Where pharmaceutical products are manufactured for clinical trials a high standard of cleanliness can be required to protect the product from corruption. Also where flammable solvents are in use buildings must be designed to accommodate the potential danger of explosive atmospheres and to protect the facility against catastrophic failures. All of these issues can add significantly to the project cost and they must be addressed seriously during scale-up to reduce the level of over-design.

20.2.5 REGULATORY REQUIREMENTS

All pharmaceutical and many other natural products are required to conform to certain regulatory standards (see Chapter 21). Validation can represent a significant proportion of total project costs and it is essential that the level of work is kept to a minimum by avoiding, where possible, repetition of work and steps that are costly to validate. The avoidance of repetition is largely encompassed by the scale-down principle, which minimizes the level of revalidation required at full-scale production. In operations which can vary from product run to product run it is important to ensure that the cost of validating the repeatability of the operation does not outweigh the process advantages of using the unit operation. This is particularly true in circumstances where very costly assay procedures are required.

20.2.6 ECONOMIC FACTORS

Across the spectrum of natural products a wide range of design standards is applied depending upon the nature of the product. For example, the value of many therapeutic protein products forces designers to invest significant over-design to ensure integrity and minimize the risk of product loss. Where the product is less valuable, say in single cellular protein manufacture, plant is often simply fit-for-purpose. In both of these instances aseptic operation is achieved for the majority of production runs yet the relative capital invested is significantly different. The driving force behind this is the need to validate the manufacture of pharmaceutical products to standards that are understandably high. The key issue is to treat each natural product extraction individually. Designers must ensure that sufficient measures are taken to ensure that a process produces a safe product of repeatable quality rather than simply adopting a high level of over-design to ensure quality rather than simply adopting a high level of over-design to ensure quality. The latter will result in severe cost penalties for the product in the market-place. To ensure that these issues can be avoided it is essential that the individual product requirements are defined as early as possible in the scale-up process.

A further important area for consideration is the balance of capital and operating costs as the scale of manufacture increases. In biopharmaceutical purifications the annual cost of disposable items such as filters and adsorption media can be considerable. For example, at production scale the annual cost of filtration can be of a similar order to the capital cost of the production equipment, especially if single-use viral removal systems are employed. It is essential that the minimization of disposable items is fully addressed. Where possible resuable filters and adsorbents should be employed for the bulk of process operations with disposable systems in place for polishing operations.

20.2.7 FINAL PROCESS

Having completed the exercises above a set of options will be available for the scale-up of the process. Each option will have its particular advantages and disadvantages and these should be compared with the original scope for the project. It will most probably be necessary to iterate around the loop to ensure that a set of options which fully satisfy all requirements can be developed.

Once the final list of options is available it will be necessary to construct a suitable facility in which to pilot the operation. The design basis for this facility will have been fully defined during the earlier stages of this exercise. Once in place the pilot facility can be used to ascertain which of the process options will best suit the product in question and hence confirm the design basis for the production facility.

20.3 Practical Aspects of Scale-up

In the earlier sections of this chapter a theoretical approach to the scale-up of natural product extractions has been outlined. In practice resources such as time, data, manpower and finance will limit the level to which this strategy can be implemented. To accommodate this resource limitation it is often necessary to rely heavily upon experience and intuition to anticipate and solve problems that may occur. In the following sections practical experience of the development design and operation of natural product recovery processes has been drawn together to provide the reader with support in such circumstances.

20.3.1 EQUIPMENT SELECTION

Although the vendor organizations supplying the biotechnology industry have made great progress in recent years there is often a shortage of equipment items and suitably experienced personnel to deal with the specific requirements of natural product recoveries. This is most noticeable when one steps outside the

Table 20.2 Critical scale-up parameters for major equipment items

Unit operation	Scale-up consideration
Centrifiguation	For total solid volumes above 10 litres, discharging machines are desirable as they avoid the use of multiple parallel machines.
	Natural product sedimentation and rheological properties are complex and must be fully studied before centrifuge type is selected.
	Serious consideration must be given to the yields of solid and liquid phases.
	Normal operation may not be compatible with clean-in-place strategies and these matters should be resolved.
Cell disruption	Mechanism of disruption must take note of downstream operations, i.e. potential chemical interactions and problems of fine particulate removal or highly viscous solutions must be addressed.
	Heat generation must be controlled to minimize product denaturation.
	Equipment should be selected to minimize or, if possible, prevent aerosol formation.
Filtration	Product loss on filter medium.
	Operating costs associated with disposable filtration systems.
	Validation of repeatable filter performance.
Extraction and precipitation	Scale-down techniques should be used to mimic full-scale heat transfer, mass transfer and fluid dynamics at pilot scale to ensure that the system is fully characterized.
	Detailed consideration should be given to chemical addition, i.e. agitator shaft, in-line mixers or directly through addition ports to eliminate localized chemical effects.
	Precipitates and extracts should be readily handled by downstream operations such as filters and centrifuges.
	Conditions should be selected to optimize the overall process yield rather than that of the extraction/precipitation stage.
Adsorption	Adsorbent life could be a significant economic factor and must be assessed as early as possible.
	Column packing becomes more complex as scale increases and the need for clean areas and potentially lifting equipment must be assessed during process development. Efforts should be made to maximize column utilization through cycling of parallel systems rather than utilizing a single large column.

reasonably mature fields of large-scale fermentation and plasma fractionation and considers smaller-scale operations such as extractions from animal and vegetable matter. The main problems arise from a lack of operating data and it is hence essential that full-scale equipment trials are undertaken before purchase is agreed. With this in mind it is clear that, for reasons of finance and flexibility, the use of purpose-built equipment should be avoided at intermediate and large scales. Where possible standard vendor equipment should be specified to ensure repeatability of operation and availability of spares. In Table 20.2 the key areas of attention in the scale-up of various unit operations have been outlined.

20.3.2 DATA TRANSFER

In most instances those responsible for the scale-up and production of a product will be different from those that have undertaken the initial design work. For effective process development it is essential that the two groups of personnel communicate effectively with each other. The terminology used by various technical groups will often be very different and it is essential that project teams can transfer data effectively between each other. To this end engineering staff must be both capable and willing to take on board the difference of approach between production and that of their scientific colleagues and vice versa. Of particular importance is the need to develop the right data during research trials such as material and solution properties which will enable early identification of problem areas. For example, adhesive precipitates, extremely viscous fluids or high agitation requirements may not be acceptable to production and should be highlighted. In addition attention should be paid to utilities' usage to ensure these are accounted for at an early stage and not dealt with as a necessary evil during detailed design.

It is also important to realize the contribution that outside assistance can bring to a project. Consultancy and engineering contractors have been forced to develop considerable in-house resources to compete in an ever tightening market. These resources are generally based around highly trained individuals with many years of experience in design. Although this experience may not be directly relevant the cross-fertilization of ideas from chemical, food and biotechnology industries can often bring significant added value to a project. External bodies can often help to eliminate the conflicts of interest that will inevitably arise between the scientific, design and financial sectors of an organization. Finally, the use of external resources may be unavoidable if managerial pressure to progress clinical trials outstrips the resources available within the production organization.

20.4 Conclusions

The scale-up of natural product recoveries is by no means a simple affair. The enormous variations from process to process necessitate attention to detail at all stages of product development and leave no room for complacency. However, the authors believe that by adopting a structured approach many pitfalls can be anticipated and it is to be hoped avoided.

The central theme of the suggested approach is a holistic assessment of the manufacturing process. Product recovery is not simply concerned with the unit operations required for purification but is centred upon the integration of all the services, personnel, equipment, waste handling and regulatory issues involved in modern manufacturing processes. If this philosophy is implemented during process development then the efficiency and cost effectiveness of scale-up and design operations should be significantly improved — factors that can only help to improve the market performance of a given product.

20.5 References

The references below are based on United Kingdom legislation. The same principles apply elsewhere but the reader should ascertain the legal position pertaining to his or her operations.

1. Advisory Committee on Dangerous Pathogens (1990) *Categorization of Pathogens According to Hazard and Categories of Containment*, 2nd edn, HMSO, London.
2. Advisory Group on Genetic Manipulation (1982) *Third Report of the Genetic Manipulation Advisory Group*, Cmnd 8665, HMSO, London.
3. The Control of Substances Hazardous to Health Regulations (1988). Statutory Instrument 1988, No. 1627.
4. R. Garside (1990) *Electrical Apparatus and Hazardous Areas*, Hexagon Technology, Aylesbury.
5. Environmental Protection Act (1990) Part 1 (Includes Integrated Pollution Control).
6. M.G. Roig (1993) in J.F. Kennedy and J.M.S. Cabral (Eds.) *Recovery Processes for Biological Materials*, John Wiley, Chichester, p. 377.

21 GMP AND QUALITY CONTROL

David Sherwood

2.1 Introduction

At the outset it is necessary to define good manufacturing practice (GMP) and quality control. Definitions are given in several documents used to govern the way in which the pharmaceutical and related industries manufacture products [1, 2]. The primary aim is to ensure that the *quality*, *safety* and *efficacy* of a product are maintained batch after batch; therefore controls must be in place to ensure that this happens. Good manufacturing practice (GMP) is the part of quality assurance that ensures that products are consistently produced and controlled to the quality standards related to their intended use.

Basic requirements of GMP relate to:

(a) a clear definition of manufacturing processes,
(b) validation of critical steps in the manufacturing process,
(c) control and validation of changes to the manufacturing process,
(d) the provision of trained personnel,
(e) adequate facilities, equipment, services, materials and approved documentation for the manufacturing process,
(f) the storage of batch history records (batch records),
(g) the documentation of product distribution together with a formal product recall system and complaints procedure.

Quality control is the part of GMP that is concerned with sampling and testing of products to defined specifications. In addition, the appropriate documentation and release procedures must be available to ensure that products are not released for use until their quality has been judged to meet the required quality standards (usually defined in the appropriate product licence). Therefore, GMP and quality control are part of an overall concept of 'quality assurance',

Downstream Processing of Natural Products. Edited by Michael S. Verrall
©1996 John Wiley & Sons Ltd

which covers all activities that influence the quality ('fitness for purpose') of product.

This chapter will review the various elements of GMP and quality control and relate them to some of the practical aspects encountered in downstream processing of natural products (with the major emphasis being given to the processing of biological products, particularly proteins).

21.2 The Production Facility

It is essential that the production facility is designed appropriately for the processes that are to be carried out. The layout and design must minimize the risk of errors (e.g. product contamination), be easily maintained and cleaned/disinfected and above all must not compromise product quality. At the facility design stage, flow diagrams for personnel, product, raw materials, waste and equipment must be thoroughly reviewed, to ensure the critical flows do not cross each other. In circumstances where flows do impact on each other it may be appropriate to provide temporal separation of activities.

The fabric of the facility (materials of construction) must be appropriate to ensure that cleaning and disinfection of rooms is easily carried out both during and between product campaigns. For example, walls, floors and ceilings should be smooth and in good repair with minimal disruption from pipework, light fittings and ventilation points which create ledges/recesses that are difficult to clean. Where ever possible, services to the manufacturing area should be accessible from outside the area to prevent disruption and impact on the manufacturing processes.

Heating, ventilation and air conditioning systems are critical to the functioning of controlled manufacturing facilities. The air flows must be organized and balanced to ensure that air from one manufacturing area does not flow into an adjacent area and hence provide a means of cross-contamination. The air quality requirements for each manufacturing area must be clearly defined. For example, in downstream processing areas where product exposure to the environment is high the facility air may be high efficiency particulate air (HEPA) filtered to Class 10 000 standard (Federal Standard 209e) and operations such as fraction collection may be carried out in Class 100 laminar air flow conditions. This minimizes microbiological contamination of the product. However, in a fermenter hall, the harvest procedure for separation of cells from growth medium (in a centrifuge) may only require facility air quality at Class 100 000, as the process is contained within a closed pipework system and hence there is no product exposure to the environment.

In addition to air quality, lighting, temperature and humidity are key factors to consider as they may affect the functioning of equipment, the stability of product and personnel comfort. Some production areas may be considered particularly hazardous which then necessitates the separation and dedication of facilities to one particular operation. For example, the production of antibiotics, hormones or highly active compounds should not be carried out in the same facility at the same time. However, the campaigned use of facilities may, in exceptional circumstances, allow the manufacture of these products in the same facility. There is then a requirement for detailed validation data to demonstrate that the previous product has been removed from the facility prior to start-up of the next process.

21.3 Equipment and Services

21.3.1 EQUIPMENT

All equipment in the manufacturing facility must be maintained to ensure safe and efficient operation in the intended processes. Equipment must not represent a risk to the processing of products; therefore they must be designed to enable efficient washing and cleaning. As discussed later, the validation and cleaning of equipment are central features of a comprehensive validation programme.

Repair and maintenance of equipment should be carried out by trained personnel and a planned preventative maintenance (PPM) system should be in operation. It is traditional in some pharmaceutical industries to implement a facility shut-down period where all essential equipment servicing, maintenance and essential repairs are carried out without impact on product manufacture. In the PPM system measuring, weighing, recording and control equipment should be calibrated and checked at defined intervals and at all times documentation must be available to confirm that calibrations and checks have been successfully completed.

The engineers' equipment must operate to defined standard operating procedures and all equipment operating and maintenance manuals (from the manufacturers) must be available on plant. Defined service schedules for equipment must be in place and a defined parts list for each piece of equipment must be available for easy reference. All equipment should bear a unique reference number (asset number) for easy and unambiguous reference in production and engineering documentation. Care must be taken with equipment that has parts directly exposed to the product, e.g. pump systems, centrifuge bowl, chromatography column, etc. In these instances attention must be paid to the cleanliness of surfaces and the maintenance of seals. It is imperative that lubricants or coolants must not come into contact with the product. Equipment that has broken down or is under maintenance must either be removed from the facility for repair/maintenance or is clearly labelled with status.

21.3.2 SERVICES

Production facilities are normally supplied with various types of water (mains water, deionized water, distilled water, water for injection, pyrogen-free water).

It is essential that pipework carrying water can be sanitized on a regular basis to ensure that the microbiological water quality is maintained within defined limits. In addition, the chemical quality of water must be kept within defined limits and thus associated resin beds, filters and distillation units associated with these systems must be regularly monitored and maintained. Clean steam is an essential service in the production facility used for steam in place of equipment and the operation of autoclaves. The steam quality must be monitored on a regular basis to ensure that chemical, microbiological and dryness amongst several parameters are maintained within required specifications.

Air handling units supplying filtered air to the facility are a critical service, particularly in downstream processing areas such as purification and bulk aseptic filling areas. Appropriate system monitoring of the supplied air must be carried out using microbiological and non-viable particulate monitoring; in addition, absolute air pressure readings in rooms and differential air pressure readings between rooms must be monitored and maintained.

Finally, the drainage system in the facility must be appropriately fitted with non-return valves, preventing back-flow of waste. Facilities dealing with biological waste from microbial or mammalian cell fermentations require a neutralization/kill tank system so that all potentially hazardous biological waste is treated before being transferred to the general sewage system.

21.4 Personnel

People are the most single important factor for the maintenance of current good manufacturing practices. It is for this reason that there must be sufficient trained, qualified personnel to carry out the tasks required in the manufacturing area.

21.4.1 TRAINING

All personnel must be trained in their assigned duties as well as in the principles of GMP. An on-going training programme must be maintained and training records must be kept, preferably in a centralized personal training log held by each person. As always, the effectiveness of training should be continually monitored by supervisors and managers. Programmes for training and retraining in new processes or procedures should be scheduled and each person should develop an annual training schedule with their supervisor or manager. Compliance to the training programmes in place can be assessed by periodic audit of training records, batch documentation (to ensure that personnel are not working on processes for which they have no training record) and observation of the task being performed.

For example, personnel involved in the aseptic filtration of a bulk liquid must be trained in aseptic manipulations, be observed carrying out the filtration procedure and their technique assessed by undertaking a media fill trial. In addition, personnel being trained in analytical techniques should be aware of the principles of the technique, be demonstrated the technique by a trainer, perform the technique under supervision and independently, and have the results assessed against the techniques carried out by a trained operator. All this training information must be recorded and held in the training files.

21.4.2 GOWNING PROCEDURES

The entry and exit of personnel into manufacturing facilities must be strictly controlled to ensure that external environmental contamination does not enter the manufacturing area. To accomplish this strict change procedures must be followed to ensure that personnel dress in factory clothing, wear over-shoes or facility shoes, beard covers and head covers. Once in the facility, further gowning requirements may be necessary. For example, in product recovery areas where liquids are handled in contained systems, e.g. harvest vessels, centrifuge, filtration rigs, additional gowning may not be required. However, where product exposure to the environment is encountered personnel gowning must be upgraded to prevent the product being exposed to contamination from personnel. In these instances personnel should re-gown over their factory clothing and wear gloves and face masks in addition to their routine factory clothing. Even though a process may be contained (as described above for the product recovery area) there will be instances where the contained system has to be broken open, e.g. for cleaning purposes or filter replacement. In these instances the operator requires protection from product residues; therefore, in addition to primary containment mechanisms (laminar flow hoods), the operator will need to wear additional protective clothing, e.g. gloves and face masks.

Thus gowning strategies must be established for personnel after consideration of what protection is required. In some cases protection of the product from personnel is necessary while in other cases protection of personnel from the product may be essential.

21.5 Raw Materials

In a production process raw materials may be diverse materials ranging from simple plastic bottles to complex protein additives or basic salts/solvents to complex medium formulations. In all instances the materials should be purchased from approved suppliers. A material specification must be in place which names the supplier and defines the following:

(a) composition of the material,
(b) packaging and labelling,
(c) safety information,
(d) storage conditions,

(e) physical inspection requirements, i.e. integrity of packaging,

(f) sampling plan,

(g) analytical and microbiological acceptance testing regime (which must include, where appropriate, identity, purity, contaminant and assay/concentration tests),

(h) expiry date assignment,

(i) suppliers certificate of analysis requirements.

On receipt of a raw material, stores personnel must check the delivery against the purchase order and then place the material in a quarantine holding area. Material inspectors then inspect the material against the material specification requirements and assign a unique number identifier. On completion of the inspection the material inspector will label each item with either an approved/released status or a reject status (the material will be moved to a reject holding area).

Raw material inspection strategies depend on a number of factors and each material must be reviewed on a case-by-case basis. Some materials may require a high level of inspection due to their critical nature (i.e. chemicals used in final product formulations); others may be inspected using guidance given in established sampling plans (i.e. British Standard 6000).

As an overall requirement of the quality assurance strategy material suppliers must be audited for compliance to their manufacturing and testing procedures. Supplier performance should be monitored and audit frequency determined after review of past performance for supply (i.e. number of rejected batches) and audit performance.

In some instances materials may be accepted using information provided in the materials certificate of analysis. If this is the case a validation strategy should be implemented whereby the test results shown on the certificate of analysis are periodically checked by third party testing. In these instances it is essential that the methods used by the third party are comparable to those used by the supplier.

As mentioned at the beginning of this section, some materials used in the manufacturing process may be very complex. For example, the manufacture of proteins by mammalian cell culture or bacterial culture will require the use of complex and in some instances poorly characterized media. The microbiological status of some of the components in these media have to be thoroughly assessed. Examples may be additives to mammalian cell culture media such as bovine serum albumin (BSA) and human transferrin. The supply of both these materials must be fully understood with respect to:

(a) *Origin of material source.* For example, the supply of BSA from animals in the United States, New Zealand or Australia is essential if potential contamination from infection agents such as bovine spongiform encephalopathy (BSE) or various bovine viruses and mycoplasma are to be excluded.

(b) *Testing of material at source.* For example, sera from human donors must be individually tested for infectious agents such as hepatitis B and C and HIV before the sera are used as a source of human transferrin.

(c) *Processing of the material.* The process by which BSA or human transferrin is made should be understood to enable evaluation of virus inactivation or virus removal during the manufacturing process.

For the future the use of 'synthetically' derived alternatives to materials such as BSA or human transferrin is desirable to avoid the problems described above.

21.6 Validation

Validation may be defined as a scientifically rigorous and well-documented study which demonstrates that a process or piece of equipment does what it is intended to do time after time. The validation sequence starts at the design of a process, facility or piece of equipment, continues through installation qualification, operational qualification and performance qualification and in many instances never stops as revalidation (or reverification) is required to ensure that the process or equipment has not changed over time or following maintenance or modification.

Detailed information on validation strategies for those interested is provided in Refs. [4], [5] and [6]; however, the essential details are described below.

21.6.1 DESIGN QUALIFICATION (DQ)

Design qualification defines the final design of a piece of equipment so that the basis of validation can be determined. For example, the definition and description of use of a piece of equipment should be outlined and a series of specifications defined including engineering drawings (piping and instrument diagrams (P and ID), process flow diagams (PFD) assembly and electrical schematics), fabrication details, bill of materials, welding requirements (procedures, certification, history log, inspection, cleaning/passivation log), material certifications (construction, surface finishes), control software and hardware definition. It should be noted that at this point the design is fixed such that if changes occur during the following validation phases, justification for change via a change control procedure should be carried out and impact on the remaining validation work needs to be assessed.

21.6.2 INSTALLATION QUALIFICATION (IQ)

This phase of validation relates to all aspects of installation of equipment and ensures that the installation follows manufacturers' recommendations, design intentions and safety regulations and that the correct equipment has been purchased and installed. IQ may include the following activities:

(a) *Equipment information summaries* specific enough to define the system (e.g. for a purification column system this may include feed tanks, tubing, pipework, pumps, filters, pressure gauges, valves, detectors and the purification column).

(b) *Utilities description* (e.g. quality of electrical sources, gasses, steam or water supplies).

(c) *Spare parts list.*

(d) *Manual listings and procedures* for the installation, operation and maintenance of the equipment.

(e) *Process instrumentation.* The type, manufacturer, range, use and calibration schedule of all process instrumentation should be listed. The list should show critical and non-critical instruments (a critical instrument is defined as one whose failure would adversely affect the product quality or safety).

(f) *Materials of construction* should be defined and those that come in contact with the product should be identified and compatibility confirmed.

21.6.3 OPERATIONAL QUALIFICATION (OQ)

This phase of validation documents whether the installed equipment, when used according to standard operating procedures, performs its intended function. Therefore the OQ will demonstrate that the user has tested the equipment and has found it to be free from mechanical defects or design defects before use in the production process.

Operational qualification must not be started before IQ is completed and in some instances calibration activities must have been completed. Calibration may be included as part of IQ or OQ (it does not matter which) as long as the activity is carried out. In addition, utilities supporting a piece of equipment must have IQ and OQ phases completed before the equipment OQ is started.

OQ may include the following activities:

(a) Definition and approval of all standard operating procedures (SOPs) associated with the operation and maintenance/calibration of the equipment.

(b) Operator training completed in the appropriate SOPs.

(c) System integrity tested at the operating pressures to establish leaks by visual detection. In some complex purification systems fluid mass balance (fluid output equals fluid input) may be an appropriate method.

(d) Series of equipment features should be tested such as:

- *Flows and pressures.* Pumps are tested to show that the required flows are achieved (within tolerances) and that pressure gauges are appropriate and recording.
- *Gradient formation.* In purification systems reproducible gradients (shape and slope) should be confirmed plus expected results if there is variation in gradient solutions.

- *Detectors*. For a variety of process control purposes (e.g. pH, conductivity, dissolved oxygen, etc.) the acceptable operating range reaction times and limits of linearity response should be demonstrated during operating conditions.
- *Filters*. Filters and filter housings should be examined to demonstrate that they are appropriate for use with the flowrates and pressures encountered (this is particularly important in broth clarification filtration systems).

(e) Alarm conditions should be mimicked either by challenge or simulation, e.g. the pressure alarm may be tested by increasing pressure in the system using pumps or valves or the sensor monitoring pressure may be simulated by sending the appropriate voltage to the alarm mechanism.

(f) Steam-in-place (SIP). Efficacy in purification and broth recovery systems should be tested using temperature distribution mapping (time, temperature and pressure) with the calculation of $F\hat{o}$ and the use of biological indicators (*Bacillus stearothermophilus* spore strips). Any 'cold spots' should be identified and corrected and the positions of thermocouples defined for routine monitoring. System sterility following SIP may be investigated during OQ but usually thorough studies are conducted in the PQ phase.

(g) Clean-in-place (CIP). This is currently one of the most controversial aspects of validation [4, 5]. During OQ the CIP regime must be confirmed with respect to optimization of cleaning agent concentrations, flow, equipment exposure time (cleaning time) and coverage (confirmation of fluid distribution to all parts of the system), temperature of cleaning fluids and removal of cleaning agent residue by final rinses. Analytical tests and limits must be determined to demonstrate cleaning agent removal. The strategies used to define the types of tests and limits are discussed in detail elsewhere [5], but include on a case-by-case basis an assessment of types of samples to be taken for analysis, e.g. surface swabs, rinse waters and analytical methods selection. e.g. pH, conductivity, detergent assay, specific ion meters. Limits of cleaning agent residues from the clean-in-place regime may be rationalized by using 'medically derived limits' (evaluation of toxicity profiles), 'arbitrary safety factor' (to be used in conjunction with medical opinion) or 'analytical sensitivity' (use of analytical methods to their ultimate sensitivity). Removal of the product or 'simulated' product by the CIP regime is addressed during performance qualification.

21.6.4 PERFORMANCE QUALIFICATION (PQ)

This is the stage at which the performance of the equipment is assessed during simulated or normal production runs. An additional aim of PQ is to establish the performance of the equipment under 'worst-case' conditions as the extreme to normal operating conditions. The decision to examine 'worst-case' situations has to be based upon simulated conditions that may actually be encountered during production operations. Unrealistic conditions do not constitue the 'worst case' and will result in an explanation of failed validation runs which would not normally be encountered in standard operations. Performance qualification is carried out to approved protocols (the same as IQ and OQ). Protocols should detail the following:

(a) a full description of methods for performing operations with relevant standard operating procedures,
(b) stages at which sampling should be taken or monitoring performed,
(c) labelling and storage of samples,
(d) tests to be performed on samples taken,
(e) instruction for recording results and monitoring tests,
(f) a statement to indicate PQ acceptance criteria,
(g) responsibilities and protocol sign-off sections.

On completion of the PQ activity a signed-off report must be available, indicating the outcome of the PQ and an assessment of data against the stated acceptance criteria defined in the protocol.

Performance qualification testing normally involves production runs or simulated production runs. Three runs are usually required to establish confidence in the consistency of the equipment. These runs can be carried out using a 'concurrent validation' approach in which equipment and the process are assessed during actual production runs. Examples of PQ evaluations to be carried out on equipment associated with broth recovery processes and downstream processing include clean-in-place (CIP) studies, steam-in-place (SIP) studies and media fills to demonstrate the effectiveness of SIP.

The CIP evaluations represent one of the most controversial areas of the validation sequence [5] and consideration must be given to the definition of what is 'clean'. During the design of a CIP protocol the following aspects need to be evaluated to demonstrate clearance of the product:

(a) What product-related materials are removed during the cleaning process? (These are active ingredients, degradation products from the active material, excipients, microbial contamination, endotoxin, particulates.)
(b) What cleaning agents are being removed? Decide which samples are to be taken (surface swabs or rinse water and examination of previous product manufactured, solvent extraction to detect residues that are poorly soluble in water).
(c) Visual assessment of cleanliness.
(d) Analytical test selection (specific product assay, total organic carbon (TOC), analysis, visible, infrared or UV scans of rinse water, pH, osmolality, conductivity, for proteins non-specific assays such as Bradford protein assay or specific assays such as those that detect functional activity).

(e) How much residue may be permitted? This can be assessed and rationalized in a variety of ways by ensuring that previous product is present at a percentage of the lowest therapeutic dose (medically derived limit) or not detectable by the analytical method (analytically derived), or demonstrate log reduction in concentration.

(f) Can validation be carried out following only one type of product or all products? (There may be a diversity of products testing handled, e.g. very potent products, insoluble products.)

21.6.5 PQ ACTIVITIES SPECIFIC TO CHROMATOGRAPHY-BASED PROCESSES

Some PQ activities specific to chromatography-based processes are impractical and uneconomical to perform at the production scale [6]. In these instances scaled-down versions of the process may be used to perform PQ studies. Examples of such studies can be given, such as scaled-down product purification runs carried out to show clearance of contaminants such as viruses, DNA or other process contaminants. The extent to which a given column-based separation can be scaled down for validation will depend upon the actual production scale and the smallest scale that can reliably reproduce the production process.

It is important to assess the quality of column packing and therefore studies must be designed to investigate packing during routine production and after prolonged storage. This may be carried out using the height equivalent to a theoretical plate value (HETP). The choice of sample material to run on a column to determine the HETP depends upon the packing material. For example, size exclusion and affinity media can be tested using 1% sodium chloride in water and ion exchange media may be tested using buffer that is 10 times more concentrated than the equilibration buffer. Column packing may also be assessed using peak shape and column performance and by comparing these parameters to previous columns or a historical column record.

An important validation activity for column-based separations is the demonstration that the overall process or process step gives a product of consistent quality which conforms to predefined specifications. Validation should demonstrate that the process will not fail when carried out within the normal operational range for critical parameters (buffer pH, ionic strength, gradient shape, product loading, temperature, flowrate, system back pressure).

Production columns are frequently used for many product batches; therefore the cleaning regeneration and useful life must be validated. Specific criteria used to evaluate production columns may be physical (cracks, channels, discoloration, deposits), performance (elution profiles, peak retention times, resolution between peaks, loss of capacity) or microbiological parameters (bioburden, endotoxin assays).

21.6.6 VALIDATION OF ASEPTIC OPERATIONS

A validation exercise is used to check the aseptic filling operation. The aim is to perform a media fill (into bulk holding vessels or final dosage form containers) rather than a product fill, including the taking of all samples (analytical, microbiological and environmental). If the filling operation is manual each operator involved in the filling operation must be validated by a media fill exercise prior to their involvement with the filling of product. The media fill validation may be repeated on a quarterly or six monthly basis or may even occur each time prior to the product fill (if the product is filled on a very infrequent basis).

21.6.7 VALIDATION OF COMPUTER SYSTEMS

Many production operations are now computer controlled; therefore the systems controlling these operations must be validated. General overviews of computer system validation are given in several texts [7–9]. However, the essential elements can be summarized as follows:

(a) System definitions should include a functional and structural specification.
(b) The system software should be developed by qualified professionals following written procedures.
(c) Installation qualifications and calibration must be carried out for the hardware with operational qualification for both hardware and software.
(d) A performance qualification plan should be implemented which includes eventful testing (system challenge) as well as uneventful testing (confirmation of system control).
(e) Periodic review and validation following changes implemented through change control procedures.
(f) Ongoing operator training and security system review.

21.6.8 ONGOING VALIDATION

Once the initial validation process of IQ, OQ and PQ have been completed there must be an ongoing assessment of equipment or processes to check that the activities demonstrated in the original validations remain accurate and routine activities such as calibrations or planned preventative maintenance have not changed the performance of the equipment or process.

It is important that an efficient change control procedure is operated as part of the quality assurance systems which will identify those changes to processes or equipment that require validation prior to implementation (i.e. planned changes). Some changes may have to be implemented in an emergency situation and therefore it may be appropriate to gather validation data for this change using a concurrent validation approach where the process is validated during actual production runs.

21.7 Documentation Systems

A good documentation system is an essential part of GMP as clear and precise documents enable unambiguous communication of specifications, procedures, manufacturing directions and overall training of batch manufacture. Specifications must be available for all raw materials used in batch manufacture of bulk products and final dosage form. They form a basis for quality evaluations and formal release requirements. Written procedures (standard operating procedures, SOPs) are required to define production, engineering and quality control activities such as cleaning, maintenance, sampling, testing, environmental monitoring, etc. Quality assurance policy documents should be included in the documentation system. These documents should define the approach to various systems such as validation strategy and standards, how to write specifications, standard operating procedures and manufacturing directions, change control policy, training programme, complaints procedure, recall procedure. Manufacturing directions provide written directions on how to make the product and together with testing records provide the batch record (the history of batch manufacture).

All documents should be defined in a standard format, circulated for review and authorized by a number of departments including production, engineering, quality assurance, regulatory, development, analytical and microbiological testing groups. Issued documents must be tightly controlled. Only authorized copies issued by the documentation group must be in circulation and identified by an authentic mark, e.g. red stamp. Documents should be kept up to date and reviewed on a regular basis. Changes to existing documents must be scrutinized by a change control system prior to approval. At the time of issue of a revised document all superseded copies must be withdrawn from circulation and the superceded master document must be kept in a historical file.

Batch documentation must be issued to the production areas and completed *at the time* actions are carried out. Any errors must be struck out using a single line through the error then initialled and dated. All entries must be clear, legible and made in permanent marker (i.e. ink *not* pencil). Each document completed during batch manufacture must bear a unique batch number such that all documentation issued for batch manufacture can be collated and reconciled to one traceable number. Batch documentation may be recorded by electronic means, and in these instances strict control and security must be maintained for the entry (and modification) of data.

Critical data entry for both electronic and manual records must be signed off by an independent checker. All batch records must be archived using either the paper copy, microfilm or magnetic tape (for electronic records) and stored in a fire-protected area. The duration of record storage is governed by regulatory requirements and company procedures. However, a system must be in place to manage archiving and efficient retrieval of records.

21.8 Quality Control

Quality control (QC) is that part of GMP which is concerned with the sampling and testing of raw materials, intermediate products, final products and the environment to defined specifications. The strategy for raw material sampling and testing has been described earlier in this chapter.

21.8.1 IN-PROCESS TESTING OF INTERMEDIATE PRODUCTS

The sampling regime for routine in-process testing during the purification of a natural product will be structured to address key critical parameters. For example, biological products such as proteins are extremely vulnerable to microbiological (bacterial) contamination throughout the process. The process must be designed to prevent bacterial contamination; however, critical steps in the process must be monitored for bioburden levels. The presence of high levels of bacteria may require further investigation by the microbiologists (e.g. determine the origin of contamination, identification of the bacteria) or action levels may be defined such that the product may be rejected if it exceeds certain levels. In addition to product testing it may be appropriate to test buffers and packed/equilibrated chromatography columns used in the purification processes for microbiological contamination (bioburden) and endotoxin levels (by *Limulus* amoebocyte lysate (LAL) assay). These analyses together with product tests provide an overall picture of the microbiological challenge experienced by the product.

Product purity assessment may be another critical parameter to monitor during purification. Purity assessments may be made (using sodium dodecyl sulfonate polyacrylamide gel electrophoresis (SDS PAGE) or HPLC analysis) at the completion of certain purification stages. The assessments may be used to analyse fractions collected from a chromatography step and data used to rationalize fraction pooling. It may be appropriate to sample and test for general chemical contaminants during the purification process. For example, complex media used in the growth of mammalian cells will contain a number of supplements which may carry through the purification process. A testing strategy may be developed to monitor the levels of certain chemical contaminants during processing to ensure that appropriate clearance (removal) of these chemicals occurs at key steps.

At the medium conditioning phase (e.g. after growth of mammalian cells in complex medium) the concentrated medium containing the desired product may be tested for adventitious viral agents (by specific virus assays or electron microscopy). The aim of this testing is to ensure that the virus loading on the purification system is not large enough to break through into the bulk purified product. To assess this, reference must be made to viral clearance studies carried out during process validation work.

Specific product assays may be carried out in-process to determine step yields achieved over each chromatography stage. Variances against established step yields may require further investigation and action to determine the cause of this variance.

At key stages during product processing aseptic filtrations are carried out on the product to ensure that bioburden levels are reduced (eliminated) and controlled. In-process filtrations may be carried out in line or at specific 'breaks' in the process, i.e. between chromatography column steps. The filtrations may be performed in laminar flow cabinets (Class 100) and settle plates are used to monitor air quality throughout the filtration process. A critical aseptic filtration step occurs at the *bulk* purified product stage where the aseptically filtered product is normally held for a period of time (justified by stability studies) prior to formulation of the dosage form. This filtration is performed in clean room facilities (Class 10 000 or 1000 by Federal Standard 209e) within a Class 100 laminar flow cabinet (LFC). During this filtration, room and LFC air monitoring are carried out using non-viable particle air samplers and viable particle air samplers. Operators may monitor their gloved hands for microbiological contamination before and after aseptic manipulations and settle plates are exposed in the Class 100 LFC. Additional personnel monitoring may occur using contact plates on gowns. During this filtration step samples are collected from the bulk purified product for extensive bulk product testing.

21.8.2 BULK PRODUCT TESTING

The strategy for bulk purified product testing should generally satisfy the following requirements:

(a) Confirm product identity.
(b) Assess product purity.
(c) Assess product concentration.
(d) Determine product contaminant profile.
(e) Assess product potency (activity).
(f) Assess product safety.
(g) Assess product sterility.

To illustrate the above requirements some of the analytical methods used in the QC analysis of a monoclonal antibody are described below.

21.8.2.1 Product Identity

Identity may be confirmed using immunological assays (radial immunodiffusion, enzyme immunoassay), isoelectric focusing or chromatographic identity by HPLC.

21.8.2.2 Product Purity

Purity may be assessed using SDS PAGE analysis or HPLC.

21.8.2.3 Product Concentration

Concentration may be measured using a product specific enzyme immunoassay, Kjeldahl, Lowry, Biuret or UV extinction coefficient at 280 nm.

21.8.2.4 Product Contamination Profile

A number of methods may be utilized to detect the variety of potential contaminants such as viruses, mycoplasmas, DNA (hybridization techniques), protein contaminants (SDS PAGE using Coomassie blue and silver stains together with Western blotting methodologies) or endotoxin (LAL assays).

21.8.2.5 Product Potency (Activity)

A variety of methods may be utilized ranging from whole animal bioassays (these are usually expensive, not precise, technique dependent and generally require statistical analyses), cell culture derived bioassays (less expensive, precise, usually can be automated, require multiple samples, difficult to trouble-shoot) or biochemical methods (demonstrate high precision, can mimic biological function, readily validated, can analyse large numbers of sample, but *in vitro* assessment). The final choice of potency assay will depend upon analytical resources available and ease of method validation.

21.8.3 METHOD VALIDATION

All analytical and microbiological methods must be validated to demonstrate a defined level of performance. For instance, many analytical methods should be evaluated for performance parameters such as accuracy, precision, linearity of response, specificity, sensitivity (limit of detection, limit of quantitation) and robustness. All these performance parameters can be investigated during validation exercises conducted under strict study conditions, i.e. good laboratory practice (GLP) studies [10] where a study protocol, study director and study report are required together with ongoing audits of the whole study from start to finish by a GLP monitoring unit.

21.8.4 STABILITY STUDIES

An ongoing stability programme must be defined to continually verify, reverify or extend existing product expiry dating periods. Analytical and microbiological methods should be selected and validated specifically for stability study purposes. Studies may be carried out on product taken from key intermediate stages in which holding times and temperatures (plus other storage conditions) need to be determined. Stability studies can be carried out in GLP or GMP environments as long as protocols and reports formally document the studies.

21.8.5 ENVIRONMENTAL MONITORING PROGRAMME

A comprehensive environmental monitoring programme must be in place to monitor viable and non-viable particulates throughout the manufacturing facility. In addition, critical service systems must be monitored, e.g. WFI, clean steam, air systems. This is achieved by implementation of a variety of monitoring programmes. The more significant monitoring activities are detailed below:

(a) *Water systems.* The microbiological and chemical quality of water is monitored at all user outlets. The frequency of monitoring depends upon the criticality of the water system, e.g. water for injection (WFI) systems are monitored every day whereas deionized water may only be monitored weekly. Alert and action limits must be defined with clear follow-up activities implemented when alert or action limits are reached.

(b) *Viable particles.* The microbiological status of air and surfaces throughout the facility must be monitored via a routine monitoring programme. Levels defined in various guidelines [11] are used to provide guidance as to what action and alert limits may be set for various types of facilities. Routine monitoring data must be trended regularly to provide a review of alert and action limits.

Sampling techniques utilizing air samplers such as slit to agar sample, single sieve to agar sampler, centrifugal sampler or surface samplers (e.g. sterile cotton swabs, agar contact plates) are commonly used. Agar substrates used in the different sampling techniques may be varied depending upon whether total counts (bacterial and fungal) are required or selective monitoring for specific micro-organisms is required.

(c) *Non-viable particles.* Non-viable particles may be measured using an optical particle counter. The sampling rate is usually defined as counts per cubic foot per minute and specifications are typically set relating to 0.5 μm size particles. United States Federal Standard 209e is widely used to define room classifications relating to non-viable particles. For example, air in a laminar flow cabinet is defined as Class 100 (less than 100 0.5 μm particles in a cubic foot of air) whereas the suite housing the laminar flow cabinet may be classified as Class 10 000 (less than 10 000 of 0.5 μm particles in a cubic foot of air).

Both viable and non-viable particle monitoring may be conducted either by manual sampling or by remote activation of an automatic sampler. Both sampling techniques have their limitations and therefore single sampling observations must be reviewed in the light of trended data.

21.9 References

1. *The Rules Governing Medicinal Products in the European Community*, Vol. IV, *Guide to Good Manufacturing Practice for Medicinal Products*, January 1989, Commission of the European Communities, Office for Official Publications of the European Communities.
2. Code of Federal Regulations of the Food and Drug Administration 21 CFR Part 210, *Current Good Manufacturing Practice in Manufacturing Processing Packaging or Holding of Drugs; General.*
3. British Standard 6000.
4. J. T. Mahar (1993) 'Scale-up and validation of sedimentation centrifuges. Part II. Validation', *Pharmaceut. Technol. Europe*, October, 32–9.
5. J. Agalloco (1992) ' "Points to Consider" in the validation of equipment cleaning procedures', *J. Parenteral Sci. and Technol.*, **46** (September–October), 163–8.
6. Anon. (1992) 'Industry perspective on the validation of column-based separation processes for the purification of proteins' *J. Parenteral Sci. Technol.*, **46**, (May–June), 87–97.
7. A. S. Clark (1988) Computer systems validation: an investigator's view', *Pharmaceut. Technol.*, January, 60–6.
8. P. J. Motise (1984) 'Validation of computerised systems in the control of drug processes: an FDA perspective', *Pharmaceut. Technol.*, March, 40–5.
9. PMA Computer Systems Validation Committee (1986) 'Validation concepts for computer systems used in the manufacture of drug products', *Pharmaceut. Technol.*, May.
10. Department of Health, United Kingdom (1989) *Good Laboratory Practice*, The United Kingdom Compliance Programme.
11. The Parenteral Society (1989) *Environmental Contamination Control Practice*, Technical Monograph 2.

INDEX

Acetonitrile 226
Acid protease 61
Acid–base interactions 95
Acrylamide 197
Actaplanin 175
Activated carbon 160
Adenyl cyclase 112
Adsorbent
    fouling 128
    life 173
    preservation 174, 202
    pretreatment 193, 202
Adsorption
    capacity 161
    chromatography 226
    isotherms 148, 151, 155, 161
Affinity
    adsorbents 4, 7, 95, 183, 195
    chromatography 98, 131, 193, 212
    emulsions 200
    ligand 196, 199
    partitioning 200
    tails 4, 205
Agarose 129, 197, 198, 209
Air quality 330, 332
Alamine™ 336 264
Alarms 337
Algae 42
Alginate 24, 128
Alcaligenes eutrophus 50
Alkaline phosphatase 116, 203
Amberlite IRA-400™ 101
Amino acids 96
Amino agarose 199
p-Aminobenzamidine 196, 202
Aminophenylboronic acid 196
    agarose 200
Ammonium sulfate 57
Amphotericin B 4
Analytical methods 4, 344
Angiotensin 110

converting enzyme 175
Annealing temperature 292
Antagonism 5
Anti-melanoma antibody 220
Antibacterial activity 2
Antibiotics 83
Antibodies 131, 196, 199
Antibody–enzyme hybrids 117
Antifoam 27, 145
Antithrombin III 203
Aqueous two-phase systems 6, 53, 114, 118, 130, 200
Arachidonic acid 177
Archiving 342
Aseptic operation 43, 118, 128, 301, 304, 323, 332, 333, 340, 343
Aspergillus sp. 46
Asset number 331
Audits 332
Autolysin 110
Automation 143, 322
Avidin 116, 205
Axially compressed columns 237

Bacillus brevis 45
Bacillus stearothermophilus 337
Bacillus subtilis 108, 126
Back extraction 61, 62, 259, 260, 267
    by esterification 99
Bacteriostatic solutions 181, 202
Baculovirus/insect cell system 110, 113, 118
Bafilomycin 229, 232
Batch adsorption 200
Batch documentation 332, 341
Bead mill 41, 49, 60, 61
Biliquid foam 85
Binodal curve 56
Bioburden 342

Biochemical assay 4
Biochemical pathways 3
Biological activity 4
BioRad P-6DG™ 221
Bioselective adsorption chromatography
    193
Biotechnology 3
Biotin 175, 205
    binding domains 116
    ligase 116
Biuret assay 344
Bovine serum albumin 221, 334
Bovine spongiform encephalopathy 335
Brominated polystyrene resin 163, 175
Broth
    clarification 11
    conditioning 21, 34
    pretreatment 22
    viscosity 22, 28, 81, 140
Bulk product testing 343
1,4-Butanedioldiglycidyl ether 199
tert-Butanol 290
Butanone 245

C.I. Reactive Blue 203
Calibration 336
Calrectulin 110
Candida utilis 50
Carbodiimide condensation 198
Carbohydrases 112
Carbohydrate binding domains 117
Carbon adsorption 100
Carbon dioxide extraction 84, 243
Carbonic anhydrase 110
Carboxylic acids 93, 95
Carboxymethyl cellulose 24
Carboxypeptidases 113, 115
Carrageenan 100
Carrier solvent 150
Carrier species 261
Cell
    breakage assessment 43
    debris 62, 124, 129, 130
    disruption 5, 7, 41, 54, 60, 124, 245
    disruption kinetics 44
    homogenate properties 48
    immobilization 100, 128
    lysis 124, 193
    permeabilization 130
    pretreatment 150
Cellulose 24, 112, 129, 197, 198, 209

Cellulose acetate membranes 18
Cellulose DE 52 183
Cellulose nitrate membranes 18
Centrifugal contactors 86
Centrifugation 20, 61, 283
    performance 35
Cephalosporins 162, 174
Ceramic membranes 18
Certificate of analysis 334
Change control procedure 340
Chaotropic agents 107
Chelating resins 114
Chelation 7, 8
Chiral chromatography 151
Chiral transport 264
Chitosan 24
Chloramphenicol 86
Chloramphenicol acetyl transferase 110,
    116
Chromatography performance qualifica-
    tion 339
ChromSpher™ 5 Biomatrix 231
ChromSpher™ UOP C18 229
Citrate 57
Citric acid 96, 98, 123
Clean in place 143, 144, 185, 186, 296,
    303, 321, 323, 337
Clean steam 321, 332
Cleanliness assessment 338
Coagulation 23, 24, 37
Coalescence 74, 82, 266, 269
Cofactors 196
Collagenase 108
Collapse inhibitors 290
Collapse temperature 280, 286
Column cleaning 197
Column packing 181, 212, 216, 339
Communications 326
Computer systems validation 340
Con A Sepharose™ 215
Concurrent validation 338
Conductivity 180, 307, 311
Contactors 83
Containment 28, 43, 139
Contaminants 199, 231, 288, 344
Continuous fermentation 27
Control of Substances Hazardous to
    Health 320
Cooling 142
Countercurrent extraction 94, 250, 268
Coupled transport 261
Coupling chemistry 197

Critical composition 244
Critical data entry 341
Critical temperature 242
Cross contamination 331
Crossflow filtration 18, 39, 139
Cryoprotectants 290
Crystallization 278
Cyanate ester groups 198
Cyanogen bromide 110, 116, 198
Cyclodextrin 176
β-Cyclodextrin silica 154
Cysteine 114, 200
Cystic fibrosis proteins 110
Cytochrome c 58, 59, 61
Cytoplasmic enzyme 43, 45

Data transfer 326
Dead end filtration 12
DEAE cellulose 215
DEAE membrane 114
Decan-1-ol 82
N-Decyl-(L)-hydroxyproline 265
Design qualification 335
Desorption 283, 295
Detergents 183, 190
Deviation factor 135
Dextran 53, 197, 290
2′,3′-Dideoxyadenosine 177
2′,3′-Dideoxyinosine 177
Di(isobutyl)octadecylsilane 228
Diafiltration 65, 145
Diatomite 23
5,10-Dideazatetrahydrofolic acid 152
Diethyl phthalate 155
Diethylaminoethyl ion exchangers 111, 183
Differential scanning calorimetry 286, 292
Diffusivity 244
N-(3,5-Dinitrobenzoyl)-7a-1-amino acid 264
Diols 98
Disc stack centrifuge 20, 140
Displacement chromatography 8, 147
Disposable adsorption media 324
Disposable filters 324
Distribution coefficient 161, 263
Divinylsulfone 199
Documentation systems 341
Drag coefficient 76
Drainage system 332

Drop behaviour 74
Droplet rigidity 78
Dual tails 111
Dye ligand adsorbents 7, 131, 196, 200
Dynamic extraction 249

Economics 28, 129, 139, 143, 144, 160, 184, 209, 276, 298, 319, 321, 323
Eddy diffusion 79
Egg-white proteins 180, 185
Elaiophyllin 224
Electrostatic effects 58
Electrostatically enhanced extraction 89
Emulsifiers 266
Emulsion liquid membrane 265
Emulsions 84
End capping 227
Endotoxin 197, 202, 303, 342, 344
Energy 94, 276
Engineering documentation 331
Enterokinase 110, 112
Enterotoxin 113
Environmental protection 137, 244, 320, 345
Enzyme inhibitors 2
Enzymes 95
Epoxides 199
Equipment
  calibration 331
  configuration 146
  selection 321
Ergot alkaloids 177
Escherichia coli 47, 108, 110, 111, 126, 135, 212
Ethanoic acid 83, 99, 196
Ethanol 82, 83, 84, 123, 137, 181, 190, 196, 244, 247, 276
Ethanolamine 199
Ethylene glycol 201
Ethylene oxide 296
Eutectic temperature 280, 286
Exclusion effect 56
Exodipeptidase 115
Expanded bed adsorption 6, 118, 130, 131, 200
Expert systems 132, 315
Express-Ion™ ion exchangers 185, 189
Extraction decanter 89

Facilitated processing 105, 119
Factor IX 218

Factor Xa   108, 110, 203
Feedstock preparation   179
Fermentation modification   26
Filter aid   23, 37
Filtration tests   32
FLAG octapeptide   112, 117
Flameproof equipment   312
Flat sheet membrane filtration   143
Flavins   60
Flocculation   23, 37, 127
Flow meters   305, 311
Fluidized bed adsorption   6, 118, 130, 131, 200
Flux decline   142
Focusing effect   149
Formaldehyde   296
Freeze drying   8, 275
Freeze-thawing of cells   49
Frontal chromatograms   154
Fumarase   66
Fusion proteases   116
Fusion protein   107, 205
Fusion tail   58, 107

Galactomannan   24
β-Galactosidase   57, 62, 108, 116, 117
Gel filtration   118, 130, 135, 212
Gibberellic acid   177
Glass transition temperature   280
Glucanase   127
Glucoamylase   113
Glucose   290
Glutamate   115
Glutathione-S-transferase   108, 205
Glycine   289
Glycols   98
Glycopeptides   175
Glycoproteins   196, 199, 200
Glycosylation   117
Good Laboratory Practice   344
Good Manufacturing Practice   3, 8, 286, 295, 302, 311, 313, 329
Gowning procedures   333
Graphitized carbon   228
Guanidine hydrochloride   114, 183, 190
Guar gum   24
Guard column   184

Haemoglobin   203
Hazardous area   237, 302, 312, 320, 323
Hazardous waste   332
Health and safety   320

Height equivalent to a theoretical plate   339
Heparin   203
Hepatitis   335
HIGEE contactor   89
High performance liquid chromatography   8, 223, 343
High resolution purification   199
Histidine   7, 114, 200
Hollow fibre membrane   129, 141, 270
Homogenizers   42, 44
Horizontal leaf pressure filter   13
Hormones   95
HPLC–mass spectroscopy   236
Human growth hormone   115
Human immunodeficiency virus   335
Human serum albumin   203
Humidity   330
Hybrid protein   107
Hydrazide agarose   199
Hydrocarbon solvents   266
Hydrogen bonding   95, 101, 130, 183
Hydrogen peroxide   296
Hydrophilic/lipophilic balance   266
Hydrophobic bonding   130
Hydrophobic interaction   107, 130, 183 chromatography   60, 63, 131, 135, 212
Hydroxyethylmethacrylate resin   197
Hydroxylamine   110, 111
N-Hydroxysuccinimide esters   198
N-Hydroxysuccinimidobiotin   205
Hypercarb™   228

Imidazole   101
Imidocarbonate bond   198
Immobilized albumin   111
Immobilized galactose   113
Immobilized immunoglobulin G   111
Immobilized metal affinity chromatography   114, 131
Immunoadsorbents   65
Immunoaffinity chromatography   112, 200
Immunoassay   343
Immunoglobulins   185, 204
Inclusion bodies   7, 105, 107
Installation qualification   335
Interfacial phenomena   74, 82, 100
Interleukins   112, 220
Internal circulation   78
Internal surface reversed phase   231

Intracellular degradation 108
Intracellular proteins 54
Integrated pollution control 320
Integrated processes 28, 48, 124, 141, 143, 301, 320
Intrinsically safe equipment 312
Ion exchange 95, 101
 chromatography 130, 135, 179, 212
 media fouling 183
 media preparation 181
 membrane 114, 118
 resins 5, 7, 129
Ion pair extraction 8, 73, 95, 97, 264
Ionic bonding 130
Isoelectric focusing 343
Isomer separations 151
Isotactic train 151

Karr column 84
Kinetics 127, 251
Kjeldahl assay 344

β-Lactamase 116
Lactate 101
Lactic acid 100, 264
Lactobacillus delbreukii 101
Lectins 112, 113, 183, 200
Level measurement 308, 311
Lewis acids 95
Ligand accessibility 196
Ligand immobilization 197
Ligand stability 196
Light meromyosin 111
Limulus amoebocyte lysate assay 342
Lipids 183
Liquid ion exchange 73, 97
Liquid membranes 259
Liquid–liquid partition 6, 7, 72, 130, 259
Lowry assay 344
Lubricants 331
Lyophilization 8, 275
Lysine 203
Lysozyme 59

Macromolecular precipitation 56
Magnesium sulfate 57
Maintenance 322, 331
Malaria antigen 111
Maltose 113
 binding protein 112
Mannitol 289, 290

Mannose 113
Mass measurement 308, 311
Mass transfer rate 73, 82, 261
Material certifications 335
Materials specification 334
Membrane bound enzyme 45
Membrane fouling 141, 143, 145
Membranes 100
2-Mercaptoethanol 199
Metal ion affinity chromatography 7, 131, 200
Metal ion complexes 114
Methanal 296
Methanol 244
Method validation 344
Microbiological contamination 330
Microbiological methods 344
Microfiltration 18, 61
Microparticle formation 255
Milbemycin 237, 255
Mimetic™ Orange 3 A6XL 202
Mimetic™ Red 2 A6XL 202
Mimetic™ Yellow 2 A6XL 202
Mitomycin C 175
Modular instrumentation 311
Molecular recognition 4
Monoclonal antibodies 112, 129, 200
Morphine dehydrogenase 202
Morphinone reductase 202
Muconic acid 177
Mycophenolic acid 1
Mycoplasma 335, 344
Myristic acid 177

1-Naphthol 154
2-Naphthol 154
(*S*)-*N*-(1-Naphthyl)leucine octadecyl ester 264
N-(1-Naphthyl)methyl-7a-1-methylbenzylamine 264
Neomycin phosphoryltransferase 116
Neurofibromatosis protein 112
Nexpert Object™ 135
Nicotine alkaloids 177
Nocardia rhodochrous 45
Non-ionic adsorbents 159
Non-selective binding 196
Non-viable particles 345
Nucleic acids 113, 130
Nucleosides 177
Nylon 198

Octadecyl trichlorosilane 226
Operational qualification 336
Operator protection 333
Operator training 336
Optimization 142, 201, 229, 236, 254, 275, 286
Organoboronate 97

Partition coefficients 58, 130
Peak symmetry 231, 339
Penicillin 2, 83, 96, 177
Penicillium brevi-compactum 1
Peptides 152, 160
Performance qualification 337
Performance tests 185
Perfusion chromatography 4, 131, 231
Perfusion cultures 129
Periplasmic products 107
Perlite 23
Permeability 23
Permeabilization 124, 130
pH measurement 306, 311
Phase separation 74
Phenols 97
Phenyl Superose™ 115
Phenylalanine 177, 265
Phosphate buffer 290
Phosphine oxides 97
Photodiode array detector 231
Piping and instrument diagrams 335
Planned preventative maintenance 340
Plasminogen 203
Plate and frame filter 17
PLRP-S™ 228
Podbielniak extractor 84, 86
Polyacrylamide 198
Poly(acrylic acid) 113
Polyarginine 113, 117
Polyaspartic acid 114, 117
Polyclonal antibodies 200
Polycysteinc 115
Polyelectrolyte precipitation 113
Poly(ethylene glycol) 53
    adsorbents 57
Poly(ethylene glycol)/potassium phosphate 115
Poly(ethyleneimine) 114
Polyhistidine 114, 117
Poly(hydroxybutyrate) 46
Polymer exclusion 62
Polymeric adsorbents 101, 147, 159

Polymethacrylate resin 152, 163, 176
Polypeptide 107
Polyphenols 60
Poly(phenylalanine) 115
Polypropylene membrane 270
Poly(styrene-divinylbenzene) resin 161, 197, 228, 231
Polysulfone membranes 18, 270
Poly(vinylpyridine) 101
Potassium phosphate 53
Predictable downstream processing 107
Preparative HPLC 236
Preparative supercritical fluid chromatography 254
Pressure measurement 306, 311
Primary drying 276, 280, 292
Process
    configuration 322
    control 8, 321
    design 107, 119, 132, 142, 184
    flow diagrams 335
    selection 319
Product
    capture 6, 132, 199, 201
    contamination 330
    potency 343
    purity 63, 132, 156
    release 6
    specification 329
    stability 320, 330
    yield 63
Production
    campaign 331
    facility 330
    rate 156
Propan-2-ol 85, 173, 190
Propanone 174, 244
Prostaglandins 177
Protease 108, 129
ompA Protease 114
Protease deficiency 108
Protein A 111, 204
Protein A Sepharose™ 220
Protein binding studies 200
Protein elution 201
Protein G 204
Protein refolding 7, 107
Protein–metal complexes 114
Proteins 152, 160
Pseudomonas putida 45, 202
    shemanii 116
PTFE membrane 270

Purification strategy 86
Pyridine 99
Pyridyl groups 99, 101
Pyrogen removal 197, 202, 303, 342, 344
Pyrogen-free water 321
Pyruvate kinase 66

Q Sepharose™ 212
QAE Sepharose™ 215
Qualified personnel 332
Quality Control 329, 342
Quarantine area 334
Quaternary amine sorbents 101
Quaternary ammonium salts 97

Radial flow chromatography 209
Rapamycin 224, 232
Rapid expansion of supercritical solvent 255
Raw materials 333, 341
Reactor design 127
Reagent recycle 66, 268
Recall procedure 341
Receptors 2
Recombinant proteins 50, 105, 129
Recovery rate 156
Recycling processes 137
Regeneration 94, 98, 173, 185, 190
Rejuvenation methods 173
Release procedures 341
Rennin 115
Retention optimization 152
Reverse micelles 97
Reversed phase chromatography 175, 212, 226
Reynolds number 76
Rheology 80, 140
Rhizopus nigricans 46
Robust processes 6
Rotary drum precoat filtration 12, 140

Saccharomyces cerevisiae 45, 118
Safety 343
Salting out 55
Sample pretreatment 224
Sampling plans 334
Sanitization 196, 197, 202, 210, 303, 332
Scale up 131, 215, 286, 317
Schistosoma japonicum 108

Screening procedures 2
Scroll decanter 20, 39
Secondary adsorption 183
Secondary drying 276, 283, 295
Secondary metabolites 2, 232
Sedimentation properties 36
Selective extraction 128, 259
Selective release 127
Selectivity 147, 152, 224, 259
Self displacement 155
Separation coefficient 135
Sephadex™ 129
Sepharose CL4B™ 215, 218
Sepharose Fast Flow™ 215
Settling velocity 36
Shear 140, 181
Silica gel 147, 198, 225
Size exclusion chromatography 118, 130, 135, 212
Slime moulds 118
Sodium azide 174
Sodium cyanoborohydride 199
Sodium hypochlorite 173
Sodium mandelate 264
Sodium periodate oxidation 199
Sodium sulfate 57
Soluble protein 43
Solute exclusion 58
Solvation 96
Somatostatin 112
    release inhibitory factor 116
Sorbitan mono-oleate 266
Soybean oil 100
Spacer arm 196
Spare parts list 331
Specific binding 107
Specific cake resistance 32, 38
Spectrophotometric assay 4
Spray column 268
Stability 275, 276, 320
    data 5, 344
Stabilizers 290
Standard Operating Procedures 336, 341
Staphylococcus aureus 111
Starch 24, 112
Starch binding domains 113
Static extraction 249
Steam in place 337
Steric hindrance 196
Sterility 276, 343
Sterilization 43, 281, 286, 296, 303
Stirred tank adsorption 200

Stoichiometric overload 96
Stoichiometry 127
Streptavidin 116, 205
Streptomyces hygroscopicus 237
Stripping phase 260, 267
Sublimation 276, 280, 292
Sucrose 290
Sugars 98, 160, 176, 290
Sulfonamide resins 110
Supercritical fluids 242
Supercritical fluid
  chromatography 8, 252
  extraction 245
Support matrix 196
Supported liquid membrane 269
Surface charge 24
Surface finish 296
Surfactants 75, 79, 80, 183, 266
System challenge 340
System integrity 336
System selection 59

Tail cleavage 116
Tangential flow filtration 6, 18
Tannins 60
Temperature
  mapping 337
  measurement 306
  optimization 142
Terminal velocity 75, 82
Tertiary amines 101
β-Tetrolone 155
Therapeutic proteins 323
Thienamycin 175
Thiocyanates 59
Thioether bonds 199
Thiogalactosidyl Sepharose™ 108
Thiopropyl Sepharose™ 115
Thrombin 110
Tissue plasminogen activator 176
Titanium dioxide 129
Toluene 224
Tosyl chloride 199
Training records 332
Transferrin 334
Tributyl phosphate 85, 244
Tridecylamine 96
Triethylamine 228
Triglycerides 252
Trimethylamine 100

Trioctylamine 93, 96
Triple point 242, 275, 280
Tris(hydroxymethyl)aminomethane 289
Trypsin 196, 202
Tryptophan 200
Tubular bowl centrifuge 39
Tyrosine kinase 110

Ubiquitin 118
Ultrafiltration 7, 60, 65
Ultrasonic treatment 224, 247
Urea 183, 190, 196, 202
Uridine phosphorylase 212
β-Urogastrone 113
Utilities 321, 326

Vacuolar enzyme 45
Validation 143, 184, 186, 286, 295, 314,
  323, 335, 341
Vancomycin 175
Vapour flux 282
Variable volume filter press 18
Vendor certification 334
Vertical leaf pressure filter 13
Viable particles 345
Viral proteins 110
Viscous fluids 326
Vitamin B_2 296
Vitamins 175
Volume measurement 308

Waste minimization 320
Water for injection 321
Water quality 332, 345
Western blotting 344
Whatman DE 52™ 219
Whatman QA 52™ 180, 185
Whatman SE cellulose™ 189
Whole broth 54
Whole broth extraction 6, 7, 73, 140
Worst case situations 338

Yeast 42, 46, 54, 60, 82, 83, 84, 116
Yeast proteins 116, 127

Zeta potential 24
Zorbax™ SB-C18 228